GLOBALIZATION AND ITS OUTCOMES

GLOBALIZATION
and Its Outcomes

Edited by
JOHN O'LOUGHLIN
LYNN STAEHELI
EDWARD GREENBERG

THE GUILFORD PRESS
New York   London

© 2004 The Guilford Press
A Division of Guilford Publications, Inc.
72 Spring Street, New York, NY 10012
www.guilford.com

Chapter 4 © 2003 United Nations Industrial Development
Organization (UNIDO)

Chapter 9 © 2004 Institute for International Economics

Printed in the United States of America

This book is printed on acid-free paper.

Last digit is print number: 9 8 7 6 5 4 3 2 1

Library of Congress Cataloging-in-Publication Data

Globalization and its outcomes / edited by John O'Loughlin, Lynn Staeheli,
Edward Greenberg.
    p. cm.
  Includes bibliographical references and index.
  ISBN 1-59385-045-X (pbk. : alk. paper) — ISBN 1-59385-046-8 (hardcover :
alk. paper)
  1. Globalization. I. O'Loughlin, John V. (John Vianney) II. Staeheli, Lynn A.
III. Greenberg, Edward S., 1942–
  JZ1318.G57915 2004
  303.48'2—dc22
                                                              2004008541

Preface

While a plethora of books on globalization have appeared in the past couple of decades, relatively few of them attempt to bring together scholars from the range of disciplines that engage globalization study. More typically, economists cluster in one set of books, sociologists in another, cultural studies scholars in a third, and so on. A student trying to get a sense of the rapidly growing literature in these disciplines must therefore scramble across the disciplinary horizons to examine the various disciplinary emphases. In this book, we attempt to pull together some of the main globalization themes and important disciplinary emphases with respect to globalization studies. As we remark in Chapter 1, no single book can cover all aspects of this burgeoning field; similarly, no single book can give equal and comprehensive attention to the foci that the individual disciplines have developed. With this caveat, we have drawn together in this book a range of perspectives from economic to cultural to political, and a range of disciplines (economics, geography, political science, sociology). We also incorporate a variety of perspectives on globalization, from highly critical ones to several that are highly supportive of further economic and political integration, though under different rules.

The genesis of this book took place a decade ago in a pair of initiatives from the National Science Foundation (NSF)—one to encourage new and innovative interdisciplinary graduate training programs, and another to promote research on democratization. Four of us in the Program on Political and Economic Change in the Institute of Behavioral Science at the University of Colorado (Edward Greenberg, John O'Loughlin, Lynn Staeheli, and Michael D. Ward, now at the University of Washington) successfully applied for a graduate training grant under the rubric of "Globalization and Democracy"(GAD) while, at the same time, John O'Loughlin and Michael D. Ward received a research grant to study the "diffusion of democracy." We are grateful to NSF for this support which, together with funding for foreign students from the University of Colorado (CU) Graduate School arranged by graduate dean Carol Lynch, and matching funding from the four cooperating departments (Economics, Geography, Political Science, and Sociology), allowed us to support over 20 graduate students. Fifteen have so far graduated from CU–Boulder, and many of them are represented as authors or coauthors in the chapters of this book.

The training program was housed in the Institute of Behavioral Science (IBS) and benefited from the continued support of many colleagues in and outside the Institute. The current and former directors of IBS, Jane Menken and Dick Jessor, have consistently offered their intellectual and financial help to the GAD program over the past decade and have also assisted us greatly in negotiating institutional and programmatic arrangements with the respective university offices. The staff of the program (Marcia Richardson, Sugandha Brooks, and Marc Lowenstein), the central IBS office (Mary Axe, Debbie Ash, and Steve Graham), and the Center for Computing and Research Services (Zeke Little, Jani Little, Richard Cook, and Tom Dickinson) provided sterling support to the GAD program, while helping the faculty and students with their research. Our colleagues in the program (Tom Mayer, Jim Scarritt, Keith Maskus, David Leblang, David Brown, Jim Bell, Paul Kubicek, and Debra Javeline) were active in the GAD program conferences and helped to shape the program and teach the GAD seminar courses.

This particular book developed from a capstone conference on the training and research held in Boulder in April 2002, with funding from NSF and the IBS. We invited individuals who had helped develop the program, others who had helped teach the program, and students in the Institute to contribute to the conference and to the book. We thank Kristal Hawkins of The Guilford Press for her interest in the book project and the reviewers of the proposal for their comments, which contributed to the book's final form. Our greatest debt is to Mike Ward, who is more than the coauthor of one of the book's chapters. As the prime mover to respond to the NSF call for proposals for Graduate Research and Training grants a decade ago, and as the first director of the training program, he continued to shape the resulting products long after he left the University of Colorado for Seattle.

Contents

GLOBALIZATION AND ITS OUTCOMES

Part I

INTRODUCTION

1

Globalization and Its Outcomes
An Introduction

JOHN O'LOUGHLIN
LYNN STAEHELI
EDWARD GREENBERG

Globalization is profoundly affecting the lives of people around the world. It is a set of processes in which capital, technology, people, goods, and information move relentlessly across the inherited map of political boundaries (Dicken, 1998; Feenstra, 1998; Mittelman, 2000), and through which the interdependence of societies over vast distances and ever-shortening time frames has been intensified. This compression of time and space across a broad range of human activities has been made possible by the dramatic decline in the costs of transportation, communication, and production, and by changes in the formal rules that once established substantial (though never entirely impermeable) barriers against flows across borders (Harvey, 1989).

Our objective in this volume is to report research that examines globalization processes and how they are affecting people and the societies in which they live. We have organized the discussion of this complex set of topics in three parts. In Part I, we present globalization as a set of economic flows, as a complex of information and knowledge linkages, and as a network of migration between countries and regions. In turn, these flows affect countries and communities in differential ways; they

3

are mediated by the actions of governments and by the level of incorporation of the communities and countries into a globalized world. While globalization has a myriad of impacts across the social, political, and cultural spectrum, we concentrate in this book on societal outcomes in the areas of health, welfare, and citizenship (i.e., how communities are reconstituting themselves as supporting or opposing new identities that have evolved out of the encounter of social groups with a globalized world). Thus we do not pretend to present a complete portrait of the processes of globalization and their diverse outcomes; such an evaluation is now beyond the scope of any one book, given the explosion of research on the subject over the past decades and the contrasting evaluations of it. Lively debates exist about the nature of the individual community experiences and about the benefits and costs of globalization; these debates are represented in the range of arguments in this book. If there is one point on which researchers of different ideological and methodological preferences can agree, it is that experiences are widely disparate across the globe.

We recognize that some of the cross-border flows that comprise globalization are not unprecedented—many have been occurring for a very long time, as argued by Immanuel Wallerstein and his world system acolytes—but we believe that their scale and their simultaneous occurrence make contemporary globalization something new and worth detailed examination. (For a skeptical view of globalization as a recent development, see Wallerstein, 2000, and Keohane & Nye, 2000.) We also recognize that globalization is neither an inevitable nor an irreversible set of processes. By this statement, we mean that the present character of globalization is not the inevitable product of scientific and technological change, or of the inexorable laws of the economy, but instead is an outcome of decisions made by individuals, groups, firms, nation-states, and transnational organizations. Moreover, globalization can be slowed, deflected, or—as some have argued—stopped in its tracks since September 11, 2001 (see Steinmetz, 2003).

Earlier trends toward global openness and diminution of trade controls at the beginning of the 20th century were soon reversed during the global slowdown of the 1920s and 1930s. Many governments reintroduced tariffs and quotas in an attempt to protect their domestic manufacturers and farmers. Recent tensions within the World Trade Organization (WTO) between the European Union and the United States over steel, bananas, genetically modified food, aircraft manufacturing subsidies, and agricultural price supports testify to the ongoing uncertainties about trade relations between the two biggest economies. Terror (real and imagined) and the war against it have constrained the pace of globalization and increased doubts among many about its efficacy, especially in the movement of people. For example, fear of terror and the numerous controls that a wide range of countries have placed on the cross-border flows of people and products to counter real or imagined terrorist threats may constrain growing globalization. The recent breakdown of the Cancun WTO trade talks on agricultural subsidies (in this case, due to the claims made by poor countries for an end to the substantial subsidies made to rich countries' farmers) is another reminder that the process of globalization is subject to challenges, setbacks, and missteps.

As editors, we want to emphasize that we are "agnostic" in a scientific sense about the outcomes of globalization. We do not simply assume, as many do, that globalization is beneficial and something to be encouraged. Nor do we simply assume, as many do, that globalization is destructive and unjust, and something to be opposed and defeated. Rather, our commitment is to the systematic examination of globalization and its impacts on people and societies, a view that is reflected in the diverse set of chapters in this volume. Individual authors clearly have their preferences, variably expressed in their chapters, and a sense of the contrasting perspectives can be gained by a comparative reading.

PROCESSES OF GLOBALIZATION

Globalization is uneven and incomplete because the pace of globalization varies over time, and not all societies are fully integrated into the global system (Hirst & Thompson, 1996; Wade, 1990; Wallerstein, 2000). Nonetheless, it remains the case that the integration of nations and societies into the global system is advancing, with increasing numbers of them subject to global flows across their boundaries year after year; states and societies are becoming more tightly integrated and interdependent as time passes (Castells, 1996; Gereffi & Korzeniewicz, 1994; Harvey, 1989; Mittelman, 2000; Strange, 1996). But it is not necessarily clear what sense we should make of these flows and whether they really constitute a challenge to our understandings of the world. James H. Mittelman raises precisely this question in Chapter 2; he asks whether globalization is a new paradigm for the field of international studies, predicated as it is on the notion of the globe as a collective of autonomous states, now numbering about 190. As globalization's economic processes continue apace and transnational corporations continue to push their operations wider into more world regions and deeper into existing communities, the issue of the "hollowing out of states" by these corporations remains a potent one (Giddens, 2000; Taylor, 1994). In the extremist perspective of Ohmae (1995), states are unable to control their own borders and seem doomed to shrink as economic partners, but caution is warranted since many commentators (e.g., Johnston, 1993) have pointed out that state functions continue to aggregate and state spending (as a proportion of gross domestic product [GDP]) is still generally growing. Mittelman concludes that globalization is a "protoparadigm" that shows every indication of undermining the assumptions of the field of international studies that has existed for about 100 years. In any case, the dialectic of states and transnational corporations is likely to evolve in an unpredictable manner, sometimes favoring one and sometimes the other.

As Mittelman argues in his chapter, globalization is a complex phenomenon that raises questions about the ability of existing frameworks to analyze, and thus it is important to clarify the kinds of processes that we are addressing. The first aspect of globalization concerns economic flows, the increasing movement across national borders of traded goods (both intermediate and final products), foreign direct investment (corporate investments in production and distribution facilities abroad), foreign portfolio investment, and financial instruments (including currencies, credit, and insurance), as well as the rules by which these economic processes are regulated (e.g., intellectual property rights, structural adjustment requirements, WTO rules).

Transnational corporations (TNCs) are the chief organizing agents of economic globalization (Gilpin, 2000). Because of TNC activity, product design, engineering, production, marketing, and sales (as well as associated financing and other business services) have become more globally dispersed, yet increasingly coordinated and integrated. These activities are supported by the operations of a highly integrated international financial system that provides the necessary liquidity for these dispersed commodity production and distribution activities. A variety of rules have also been enacted over the years to bring a measure of order and predictability into global economic relations. Thus transnational financial, legal, and intellectual property rights regimes (see Leblang & Bernhard, 1999; Maskus, 2000) have taken shape, often in formal organizations like the WTO and the International Monetary Fund (IMF), and sometimes in informal arrangements between economically powerful actors (e.g., the Basel Committee on Banking Supervision). These rules are often the outcome of negotiations among nations and groups with substantial differences in power.

One result of these complex processes is a significant and continuous restructuring of the location of economic activities, both within nations and across them. In Chapter 3, Richard Grant and Jan Nijman examine globalization as a geographical–spatial phenomenon in which economic activities are continuously reconfigured and negotiated within countries. Much of the research on globalization has taken the country as the unit of analysis and compared the 190 or so countries in terms of general indices of gain and loss. A recent recognition by economists (e.g., Gallup, Sachs & Mellinger, 1998) of the (often) dramatic differences within countries is helping to draw attention to the geographical perspectives that have long stressed that globalization's impacts need to be measured in small-areal terms, as well as by class, ethnicity, and gender. In their study of India and Ghana, Grant and Nijman show how a "hyperdifferentiation" of space at regional levels and within metropolitan areas can be identified in terms of capital accumulation and levels of prosperity. Winners and losers are easily identified by geographic location.

International trade is an extremely important part of the process of globalization. It is also the subject of contentious views in both the journalistic and the scholarly literatures about the outcomes of the trading system for poor countries, for the well-being of people in poor countries, and for wage earners in the rich countries. (See, e.g., Dollar & Kraay, 2001; Rodrik, 1997; and Stiglitz, 2002.) All too often extreme views dominate the discussion, with one side claiming that the global trading system is and must be a disaster, and the other side claiming that the global trading system is generally a good thing for all. Clearly, a more nuanced position makes more sense, one that recognizes that participation in the global trading system is a precondition for national economic growth, but that the existing rules do not allow the global trading system to benefit all who are involved in it. Three chapters in this section examine the nature of trade, the influence of trade rules on the trading system, and some of the outcomes of the trading systems for societies at different places in the global division of labor. In Chapter 4, Gary Gereffi and Olga Memedovic look at how commodity chains in the apparel industry are organized at the global level, where different countries find themselves in this commodity chain, and how such placements affect the development strategies and possibilities of different nations. Their purpose is to ask whether developing countries ought to try to emulate the successful East Asian development strategy in which production for the clothing industry played a leading role in its early growth strategies for development. They conclude that under certain conditions, such a development strategy makes sense, even in light of common criticisms about the efficacy and equity of low-wage, low-skill, assembly-oriented export production. Gereffi and Memedovic suggest that the secret for sustained economic development for countries at the low end of the apparel commodity chain is to make the transition from simple assembly to full-package production, which involves a move from subcontracting to the design, production, and sale of their own commodities.

In Chapter 5, Keith E. Maskus looks carefully at the world trading system and the rules under which it operates, and acknowledges that it is wanting in important respects. He argues that in its present form the trade regime is inconsistent with the needs of poor countries for sustained economic development. He suggests that many trade economists, who accept the underlying promise of greater international integration, are nonetheless vitally concerned about the forms in which this integration is taking place. Their dissatisfaction arises from a recognition that WTO agreements contain a far more comprehensive set of rules that limit the ability of governments in poor societies to interfere with trade than did the previous General Agreement on Tariffs and Trade (GATT), and that these new rules are highly advantageous to powerful economic interests in the rich countries. Maskus argues that trade rules must be made more consistent with the economic development needs of poor countries by making allowances for the special circumstances of each country, by recognizing the costs that these countries are being asked to shoulder, and by rectifying the asymmetry of benefits that flow from existing rules that substantially favor rich countries.

In Chapter 6, Michael W Nicholson considers the impact of the global intellectual property rights regime as it pertains to technology transfers on the development possibilities for poor nations. Like Yang and Maskus (2001), he finds much in the rules of the game that have asymmetrical effects on the economic growth of rich and poor countries, much to the benefit of the former. In particular, he finds that while the existing intellectual property rights (IPRs) regime encourages direct foreign investment by multinational firms in poor countries, it also leaves proprietary control in the hands of the technology innovators in the rich countries. Consequently, the ultimate impact of stronger international IPRs may be a decrease in the transfer of *usable* technology. That is, IPRs provide an incentive for the physical transfer of technology, but with proprietary control (and the associated rents) remaining with multinationals based in the North.

As well as reflecting economic flows, globalization is also about the flow of knowledge and information across national borders (Tomlinson, 1999). Increasingly, knowledge and information are being carried across such borders by mass media and popular entertainment, the Internet, telephony, and educational institutions. Such knowledge and information flows are commonly recognized as a part of life in the rich democracies, but they happen as well in other places. There is mounting evidence, for example, that a massive increase has occurred over the past decade or so in the volume of international telephone traffic, international travel, access to first world-provided television programming and movies, and Internet access in developing and transitional societies (Kearney, 2001). What makes this important is the general agreement among scholars that knowledge and information carry embedded cultural messages (Calabrese, 1999).

Given the general paucity of systematic empirical research in this area, it is less clear what this flow means for people in different societies (Holton, 2000). Some researchers argue that the mass flow of messages leads to cultural homogenization. Homogenizers suggest that the most important outcome of globalization is the creation of global consumers tied to dreams of affluence, personal success, and sensual gratification as encouraged by advertising, Western television programs and Hollywood films, and growing belief in the superiority of market economies. For those who lament cultural homogenization of this sort, enticement into the consumer culture of the West all too often draws people in developing and transitional societies into lifestyle choices that negatively affect health and human capital development. Many scholars disagree with the homogenizers, suggesting that the flow of information and knowledge more often leads to a clash of cultural values between rich Western countries and poorer societies (Barber, 1995; Huntington, 1996; Said, 1978). Yet a third group of scholars argues for hybridization or synthesis, believing that the flow of knowledge and information from the West mingles with local cultures to produce a synthesis or hybrid form of culture (Hannerz, 1992). In Chapter 7, Andrew Kirby addresses the claim that transborder information flows carry cultural messages that put receiving societies at risk and the notion that cultural globalization is in the hands of a few corporations. He shows that corporations from several countries (especially India, China, and Japan) have joined longer established U.S. companies in producing cultural goods that include movies, videos, music, electronic games, and other entertainment products. The cumulative result is a complex networking of these corporations and a popular culture that remains strongly rooted in the diverse mosaic of the world's populations. Due to the falling price of producing consumer cultural products, there is the likelihood that both a uniformity and a diversity of "global culture" will emerge, perhaps producing a hybridization of cultural products in the world's markets.

GLOBALIZATION'S OUTCOMES: HUMAN WELL-BEING

Globalization is enormously consequential for people and societies at all levels of economic and social development, and no single volume can be expected to cover its entire range. Among the most important outcomes, in our view, are globalization's impact on societal development; patterns in the distribution of income, wealth, and poverty in societies at various levels of economic development; the possibilities for the development of vibrant civil societies; and the formation of identities, including those associated with opposition to and rejection of globalization itself. In Part III of this book, the authors focus on the effects on individual welfare of globalization processes, while in Part IV, attention is directed to issues of group identity and mobilization, and to evolving definitions of citizenship in a world with increasing numbers of people who move back and forth across borders and who develop a sense of belonging to different governmental units.

Ideology also flows across national boundaries; particularly important in recent years has been the growing influence of the ideologies that shape liberal capitalism. The broad acceptance of these norms—in contradistinction to alternative but discredited systems of ideas such as socialism and the "import substitution" model of development—has had consequences for the shape of social and economic policies in many societies. Many, for example, have reduced the role of government in the economy by moving toward deregulation, privatization, and open national economies (Gilpin, 2000). Though there has been a substantial backlash against this ideology in the midst of the current global economic malaise (as of mid-2003), the core ideas of liberal capitalism continue to guide the actions of the U.S. government and its agencies, as well as the global financial institutions it dominates, including the IMF and the World Bank. Two of the more dramatic trends since the fall of the communist regimes in 1989–1991 have been the perceived lack of a viable alternative to this "Washington Consensus" and the eagerness of successive U.S. administrations to promote worldwide democracy (O'Loughlin, in press).

Globalization has numerous complex and interrelated effects on the economic welfare of individuals and communities in those countries that embrace it (Ranis, Stewart, & Ramirez, 2000). By forcing a reallocation of resources and providing access to newer technologies from abroad, aggregate (national) incomes ordinarily would be expected to increase, at least over time. This substantially affects the ability of governments in the poorer parts of the world to meet the welfare, health, and educational needs of their citizens. Key political choices must be made about how open the economy should be to foreign investment and imported goods; to the relative spending levels on welfare, which in turn dictates a certain level of taxation; to the balance between business and consumer interests; and to long-term (e.g., education, infrastructure) versus short-term (e.g., police, telecommunications) investment.

Though some economists disagree (e.g., Rodrik, 1997), considerable evidence to date supports theoretical expectations that integration into global economic networks contributes to aggregate gains in a country's GDP over time (Bhagwati, 2002; Bhalla, 2002; Dollar & Kraay,

2000). However, the pace at which a country would see its incomes go up depends on specific circumstances and the sequencing of its liberalization. For example, it is unlikely that a society with weak financial institutions would find it advantageous to open itself fully to international financial flows because the potential volatility of exchange rates and interest rates could be devastating in the short run (Stiglitz, 2002). In Chapter 8, Michael D. Ward and Kristian Skrede Gleditsch show that a society's degree of integration into the global economy—in this case, measured by trade and economic openness—has significant positive impacts on its economic growth rate and on the reduction of aggregate poverty. They recognize that the benefits of globalization vary by global region (former Soviet and African countries lag), but they conclude that the advantages of globalization are evident in the example of Peru's experience and in their statistical evidence that globalization promotes democratization. In turn, they suggest that democratization adds to the number of countries that are less likely to fight international conflicts with other democratic states.

However, increases in aggregate measures, such as GDP, tell us little about the distribution of income, wealth, and other advantages in any particular society. It may be the case, for example, that even income-enhancing globalization can have significant adverse impacts on the distribution of income and wealth, and on the levels of poverty. Standard international trade theory argues that poor countries should observe a decline in poverty after opening themselves up to trade and investment precisely because it is the abundance of unskilled (poor) workers that provides them with a comparative advantage. However, this statement is logically true only under particular circumstances of competition, technologies, and preferences. A more trenchant claim is that individuals (and communities) that offer skills that are poorly suited to compete in the global economy will suffer reductions in real income. Similarly, if workers' jobs are tied to the use of natural resources, the price of which falls with trade liberalization, they can experience large income declines. In Chapter 9, J. David Richardson examines globalization's impact in the United States with reference to individual welfare in the form of employment and income. Adopting a balance-sheet approach that is designed to counter impressionistic accounts of the effects of globalization, he shows that marginal groups do not benefit from the increasing globalization of the U.S. economy as much as do skilled and managerial workers. However, globalization cannot be blamed for the stagnation or even the falling standard of living of poor Americans. Instead, it is technological change that is more responsible for job dislocation. Richardson, however, is well aware of the significant differences by job sector and skill/educational level and proposes that government insurance programs be instituted to help displaced workers cope with job losses. He is cognizant of the variable effects of globalization within U.S. society and argues that careful calculations of the losses and gains by each group and region should allow for more careful consideration in the political debate of balancing interests. Whether this can be accomplished in the increasingly partisan and vociferous domestic political debate, however, is questionable.

It appears increasingly that globalization has consequences for the health of a society's population. According to *The World Health Report 2002*, "it should be recognized that health itself has become globalized" (World Health Organization, 2002, p. 5). Certain key features of globalization—the liberalization of international trade, increased international travel and migration, and pervasive exposure to Western mass media—have had reverberating implications for health, especially in developing countries. Infectious diseases can now be spread from country to country in a matter of days or even hours. And as that same report noted, the lifestyles of whole populations are changing around the world in ways that are consequential for global health—lifestyle changes having to do with food consumption, tobacco and alcohol use, and diminished physical activity, among others. Together, these trends are seen as constituting a "risk transition," one that now places "a double burden of disease on developing countries—the combination of long-established infectious diseases [e.g., malaria, tuberculosis, and HIV/AIDS], and the greater relative importance of chronic, non-communicable diseases [e.g., cancer, cardiovascular disease, and diabetes]" (World Health Organization, 2002, p. xviii).

Although there has been a salutary global decrease in infant mortality and an increase in life expectancy, physical health continues to be compromised by poverty and malnutrition, poor sanitation, substance abuse (especially tobacco and alcohol use), violence, and sexually transmitted diseases, including HIV/AIDS. The World Health Organization report designates undernutrition as the leading cause of disease burden and a major cause of death among children: " It was a contributing factor in 60% of all child deaths in developing countries" (2002, p. xv). Michael E. Shin addresses several of these issues in Chapter 10. Dramatic improvements in overall global health in the past half-century have masked some strong regional differences and new challenges in the forms of infectious disease diffusion and the chronic health problems of the growing elderly segment of rich countries. Shin shows that globalization (increasing trade openness) leads to increased wealth, which in turn leads to increased health and longer life expectancy, although the benefits are not evenly distributed between or within countries. While the relationship between globalization and health has evolved over a long time and shows macrostructural developments, the recent outbreak of SARS (severe acute respiratory syndrome) in East Asia in 2003 illustrates well the rapidity with which infectious diseases can now spread in an age of cheap long-distance air travel. In turn, endemic disease has a deleterious effect on regional economic output (the dramatic drops in output in parts of southern Africa as a result of the high rates of HIV-infected adult males is well known), and panic in the face of new diseases can also set back globalization's advances, as Shin's example of Hong Kong shows.

Globalization may also have an impact on a society's ability to provide educational, medical, and income support benefits if, in the effort to remain globally competitive, societies must cut taxes on business firms, reduce the regulatory load on businesses, and reduce government's role in society and economy. In Chapter 11, George Avelino, David S. Brown, and Wendy Hunter address these issues with respect to the Latin American countries. They argue that most of the work on globalization and welfare spending has reflected the situation in rich countries and that work on poor regions like Latin America is needed in order to complete the picture. The "Washington Consensus" posits that economic growth follows on neoliberal policies that promote government efficiency—especially by reducing taxes and in keeping government small. But citizens, especially in increasingly democratic societies, demand that their governments provide welfare, often in the form of food subsidies, free education, free health care, and adequate public transportation. The authors stress the careful use of statistics, since the debate about the relative advantages of the equity (welfare) versus efficiency (trade openness and low taxes) model for societies is replete with authors using widely fluctuating numbers and definitions. Without resolving the efficiency/ compensation debate, Avelino, Brown, and Hunter conclude that intermediate factors such as the relative importance of trade unionism, the democratic nature of the regime, and the regional setting are important determinants of the outcome of globalization on social welfare.

GLOBALIZATION'S OUTCOMES: CITIZENSHIP AND CIVIL SOCIETY

By the turn of the 21st century, few political leaders—even in authoritarian states—were willing to argue aloud against democracy, since its virtues were almost universally accepted (Diamond, 1999, p. 4). Moreover, global norms and conventions supportive of key human and political freedoms associated with democracy have been coalescing, starting with the United Nations' "Universal Declaration of Human Rights" (1948), and extending to the 1993 World Conference on Human Rights in Vienna. The issue, however, is whether human rights protections and democratic political arrangements are becoming more common as societies become more integrated into the global system, and further if the institutional and cultural foundations for human rights and democracy are firmly in place. In short, the debate has moved beyond the procedural aspects of democratization to consider more of the substantive questions of what it means to be a democratic citizen and the contexts and conditions that are necessary to support citizenship in varied settings.

Similar to the growing concern about globalization in many circles, some actors and groups in developing and transitional societies have begun to question the efficacy of democratic forms of governance, or at least the attenuated and distorted forms that democracy has taken in some of these societies (e.g., Uzbekistan, Russia, Ukraine, and the Philippines [Zakaria, 2003; O'Loughlin, in press]). Democracy, at least in its liberal democratic form, is being looked at more skeptically. New political voices are being raised, questioning the models of democracy on offer, asking who they benefit, and querying what they assume about political subjects. They ask who is promoting liberal, procedural versions of democracy and for what purposes? Related to these topics is the specific issue of the rights of people who are recent residents of the receiving country. The movement of people, both within societies and across national boundaries, is a phenomenon that is both stimulated by other globalization processes and an essential part of the globalization phenomenon.

Many of the authors of this book have stressed the differential impacts of globalization within the states of the world system, either by class, region, or by involvement with the global economy. If a society finds that inequality grows as a function of differential incorporation into the world economy along ethnic lines, existing polarization can be magnified and in turn threaten the stability of the state itself. Theories of nationalist mobilization have increasingly turned to explanations that rely on social constructionist arguments—while it is helpful to a nationalist movement to have predefined lines in terms of race or ethnicity in order to build a movement, nationalist activists can point to inequalities. In this part, we include seven chapters that address issues of cultural and political identity, grassroots mobilization, and the intersection of the interests of national states and their new residents in locations as diverse as Ecuador, Mexico, Okinawa, Turkey, and the United States. Of particular concern are the changes associated with globalization and their effects on the ways in which people are able to claim—and conceptualize—citizenship. By this, we do not only mean how individuals understand the legal category of citizen, though this is clearly important. Rather, we intend a broader conceptualization of how individuals and social groups understand their rights and identities as political subjects. If the role of the nation-state is changing in response to globalization—as Mittelman suggests in Chapter 2 —then we should anticipate that new forms of—or claims to—political identity may emerge. The chapters in this part of the book address this possibility.

A theoretical starting point shared by many of the authors is that the effects of globalization are likely to be differentiated. Since social groups are variously positioned within and between countries, it makes sense to expect that the *experience* of globalization on subjectivity is likely to be uneven as local and global forces and identities interact; in this perspective, most of the authors are sympathetic to the kinds of arguments put forward by Grant and Nijman (Chapter 3). Collectively, the authors demonstrate the ways in which specific effects of globalization interact with institutions at the national and local levels to shape the ways in which globalization is experienced, made sense of, and sometimes acted upon by social groups in different circumstances. If globalization is a protoparadigm, then one element of this paradigm must be the effects on political identities and citizens.

In Chapter 12, Caroline Nagel examines the effects of migration on debates about citizenship. While citizenship is at one level a legal category, debates about citizenship can be used as a marker of how a society defines itself and the people who live within its territory. Nagel argues that migration is only one of many flows associated with globalization, but it is the one in which questions of identity at a societal scale have been crystallized. Using both theoretical arguments and public debates within Western countries, Nagel demonstrates the ways in which various narratives of citizenship and belonging have been used to frame normative expectations of how migrants *should* fit into the societies to which they have moved. But she also shows that in most of these debates the perspectives of the migrants themselves are often absent. Drawing from interviews with Arab migrants to the United States and the United Kingdom, she argues that migrants use other forms of globalization—specifically the flows of information—to express ideas about citizenship based on the complex relations between host and sending societies. Thus globalization may facilitate complex political identities not adequately captured through statecentric models of citizenship.

The dislocations of globalization are not only experienced by international migrants, but also by populations who move within their countries and by people whose worlds seem to shift around them due to institutional changes mandated by international actors. In Chapter 13, Victoria A. Lawson traces the effects of international political and economic changes on the national government, on firms, and on workers in

Ecuador. In this, her analysis stands in contrast to that provided in Ward and Gleditsch's laudatory account of economic growth and the effects of inequality in Peru (Chapter 8). Lawson documents changes in the Ecuadorian state's fiscal and social welfare policies in response to pressures from the IMF and the World Bank; she then uses the accounts of labor migrants who left rural areas of Ecuador in order to understand how the dislocations of globalization affected this already impoverished sector of society. Lawson notes that many of these migrants express faith in the project of modernization, but that they also express a sense of exclusion from it that is both economic and political. In response, many of these migrants are nostalgic for the rural communities they left. Rather than a global or even a national identity, these workers articulate a spatial and political identity connected with their locality, but that remains isolated from efforts to build new political identities from the grass roots (see Mittelman, Chapter 2, this volume).

As Lawson suggests, identity formation in the context of globalization is a complex process that is expressed in many different ways and that reflects tensions between different globalization processes. In Chapter 14, Anna J. Secor further examines this argument through an analysis of the sources of support for Islamist politics among middle- and working-class men and women in Istanbul. Like Lawson, Secor believes that the political, economic, and informational flows associated with globalization are implicated in new political movements and identities. Through surveys and focus groups, Secor questions the ways in which Islamists use narratives of globalization in their political campaigns and the ways in which these are incorporated into residents' political outlooks and identification. Secor demonstrates that the role of globalization with regard to political support for Islamist politics varies by gender, reflecting the differentiated ways in which globalization is experienced—sometimes creating openings and sometimes creating conditions supporters find alarming or dangerous. While both women and men who voted for Islamist candidates express a religious identity, the influences of globalization on support are minimal at an aggregate level, although there are strong gender differences that are further cross-cut by class. Most importantly, support for Islamist candidates seems to be affected by the different engagements with globalization for women and men. Thus Secor concludes that the effects of globalization on political identities are not easily predicted and that they do not easily fit within the narratives of globalization theorists.

While the first three chapters of this part consider some of the dislocations of globalization, the next two chapters consider the ways in which emerging institutional structures may shape people's abilities to forge identities and to act politically. These chapters take up the pressing questions of how procedures instituted as part of a move to greater democratization can be made more effective.

In many countries around the world institutions have been built in the name of providing better channels for the citizenry to engage in debate and to force a measure of accountability onto their leaders. These institutional reforms have often been mandated by international donor agencies as a condition of receiving aid. Other institutions and practices have been established by nongovernmental organizations in an attempt to involve community members in decision making. Yet Lawson's analysis shows that these efforts are not always successful, as the migrants she interviewed remained unconnected to these structures and institutions.

In Chapter 15, Patricia M. Martin examines the recent trends toward greater democracy and participation in Mexican politics. While she notes the important steps being made toward greater electoral openness, she questions whether these have been complemented by opportunities for debate, for opinion formation, and for the development of social movements. Drawing on arguments from Habermas (1989), Fraser (1995), and feminist theorists, Martin explores the opportunities for an active citizenship; she is particularly interested in the scales at which new public spheres may operate, including the local, national, and global. She finds that activists engage in debate and mobilization at various scales, but that their sense of the scales at which public spheres operate are primarily local and global, not national. Although these emerging public spheres do not converge with the formal bounds of the Mexican state, activists believe—or perhaps hope—that they will provide leverage for rethinking political life and the paths that further democratization may follow. Thus these new public spheres may prove influential in the deepening of democracy in Mexico, and give greater meaning and substance to political life than procedures alone can do.

As Martin demonstrates, civil society has in many ways become a central arena from which new ideas about democracy, citizenship, and political identity may emerge. In the best of circumstances, civil society is believed to be a wellspring of trust—of social capital—that can be mobilized to solve problems that communities face. It is this willingness to work together, to balance collective and individual interests, that is said to make governments and other institutions solve collective needs—or in Putnam's (1993) term, to "make democracy work."

In an effort to "speed up" the processes of democratization and economic development, many organizations have attempted to produce social capital within civil society. In Chapter 16, William Muck examines the efforts and efficacy of intergovernmental organizations (IGOs) and nongovernmental organizations (NGOs) in artificially inducing the development of social capital. These organizations attempt to stimulate intracommunity, extracommunity, and state–society networks in order to bring isolated or marginalized populations into the globalized system. Muck argues that because social capital is multidimensional, efforts should be directed at engaging all aspects of social capital in order to be successful. He concludes that concerted efforts on the part of state, IGO, and NGO agents are required to enhance civil society and to improve prospects for development.

As noted, many of the authors in this book have stressed the differential impacts of globalization, either by class, gender, region, or involvement with the global economy. If a society finds that inequality grows as a function of differential incorporation into the world economy along ethnic lines, existing polarization within civil society can be magnified and in turn threaten the stability of the state itself; in some cases, social capital can be turned against the more laudatory ends discussed by Muck. Theories of nationalist mobilization have increasingly turned to explanations that rely on social constructionist arguments—while it is helpful to a nationalist movement to have predefined lines in terms of race or ethnicity in order to build a movement, nationalist activists can point to inequalities in income, welfare, government support, or educational opportunity to build a protest movement. Additionally, in the face of a challenge to traditional state identity and definition from cultural products and changing consumer preferences allied to materialist interests, traditionalists often appeal to historic identities to rally support against globalization or to use its impacts to redefine their community, including drawing the lines of who qualifies for membership and who does not.

Benjamin Barber (1995) contrasted these tensions as a choice between "Jihad" (traditionalism) and "McWorld (globalization) and Thomas Friedman (2001) chose the "Lexus" (globalization) and the "olive tree" (traditional community) as his analogy. Two examples of the globalized–localized tension close out the book. In Chapter 17, Takashi Yamazaki examines the experience of Okinawa, an island chain in the south of Japan, in the face of an economic crisis beginning in the 1970s at a time when Japan was fully globalizing. As the site for massive U.S. military bases, Okinawans chafed at U.S. control. After Japan regained autonomy over the island chain, the key choice was to try to "go it alone" as an offshore node in the global economic network, emulating the example of Hong Kong, or to integrate fully into Japan in expectations of Tokyo government largesse to improve the quality of life in Japan's poorest province. Within the Okinawan population, the political debate became rancorous as local activists tried to use the recently mobilized ethnic identity to push the separation option by "riding on globalization," while using the widespread opposition to the U.S. military presence as a vehicle for promotion of their strategy. After initial success, the effort failed in the face of concerted Japanese government aims to persuade Okinawans of the long-term economic benefits of integration into the national economy. The power of government and the ability of the state to use its resources to pursue its policies, coupled with the competition from other East Asian locations and the changing economic fortunes of Japanese corporations, undermined regional attempts to renegotiate relations with the national, regional, and world economies.

It is only appropriate that we return to the situation in the global hegemon, the United States, to end the book. In the decade after the collapse of the communist alternative, the United States had seen enormous growth in its global power, presence, and prestige, as well as dramatic economic growth at home spurred by investment in a high-tech economy. Commentators crowed about the fate of Japan, once a putative challenger to U.S. dominance, but now mired in recession and stagnation since 1993. The twin shocks of the stock market collapse in 2000 and the terrorist attacks on U.S. soil in September 2001 forced a debate in American society about the U.S. role in a globalized world order. In Chapter 18, Colin Flint uses a world systems approach to examine the "hegemonic dilemma," defined as the choice facing the United States as the most important (and in many ways, the most open) of the globalized states of pushing the globalization agenda as it had while defending its obviously porous borders. The United States is an "extraterritorial" power with global interests, evidently present in all corners of the world. Whether the United States is an empire, now widely discussed (Ferguson, 2003), hardly matters; the tensions among Americans are showing signs of a growing cleavage around this balance of global and domestic interests. Groups that have traditionally been part of the social compact, such as trade unionists and rural Southerners who have bought into the prime modernity of the U.S. project of global leadership and worldwide presence, are now becoming increasingly uneasy with this project despite the strenuous attempts by the Bush administration to push a globalization strategy as the best economic option for the United States. Whether the balance can be maintained in favor of the "extraterritorial" strategy is now problematic; key political decisions need to be made starting with the 2004 presidential election about where to fix the level of immigration, which parts of the globe demand most attention, where to set the government spending ratio of guns to butter, and where to assert domestic welfare interests of workers and communities against the interest of U.S. transnational corporations. The outcome of these political choices will shape the face of U.S. globalization for decades.

CONCLUSIONS

As the chapters in this book demonstrate, globalization is a complex, interlocking set of processes that is not reducible to one trend or manifested in one outcome. As the individual authors of the chapters demonstrate, the outcomes of globalization are multiple, apparently taking different forms and holding different implications depending on the aspect one examines, the level of analysis, and the groups and settings under consideration. Furthermore, and as some of the authors have suggested, the analysis of globalization benefits from an approach that historicizes processes, such that an effect at one point in time is not solidified or finalized as an "outcome." Rather, processes of globalization continuously shape and reshape the societies in which we live.

It is in this spirit that we refuse to make a final assessment of globalization, other than to say it is complicated and ongoing. Any other kind of assessment would not reflect the spirited debates that are reflected in these chapters, or even the diverging assessments between us as editors. All we can conclude is that globalization is complex, unpredictable, subject to shocks, but undeniably important to the ways in which people live their lives and participate in societies around the world.

REFERENCES

Barber, B. (1995). *Jihad versus McWorld*: New York: Random House.

Bhagwati, J. (2002). *Free trade today*. Princeton, NJ: Princeton University Press.

Bhalla, S. (2002). *Imagine there's no country: Poverty, inequality, and growth in the era of globalization*. Washington, DC: Institute for International Economics.

Calabrese, A. (1999). The welfare state, the information society, and the ambivalence of social movements. In A. Calabrese & J. C. Burgelman (Eds.), *Communication, citizenship, and social policy: Re-thinking the limits of the welfare state* (pp. 257–277). Lanham, MD: Rowman & Littlefield

Castells, M. (1996). *The rise of the network society*. Oxford, UK: Blackwell.

Diamond, L. J. (1999). *Developing democracy: Toward consolidation*. Baltimore: Johns Hopkins University Press.

Dicken, P. (2003). *Global shift: Reshaping the global economic map in the 21st century* (4th ed.). New York: Guilford Press.

Dollar, D., & Kraay, A. (2001). Trade, growth and poverty. *Finance and Development, 38*(3), 16–19.

Feenstra, R. C. (1998). Integration of trade and disintegration of production in the global economy. *Journal of Economic Perspectives, 12*(4), 31–50.

Ferguson, N. (2003) *Empire: The rise and demise of British world order and the lessons for global power*. New York: Basic Books.

Fraser, N. (1995). Politics, culture, and the public sphere: Towards a postmodern conception. In L. Nicholson & S. Seidman (Eds.), *Postmodernism: Beyond identity politics* (pp. 287–312). Cambridge, UK: Cambridge University Press.

Friedman, T. L. (2000) *The Lexus and the olive tree.* New York: Anchor Books.

Gallup, J. L., Sachs, J. D., & Mellinger, A. D. (1998) *Geography and economic development* (NBER Working Papers No. 6849). Cambridge, MA: National Bureau of Economic Research.

Gereffi, G., & Korzeniewicz, M. (Eds.). (1994). *Commodity chains and global capitalism.* Westport, CT: Praeger.

Giddens, A. (2000). *Runaway world: How globalization is reshaping our lives.* New York: Routledge.

Gilpin, R. (2000). *The challenge of global capitalism.* Princeton, NJ: Princeton University Press.

Habermas, J. (1989). *The structural transformation of the public sphere.* Cambridge, MA: MIT Press.

Hannerz, U. (1992). *Cultural complexity.* New York: Columbia University Press.

Harvey, D. (1989). *The condition of postmodernity.* Oxford, UK: Blackwell.

Hirst, P., & Thompson, G. (1996). *Globalization in question.* Cambridge, UK: Polity Press.

Holton, R. (2000). Globalization's cultural consequences. *Annals of the American Academy of Political and Social Science 570,* 140–152.

Huntington, S. P. (1996). *The clash of civilizations and the remaking of world order.* New York: Free Press.

Johnston. R. J. (1993). The rise and decline of the corporate-welfare state: A comparative analysis in global context. In P. J. Taylor (Ed.), *The political geography of the twentieth-century* (pp. 115–169). London: Belhaven Press.

Kearney, A. T. (2001). Measuring globalization. *Foreign Policy, 122,* 56–65.

Keohane, R., & Nye, J., Jr. (2000). *Power and interdependence* (3rd ed.). New York: Pearson Addison Wesley.

Leblang, D., & Bernhard, W. (2000). The politics of speculative attacks in industrial societies. *International Organization, 54,* 291–324.

Maskus, K. E. (2000). *Intellectual property rights in the global economy.* Washington, DC: Institute for International Economics.

Mittelman, J. H. (2000). *The globalization syndrome.* Princeton, NJ: Princeton University Press.

Ohmae, K. (1995). *The end of the nation-state: The rise of regional economies.* New York: Free Press.

O'Loughlin, J. (in press). Global democratization. In M. Low & C. Barnett (Eds.), *Spaces of democracy.* Thousand Oaks, CA: Sage.

Putnam, R. (1993). *Making democracy work: Civic tradition in modern Italy.* Princeton, NJ: Princeton University Press.

Ranis, G., Stewart, F., & Ramirez, A. (2000). Economic growth and human development. *World Development, 28,* 197–217.

Rodrik, D. (1997). *Has globalization gone too far?* Washington, DC: Institute for International Economics.

Said, E. W. (1978). *Orientalism.* New York: Vintage Books.

Steinmetz, G. (2003). The state of emergency and the revival of American imperialism: toward an authoritarian post-Fordism. *Public Culture, 15*, 323–345.

Stiglitz, J. (2002). *Globalization and its discontents*. New York: Norton.

Strange, S. (1996). *The retreat of the state: The diffusion of power in the world economy*. Cambridge, UK: Cambridge University Press.

Taylor, P. J. (1994). The state as container: Territoriality in the modern world-system. *Progress in Human Geography, 18*, 151–162.

Tomlinson, J. (1999). *Globalization and culture*. Chicago: University of Chicago Press.

Wade, R. (1990). *Governing the market*. Princeton, NJ: Princeton University Press.

Wallerstein, I. (2000). Globalization or the age of transition?: A long-term view of the trajectory of the world system. *International Sociology, 15*, 249–265.

World Health Organization. (2002). *The world health report, 2002: Reducing risks, promoting healthy life*. Geneva, Switzerland: Author.

Yang, G., & Maskus, K. E. (2001). Intellectual property rights, licensing and innovation in an endogenous product cycle model. *Journal of International Economics, 53*, 169–187.

Zakaria, F. (2003). *The future of freedom: Illiberal democracy at home and abroad*. New York: Norton.

Part II

PROCESSES OF GLOBALIZATION

2

Globalization
An Ascendant Paradigm?

JAMES H. MITTELMAN

This chapter explores the question, Does globalization constitute an ascendant paradigm in international studies? Put in perspective,
this question goes beyond our field's three "great debates" over ontology, methodology, and epistemology. Now another debate, which
focuses on globalization as a paradigmatic challenge, is heating up, kind-
ling theoretical controversies, and fusing the issues vetted in earlier
rounds. The first debate was waged between "realists" and "idealists";
the second, "traditionalists" and "scientists"; the third, "positivists"
versus "postpositivists," or "mainstreamers" versus "dissidents" (in the
terms of Lapid, 1989; Wendt, 1999, p. 39; and Puchala, 2000, p. 136).

Now it is time to move on. International studies is on the cusp of a
debate between those whom I will call *para-keepers*, observers who are
steadfast about maintaining the prevailing paradigms and deny that
globalization offers a fresh way of thinking about the world, and *para-
makers*, who bring into question what they regard as outmoded categories and claim to have shifted to an innovatory paradigm. This distinction is a heuristic for examining multiple theses. The ensuing heuristic
argument does not posit a relation between two positions such that one
is the absence of the other. Rather, between the keepers and the makers

there are many gradations and dynamic interactions. These are tendencies, not absolutes.

In our field, ascendancy to a new *paradigm* would mark something other, or more, than the fourth, a successor, in a sequential progression of debates. True, building new knowledge may be a cumulative process, but it is not necessarily a linear one, and only occasionally involves paradigmatic rupture. To be sure, paradigms do not shift frequently, quickly, or easily. International studies specialists are supposed to be the knowers, but, frankly speaking, often follow the doers in the sense that we trail events, even massive ones, as with our failure to anticipate the end of the cold war, and still resist changing the paradigms in which many of us are invested.

If a *paradigm* in Kuhn's (1970) sense is understood to mean a common framework, a shared worldview that helps to define problems, a set of tools and methods, and modes of resolving those research questions deemed askable, then globalization studies makes for strange bedfellows. Perhaps constituting an up-and-coming subfield within international studies, globalization research brings together different types of theorists, with varied commitments and stakes.

No one would deny that globalization is the subject of a rapidly proliferating theoretical literature. Notwithstanding its antecedents, primarily studies in classical social theory, world history, and the rise of capitalism, a scholarly literature on globalization per se did not really exist before the 1990s. To a certain extent, globalization is a synthetic concept—a reconstruction of precursor concepts through which analysts seek to comprehend reality. Clearly, this reconstruction is of recent vintage, and the literature and contestation over its importance go to the heart of our field: What is the fundamental problematic in international studies? Primarily peace and war? Mainly what states do to each other? Rather, states and markets, a binary in much teaching and research on international political economy (even though Strange, 1996, 1998, and others exploded it to include a wide variety of nonstate actors)? Or, if globalization really strikes a new chord, how does it change the problematic, and what are the implications for the ways in which disciplinary, cross-national, development, and area studies relate to our field?

For the purpose of addressing these issues, globalization may be best understood as a syndrome of political and material processes, including historical transformations in time and space and the social relations attendant to them. It is also about ways of thinking about the world. Globalization thus constitutes a set of ideas centered on heightened market integration, which, in its dominant form, neoliberalism, is embodied in a policy framework of deregulation, liberalization, and privatization.[1]

In this chapter, then, the objective is to pull together the divergent positions, which heretofore are fragmented and may be found in many scattered sources, on the question of the ascendancy of these ideas and the formation of a new paradigm. I want to frame and sharpen the debate, and seek to strike a balance, though not necessarily midway, along a continuum, marked on either end by the resolute arguments put forward by the para-keepers and the more grandiose claims of the para-makers. In so doing, I will stake out postulates in globalization studies, disclose its inadequacies, and note its explanatory potential.

AN EMERGING DEBATE

In the evolving debate, it is worth repeating, there are different shadings on a spectrum, not a sharp dichotomy, between para-keepers and para-makers. Indeed, in time, the para-makers may become wedded to keeping their paradigm and experience attacks by other para-makers. To discern their positions in respect to globalization, one can illustrate—not provide comprehensive coverage—by invoking explicit statements expressing the commitments of scholars and by examining logical extensions of their arguments, while taking care, of course, not to do injustice to them.

The keepers are naysayers who doubt or deny that globalization constitutes an ascendant paradigm. They include realists, interdependence theorists, social democrats, and some world-system theorists. Regarding globalization as "the fad of the 1990s" and as a model lacking evidence, Waltz declares that contrary to the claims of theorists whom he calls "globalizers"—what I take to be his shorthand for globalization researchers—"politics, as usual, prevails over economics" (1999, pp. 694, 696, 700). Clinging to the neorealist position that "national interests" continue to drive the "interstate system"—a position he advanced two decades earlier (Waltz, 1979)—he does not examine the foundational theoretical literature by "globalizers" who worry about the same problems that concern him. Surprisingly, Waltz fails to identify major pioneering theoreticians (such as Giddens, 1990; Harvey, 1990; and Robertson, 1992), opposing points of view, and different schools of globalization studies. Waltz would probably find much to respect and much to correct in this work. Recalling Keohane and Nye's 1977 book, *Power and Interdependence*, Waltz's point (1999) is that the globalizers' contention about interdependence reaching a new level is not unlike the earlier claim that simple interdependence had become complex interdependence—that is, countries are increasingly connected by varied social and political relationships and to a lesser degree by matters of security and force.

In fact, more recently, Keohane and Nye maintained that contemporary globalization is not entirely new: "Our characterization of interdependence more than 20 years ago now applies to globalization at the turn of the millennium" (2000, p. 104). Thus, like complex interdependence, the concept of globalization can be fruitfully extended to take into account networks that operate at "multicontinental distances," the greater density of these networks, and the increased number of actors participating in them (Keohane & Nye, 2000). In comparison to Waltz, Keohane and Nye reach beyond classic themes in politics to allow for more changes and build transnational issues into their framework. However, like Waltz, Keohane and Nye (1998) posit that the system of state sovereignty is resilient and remains the dominant structure in the world. Implicit in their formulation is that the state-centered paradigm is the best-suited approach to globalization; by inference, it can be adjusted so long as it is utilized in an additive manner—that is, incorporates more dimensions into the analysis.

Not only do interdependence theorists (and neoliberal institutionalists, in Keohane's [1984] sense of the term) seek to assimilate globalization to tried-and-tested approaches in international studies, but social democrats have similarly argued that there is nothing really "new" about globalization. By extension, from this standpoint, a new theoretical departure is unwarranted. In an influential study, Hirst and Thompson (1999), echoing Gordon (1988), claim that the world economy is not really global, but centered on the triad of Europe, Japan, and North America, as empirically demonstrated by flows of trade, foreign direct investment, and finance. They argue that the current level of internationalized activities is not unprecedented; the world economy is not as open and integrated as it was in the period from 1870 to 1914; and today the major powers continue to harmonize policy, as they did before. Leaving aside methodological questions about the adequacy of their empirical measures and the matter of alternative indicators (see Mittelman, 2000, pp. 19–24), clearly Hirst and Thompson adhere to a Weberian mode of analysis consisting of a dichotomy between two ideal types, an international economy based on exchange between separate national economies versus a full-fledged global economy. Taking issue with advocates of free markets whom, the authors believe, exaggerate globalizing tendencies and want to diminish regulation, Hirst and Thompson, on the contrary, favor more extensive political control of markets—that is, greater regulation.

World-system theorists also contend that there is nothing new about globalization, a phenomenon that can be traced back many centuries to the origins of capitalism (Wallerstein, 2000) or even earlier. From this perspective, it is argued that the basic conflict is between a capitalist world-system and a socialist world-system. However, as will be discussed, the point of much globalization research is to expand binaries such as the inter-national versus the global and capitalism versus socialism so as to allow for multiple *globalizing* processes, including those at the macroregional, subregional, and microregional levels, as well as in localities. If anything, globalization blurs many dualities—state and nonstate, legal and illegal, public and private, and so on—that are customary in our field.

Coming down differently on the debate over globalization qua paradigm are diverse theorists who resist pigeonholing into any particular tradition or traditions, yet all of whom support the proposition that globalization constitutes a distinctive theoretical innovation. However difficult to categorize collectively, this transatlantic group of authors signals the stirrings of a paradigmatic challenge to international studies. Emblematic of this position are the writings of four scholars with different commitments but whose position on new knowledge converges.

Representative of the innovatory stance is Cerny's assertion that theorists are seeking an alternative to realism and that "the chief contender for that honour has been the concept of globalization" (1996, p. 618). Similarly, Clark's *Globalization and International Relations Theory* makes the unequivocal argument that "globalization offers a framework within which political change can be understood" and that "if globalization does anything, it makes possible a theory of change" (1999, p. 174). Joining Cerny and Clark, Scholte holds that "contemporary globalisation gives ample cause for a paradigm shift" (1999, p. 9), or, in another formulation, "the case that globalism warrants a paradigm shift would seem to be incontrovertible" (1999, p. 22). Although Scholte does fill in some of the blanks, the question still is, What are the characteristics of this new paradigm?

While globalization theorists have tentatively, but not systematically, responded to this question (an issue to which I will return), there is also a more moderate intervention in the debate over globalization's status as a paradigm. Noting "parametric transformations" in world order, Rosenau (1997) clearly sides with those who affirm that globalization forms a new point of paradigmatic departure; however, he holds that his concept of globalization is "narrower in scope and more specific in content" than are many other concepts associated with changing global structures. According to Rosenau, globalization refers to "processes, to sequences that unfold either *in the mind* or in behavior" as people and organizations attempt to achieve their goals (1997, p. 80; emphasis added). In other words, globalization is not only an objective trend, but also constitutes, or is constituted by, subjective processes. It is a mental, or intersubjective, framework that is implicated both in the exercise of power and in scholarship that informs, or is critical of, public policy. Certainly, because of the need for greater theoretical, as well as empirical, precision, a qualified response to the question of the rise of a new paradigm is worthy of consideration. The route to this response will be a Kuhnian notion of what sparks paradigmatic transformations.

THE QUESTION OF NEW KNOWLEDGE

In his study of the history of the natural sciences, Kuhn (1970) famously argued that new paradigms appear through ruptures rather than through a linear accumulation of facts or hypotheses.[2] Normal science, he claimed, is a means of confirming the type of knowledge already established and legitimized by the paradigm in which it arises. According to Kuhn, normal science often suppresses innovations because they are subversive of a discipline's fundamental commitments:

> No part of the aim of normal science is to call forth new sorts of phenomena; indeed those that will not fit the box are often not seen at all. Nor do scientists normally aim to invent new theories, and they are often intolerant of those invented by others. Instead, normal-scientific research is directed to the articulation of those phenomena and theories that the paradigm already supplies. (1970, p. 24)

Or, to extrapolate, one might say that members of a shared knowledge community not only normalize certain types of questions, but also suppress the ability to raise other types of questions. Most important, Kuhn's insight is that only rarely do intellectuals refuse to accept the evasion of *anomalies*, observations at odds with expectations derived from prior theoretical understandings. A new paradigm emerges when the burden of anomalous phenomena grows too great and when there is incommensurability between competing paradigms to the extent that proponents of alternative frameworks cannot accept a common ground of assumptions.

Some observers dispute whether Kuhn's thesis, derived from the natural sciences, can be imported into the social sciences—and, I might add, into a field like international studies, which is far more heterogeneous than disciplines such as physics. My concern here, however, is not the epistemological debate over the disparate means of discovery in respective branches of knowledge (see Lakatos, 1970; Ball, 1976; and Barnes, 1982). Rather, my contention—that globalization is not only about "real" phenomena, but also a way of interpreting the world—is more pragmatic.

To be sure, a Kuhnian perspective on the generation of knowledge is vulnerable insofar as it is limited to social and psychological conditions *within* the scientific community and does not give sufficient credence to socially constructed knowledge *outside* this community. The factors internal to the social sciences cannot be fully explained without reference to the external elements. There is nothing, however, to prevent joining Kuhn's insight about theoretical innovation with a broader analysis of social conditions. Moreover, unless one believes that international studies is rapidly approaching a Kuhnian crisis, that is, the overthrow of a reigning paradigm or paradigms—and I do not—then it is important to grasp the dynamic interface between established knowledge sets, including the structures (e.g., curricula, professional journals, funding agencies, etc.) that maintain and undermine them and a potentially new paradigm. It would appear that even without a paradigm crisis, an ascendant paradigm could emerge.

For Kuhn, the transition to a new paradigm is all or nothing: "Like the gestalt switch, it must occur all at once (though not necessarily in an instant) or not at all" (Kuhn, 1970, p. 150; also pertinent are the nuances in his subsequent work, 1977a, 1977b). In explaining transformations in this manner, Kuhn falls short insofar as he underestimates the tenacity of forerunner paradigms and their ability to modify themselves. By all indications, in the social sciences, they fight back, usually with gusto. Nevertheless, by identifying the propellant of a new paradigm as the refusal to accept the evasion of anomalies in conjunction with the quest for an alternative, Kuhn has contributed powerfully to understanding theoretical innovation.

In this vein, it is well to recall Weber's (1949) " 'Objectivity' in Social Science and Social Policy." Like Kuhn, Weber indicated that the prevailing intellectual apparatus is in constant tension with new knowledge. According to Weber, this conflict is a propellant for creativity and discovery: concepts are and should be subject to change. However, there should also be a certain staying power in the intellectual apparatus that enables one to ferret out what is worth knowing. In other words, there is nothing worse than the fads and fashions that come in and go out of vogue. In the end, Weber called for a midcourse between unyielding old concepts and unceasing shifts in paradigms.

Following Kuhn and Weber in the chase for paradigmatic advance, what are the anomalies in our field, and is globalization a viable contender for fixing these imperfections?

DISCOMFORT WITH INTERNATIONAL STUDIES

A discipline without complaints would be a non sequitur. After all, scholars are trained in the art of debate; the skills of nuance are our stock-in-trade. That said, it is important to consider the specific anomalies within international studies. Although some of these anomalies are perennial, it is no wonder that others have recently appeared, given monumental changes after the cold war, and with the distinctive mix of global integration and disintegration at the dawn of a new millennium. While others could be cited, five anomalies seem most important, but can be considered only succinctly here.[3]

First, the term *international studies* suggests a focus on relations *between* nations. But this is not so. The discipline has primarily concerned relations between states, the nation being only one of many principles of social organization (Shaw, 1994, p. 25; also see Shaw, 1999). Closely related, observers (e.g., Rosenau, 1997; Baker, 2000, p. 366) have long argued that the conventional distinction between separate national and international spheres of activity is misleading. Nowadays, it is increasingly difficult to maintain the lines of demarcation between the domestic and the foreign realms, or between comparative politics and international politics. Globalization means that the distinction between them is hard to enforce. Increasingly evident are myriad forms of interpenetration between the global and the national—global economic actors even exist within the state, as with global crime groups in Russia or the International Monetary Fund/World Bank's structural adjustment programs in developing countries.

Thus, a third discontent is opposition to the persistence of state-centrism. From this angle, the case for an ontological shift springs from the anomaly between the objects of study seen through a realist or neorealist lens and globalists' vision of a polycentric, or multilevel, world order. New ontological priorities—an issue to which I will return—would consist of a series of linked processes. Toward this end, globalization researchers are attempting to design a framework for interrelating economics, politics, culture, and society in a seamless web. Hence, in large measure as a response to globalization, some scholars have shifted their attention to global governance in an effort to incorporate a broader ontology of structures and agents. The state is treated as one among several actors. It is not that state sovereignty is losing meaning, but that the multilevel environment in which it operates, and hence the meaning of the concept, is changing.

Methodologically, the field of *international* studies is based on the premise of territoriality, reflected in central concepts such as state-centered nationalism, state borders, and state sovereignty. Yet, with the development of new technologies, especially in communications and transportation, the advent of a "network society" (Castells, 1996), and the emergence of a "nonterritorial region" (Ruggie, 1993), there has been a marked shift toward a more deterritorialized world. Hence,

Scholte (1999, 2000) has challenged "methodological territorialism" (1999, p. 17), the ingrained practice of formulating questions, gathering data, and arriving at conclusions all through the prism of a territorial framework. Without swinging to the opposite extreme of adopting a "globalist methodology" by totally rejecting the importance of the principle of territoriality, Scholte calls for a "full-scale methodological reorientation," and concludes "That globalisation warrants a paradigm shift would seem to be incontrovertible" (1999, pp. 21–22).

Finally, there is the postmodernist complaint, which, arguably, has not really registered in our field.[4] As Said (1979) contends in regard to orientalism, it is hard to erase certain representations of reality, for, in Foucaultian terms, they take on the aura of authoritative expressions and are implicated in the exercise of power. Knowledge sets may thus operate as closed systems—what Caton (1999, p. 8) terms "endless cycles of self-referring statements"—thwarting counterrepresentations that might have the power to challenge normal knowledge. As scholars in international studies, perhaps we should reflect on this allegation about collectively self-referential work, for we spend an enormous amount of time engaging in intramural debates over concepts, often without paying sufficient attention to the phenomena themselves. Still, it would be wrong to gloss over Said's insight that representations made manifest as knowledge are tied to the establishment, maintenance, and exercise of power. In international studies, probing Said's point about reflexivity involves shifting explanatory levels above and below the state—a characteristic of globalization research.

CHARACTERISTICS OF GLOBALIZATION STUDIES

Globalization theorists, of course, are not univocal. Inasmuch as their writings abound, there are different interpretations and considerable contestation. As Puchala aptly put it, "Conventional theories all have a table of outcomes that inventory what needs to be explained" (personal communication, January 30, 2001). For example, the realist table of outcomes is chiefly wars, alliances, balances of power, and arms races. For liberals, the outcomes are regimes, integration, cooperation, and hegemons (Puchala, 2001). By contrast, the problematic that globalization theorists seek to explain, while dynamic and open-ended, not invariant, may be gleaned from an emerging series of core linked propositions. I will highlight six of them.

1. Many contemporary problems cannot be explained as interactions among nation states, that is, as international studies, but must be

construed as global problems. Although this claim is not unique to globalization studies, at issue is a series of problems—for example, the rise of
organized crime, global warming, and the spread of infectious diseases—
partly within and partly across borders, partially addressed by states and
partially beyond their regulatory framework.

2. Globalization constitutes a structural transformation in world
order. As such, it is about not only the here-and-now, but also warrants
a long perspective of time and revives the study of space. A preoccupa-
tion with what Braudel (1980, pp. 3, 27), and François Simiand before
him, called "the history of events"—the immediate moment—focuses attention on a frame that differs from the *longue durée*, an observation
point that some researchers find advantageous for viewing the spatial reorganization of the global economy.

3. As a transformation, globalization involves a series of continui-
ties and discontinuities with the past. In other words, the globalization
tendency is by no means a total break—as noted, there is considerable
disagreement about how much is new—but the contemporary period is
punctuated by large-scale acceleration in globalizing processes such as
the integration of financial markets, technological development, and
intercultural contact.

4. New ontological priorities are warranted because of the emergence of a dialectic of suprastate and substate forces, pressures from
above and from below. The advent of an ontology of globalization is
fluid, by no means fixed. It includes the global economy as an actor in its
own right (as embodied, e.g., in transnational corporations), states and
interstate organizations, regionalist processes (at the macro-, sub-, and
microlevels), world cities, and civil society, and sometimes is made manifest as social movements.

5. Given shifting parameters, the state, in turn, seeks to adjust to
evolving global structures. States, however, are in varied positions vis-à-
vis globalizing structures, and reinvent themselves differently, the gamut
of policies running from a full embrace, as with New Zealand's extreme
neoliberal policies from 1984 to 1999, to resistance, illustrated by Malaysia's capital controls in 1998.

6. Underpinning such differences is a set of new, or deeper, tensions
in world order, especially the disjuncture between the principle of
territoriality, fundamental to the concept of state sovereignty, and the
patent trend toward deterritorialization, especially, but not only, apparent in regard to transborder economic flows. The horizontal connections
forged in the world economy and the vertical dimensions of state politics
are two dissimilar vectors of social organization, with the latter seeking
to accommodate the changing global matrix.

However schematically presented, the aforementioned, interrelated propositions put into question some of international studies's ingrained ways of conceptualizing the world. At present, although the attempts at reconceptualization are in a preliminary stage of formulation, it is worth identifying the traps and confusions.

DISCOMFORT WITH PARADIGMATIC PRETENSION

Barring caricatures of the concept *and* phenomena of globalization—for example, it is totalizing, inevitable, and homogenizing, rather than, as many scholars maintain, partial, open-ended, and hybrid—surely there are grounds for discontent. For one thing, globalization may be seen as a promiscuous concept, variously referring to a historical scenario, interconnections, movements of capital, new technologies and information, an ideology of competitiveness, and a political response to the spread and deepening of the market. Hence, the complaint lodged earlier in this chapter that observers (e.g., Kearney, 2001) are crying out and striving for more analytical precision.

Moreover, globalization is sometimes deemed overdetermined— too abstract, too structural, and insufficiently attentive to agency. From this perspective, it is thought that the logic is mechanically specified or misspecified in that it is too reductive. For some, especially scholars carrying out contextualized, fine-grained research on particular issues and distinct areas, globalization is regarded as too blunt a tool. After all, what does it leave out? What is not globalization? In response, it may be argued that globalization is mediated by other processes and actors, including the state. Furthermore, globalization has a direct or indirect impact on various levels of social organization, and becomes inserted into the local, thus complicating the distinction between the global and the local.

Another problem, then, is that the globalization literature has spawned its own binary oppositions. On the one hand, as indicated, the phenomena of globalization blur the dichotomous distinctions to which international studies has grown accustomed. For example, civil society now penetrates the state (as with members of environmental movements assuming important portfolios in government in the Philippines; and in several African countries, state substitution is abundantly evident—some so-called nongovernmental organizations are sustained by state funding or, arguably, their agendas are driven by the state or interstate organizations). On the other hand, globalization research itself presents new binary choices—"globalization from above" and "globalization from below," top-down and bottom-up globalization, and so on—that certainly have heuristic value but must be exploded in order to capture the range of empirical phenomena.

HOW FAR HAVE WE COME?

It would be remiss not to join a discussion of the drawbacks to globalization as an avenue of inquiry with its real gains, even if the nature of a new paradigm is tentative and being contested.

In the main, globalization studies emphasizes the historicity of all social phenomena. There is no escaping historiography: What are the driving forces behind globalization, and when did it originate? With the beginnings of intercultural contact, the dawn of capitalism in Western Europe in the long 16th century, or in a distinct conjuncture after World War II? Research has thus opened new questions for investigation and debate. And even if one returns to old issues, such as theories of the state, there are opposing views and vexing questions, especially in the face of public representations, such as Margaret Thatcher's attack on the "nanny state." Should the state be construed as in retreat (Strange, 1996), as an agent of globalization (Cox, 1987), or, in an even more activist role, as the author of globalization (Panitch, 1996; or from another perspective, Weiss, 1998)? Taken together, the writings on these issues combat the fragmentation of knowledge. Not surprisingly, given the themes that globalization embraces—technology, ecology, films, health, fast-food and other consumer goods, and so on—it is transdisciplinary, involving not only the social sciences, but also the natural sciences, the humanities, and professional fields such as architecture, law, and medicine.

Arguably, within the social sciences, economic and political geographers (including Dicken, 2003; Harvey, 1999; Knox & Agnew, 1998; Olds, 2001; Taylor, 1993; Taylor, Johnson, & Watts, 1995; and Thrift, 1996) have carried out some of the most sophisticated research on globalization. Even though the importance of spatial concerns is increasingly apparent, many international studies specialists have not noticed the work of economic and political geographers.

For the purposes of teaching globalization, one way to draw students into a subject that, after all, involves thinking about big, abstract structures, is to focus on spatial issues as they relate to the changes in one's own locale. Reading a collection of essays consisting of anthropological fieldwork carried out at McDonald's restaurants in different Asian countries (Watson, 1997), and then comparing the findings in the literature to their own fieldwork, including interviewing employees and customers at a nearby McDonald's, my students are asked to analyze the cultural political economy of globalization: a production system, the composition of the labor force (largely immigrants and members of minority groups in our locale), social technologies, and the representations conveyed by symbols. The students pursue the question of meanings—the intersubjective dimensions of globalization—in the writings of architects, for example, on shopping malls and theme parks (Sorkin, 1992) and by visits to local sites.

Time permitting, consideration is also given to the legal and medical spheres. Cybergangs and some novel types of crime do not neatly fit into the jurisdiction of national or international law (see, e.g., Sassen, 1998). The field of public health has called attention to the nexus of social *and* medical problems, especially with the spread of AIDS. The tangible consequences of a changing global division of labor and power include new flows and directions of migration, the separation of families, a generation of orphans, and the introduction of the HIV virus into rural areas by returning emigrants. As these topics suggest, globalization studies identifies silences and establishes new intellectual space—certainly one criterion by which to gauge an ascendant paradigm.

PUSHING THE AGENDA

Notwithstanding important innovations, as a paradigm, globalization is more of a potential than a refined framework, worldview, kit of tools and methods, and mode of resolving questions. Where, then, to go from here? Although these are not the only issues, the following challenges stand out as central to developing globalization studies:

1. Just as with capitalism, which has identifiable variants, there is no single, unified form of globalization. Researchers have not yet really mapped the different forms of globalization, which in the literature is sometimes preceded by adjectival designations such as "neoliberal," "disembedded," "centralizing," "Islamic," "inner and outer," or "democratic." The adjectival labels are but hints at the need for systematic study of the varieties. Or should the object of study be globalization*s*?

2. Closely related is the problem of how to depict the genres of globalization research. What are the leading schools of thought? How to classify them so as to organize this massive literature and advance investigation? To catalog globalization studies according to national traditions of scholarship, by disciplinary perspective, or on single issues risks mistaking the parts for the whole. Avoiding this trap, Guillén (2001) decongests the burgeoning globalization research by organizing it into

key debates: Is globalization really happening, does it produce convergence, does it undermine the authority of nation-states, is globality different from modernity, and is a global culture in the making? In another
stocktaking, Held, McGrew, Goldblatt, and Perraton (1999) sort the
field into *hyperglobalizers* who believe that the growth of world markets
diminishes the role of states, *skeptics* who maintain that international interactions are not novel and that states have the power to regulate international economic flows, and *transformationalists* who claim that new
patterns and an unprecedented configuration of global power relations
have emerged. But there are other debates, major differences among policy research (Rodrik, 1997), structural approaches (Falk, 1999), and
critical/poststructural accounts (Hardt & Negri, 2000).

3. What are the implications of globalization for disciplinary and
cross-national studies? How should these domains of knowledge respond to the globalization challenge? It would seem that in light of the
distinctive combinations of evolving global structures and local conditions in various regions, globalization enhances, not reduces, the importance of the comparative method. However, there is the matter of
exploring disciplinary and comparative themes within changing parameters and examining the interactions between these parameters and the localities.

4. Similarly, what does globalization mean for development and
area studies? McMichael (2000, p. 149) holds that "the globalization
project succeeds the development project." Surely development theory
emerged in response to a particular historical moment: the inception of
the cold war, which, if anything, was an ordering principle in world affairs. After the sudden demise of this structure, development studies
reached a conceptual cul-de-sac. Put more delicately, it may be worth re-
visiting development studies' basic tenets, especially apropos the dynamics of economic growth and the mechanisms of political power in the
poorest countries, which have experienced a fundamental erosion of the
extent of control that they had maintained—however little to begin
with. This loss has been accompanied by changing priorities and reorga-
nizations within funding agencies, a crucial consideration in terms of
support for training the next generation of scholars, particularly apparent with regard to fieldwork for dissertations. Although some para-
keeper area specialists have dug in their heels and have fought to protect
normal knowledge in their domain, the task is to reinvent and thereby
strengthen area studies.

5. Insufficient scholarly attention has centered on the ethics of
globalization. The telling question is, What and whose values are inscribed in globalization? In light of the unevenness of globalization, with
large zones of marginalization (not only in a spatial sense, but also in

terms of race, ethnicity, gender, and who is or is not networked), there is another searching question, Is globalization ethically sustainable? What is the relationship between spirituality and globalization, an issue posed by different religious movements? Which contemporary Weberian will step forward to write *The Neoliberal Ethic and the Spirit of Globalization*?

6. Emanating mostly from the West, globalization studies is not really "global." In terms of participating researchers and the focus of inquiry, there is a need for decentering. Moreover, the literature on globalization unavailable in the English language (e.g., Ferrer, 1997; Gómez, 2000; Kaneko, 1999; Norani & Mandal, 2000; Podestà, Gómez Galán, Jácome, & Grandi, 2000) is rarely taken into account in the English-speaking world. Still, only limited work has thus far emerged in the developing world, including studies undertaken by the Council for the Development of Economic and Social Research in Africa (1998), the National University of Singapore (Olds, Dicken, Kelly, Kong, & Yeung, 1999), the Latin American Social Sciences Council (Seoane & Taddei, 2001), and the Institute of Malaysian and International Studies at the National University of Malaysia (Mittelman & Norani, 2001).

7. Apart from the development of individual courses, there is a lack of systematic thought about the programmatic implications of globalization for the academy. Does global restructuring warrant academic restructuring in the ways in which knowledge is organized for students? If a new paradigm is emerging, then what does this mean in terms of pedagogy and curriculum? Will universities—and their international studies specialists—be in the forefront of or trail behind changes in world order? Will they really open to the innovation of globalization studies?

To sum up, it is worth recalling that on more than one occasion Susan Strange held that international studies is like an open range, home to many different types of research. Today, there is diversity, but surely one should not overlook the fences that hold back the strays. Mavericks who work in non-Western discourses, economic and political geographers, postmodernists and poststructuralists, not to mention humanists (whose contributions are emphasized by Alker, 1996; Puchala, 2000; and others), have faced real barriers.

It is in this context that globalization studies has emerged as a means to explain the intricacy and variability of the ways in which the world is restructuring and, by extension, to assess reflexively the categories used by social scientists to analyze these phenomena. The parakeepers, to varying degrees, are reluctant to embrace globalization as a knowledge set because some of its core propositions challenge predominant ontological, methodological, and epistemological commitments—

what Kuhn (1970) referred to as "normal science." Again, not to dichotomize positions, but to look to the other end of the spectrum, para-makers advance a strong thesis about the extent to which a new paradigm is gaining ascendancy. The debate is fruitful in that it engages in theoretical stocktaking, locates important problem areas, and points to possible avenues of inquiry. It also helps to delimit space for investigation and to identify venues of intellectual activity. But, in the near term, there is no looming Kuhnian crisis in the sense of an impending overthrow that would quickly sweep away reigning paradigms. Given that systematic research on globalization is only slightly more than a decade in the making, it is more likely that international studies has entered an interregnum between the old and the new.

Although globalization studies entails a putting together of bold efforts to theorize structural change, it would be wrong to either underestimate or to exaggerate its achievements. Judging the arguments in the debate, on balance, a modest thesis is in order. The efforts to theorize globalization have produced a patchwork, an intellectual move rather than a movement, and more of a potential than worked-out alternatives to accepted ways of thinking in international studies. In sum, this fledgling may be regarded as a protoparadigm.

ACKNOWLEDGMENTS

This chapter appeared in *International Studies Perspectives* (2002). It is reproduced with the permission of the journal. I owe a debt of gratitude to Donald J. Puchala, three anonymous reviewers, and the ISP editors for critical comments on drafts of this article. Thanks too to Patrick Jackson, who generously shared materials and insights—too numerous to pick up on entirely here.

NOTES

1. The literature (e.g., Beck, 2000; Giddens, 2000) suggests a number of other ways to come to grips with what constitutes globalization.
2. This section draws from, and builds on, Mittelman (1997).
3. The question of the meaning of power and counterpower under globalization is a topic too broad to examine here. I am exploring this theme elsewhere.
4. I have the strong impression, but cannot "prove," that international studies scholars, with notable exceptions (e.g., Der Derian, 1994; Peterson, 1992; Sylvester, 1994; Walker, 1993), have been more insular in the face of incursions from postmodernism and poststructuralism than have those in the other social sciences.

REFERENCES

Alker, H. R. (1996). *Rediscoveries and reformulations: Humanistic methods for international studies.* Cambridge, UK, and New York: Cambridge University Press.

Baker, A. (2000). Globalization and the British "residual state." In R. Stubbs & G. R. D. Underhill (Eds.), *Political economy and the changing global order* (2nd ed., pp. 362–372). Don Mills, Ontario, Canada: Oxford University Press.

Ball, T. (1976). From paradigms to research programs: Toward a post-Kuhnian political science. *American Journal of Political Science, 20,* 151–177.

Barnes, B. (1982). *T. S. Kuhn and social science.* New York: Columbia University Press.

Beck, U. (2000). *What is globalization?* (P. Camiller, Trans.). Cambridge, UK: Polity Press.

Braudel, F. (1989). *On history* (S. Matthews, Trans.). Chicago, IL: University of Chicago Press.

Castells, M. (1996). *The rise of the network society.* Oxford, UK: Blackwell.

Caton, S. C. (1999). *Lawrence of Arabia: A film's anthropology.* Berkeley and Los Angeles: University of California Press.

Cerny, P. G. (1996). Globalization and other stories: The search for a new paradigm for international relations. *International Journal, 50,* 617–637.

Clark, I. (1999). *Globalization and international relations theory.* Oxford, UK, and New York: Oxford University Press.

Council for the Development of Economic and Social Research in Africa. (1998). Social sciences and globalisation in Africa. *CODESRIA Bulletin, 2,* 3–6.

Cox, R. W. (1987). *Production, power and world order: Social forces in the making of history.* New York: Columbia University Press.

Der Derian, J. (Ed.). (1994). *International theory: Critical investigations.* New York: New York University Press.

Dicken, P. (2003). *Global shift: Reshaping the global economic map in the 21st century* (4th ed.). New York and London: Guilford Press.

Falk, R. (1999). *Predatory globalization: A critique.* Cambridge, UK: Polity Press.

Ferrer, A. (1997). *Hechos y ficciones de la globalización* [Facts and fictions of globalization]. Buenos Aires, Argentina: Fondo de Cultura Economica [Collection of economic writings].

Giddens, A. (1990). *The consequences of modernity.* Cambridge, UK: Polity Press.

Giddens, A. (2000). *Runaway world: How globalization is reshaping our lives.* New York: Routledge.

Gómez, J. M. (2000). *Política e democracia em tempos de globalização.* Petrópolis, RJ, Brazil: Editora Vozes.

Gordon, D. (1988). The global economy: New edifice or crumbling foundations? *New Left Review, 68,* 24–64.

Guillén, M. F. (2001). Is globalization civilizing, destructive or feeble? A critique of five key debates in the social-science literature. *Annual Review of Sociology, 27,* 235–260.

Hardt, M., & Negri, A. (2000). *Empire.* Cambridge, MA: Harvard University Press.

Harvey, D. (1990). *The condition of postmodernity.* Oxford, UK: Blackwell.

Harvey, D. (1999). *Limits to capital.* London: Verso Press.

Held, D., McGrew, A. G., Goldblatt, D., & Perraton, J. (1999). *Global transformations: Politics, economics and culture.* Cambridge, UK: Polity Press.

Hirst, P., & Thompson, G. (1999). *Globalization in question: The international economy and the possibilities of governance* (2nd ed.). Cambridge, UK: Polity Press.

Kaneko, M. (1999). *Han Gurouburizumu: Shijou Kaiku no Senryakuteki Shikou* [Antiglobalism: Strategic thinking on market reforms]. Tokyo: Iwanami Shoten.

Kearney, A. T. (2001). Measuring globalization. *Foreign Policy, 122,* 56–65.

Keohane, R. O. (1984). *After hegemony: Cooperation and discord in the world political economy.* Princeton, NJ: Princeton University Press.

Keohane, R. O., & Nye, J. S., Jr. (1977). *Power and interdependence: World politics in transition.* Boston and Toronto: Little, Brown.

Keohane, R. O., & Nye, J. S., Jr. (1998). *Power and interdependence in the information age.* New York: Council on Foreign Relations. Available online at *web.lexis-nexis.com/universe/printdoc/.*

Keohane, R. O., & Nye, J. S., Jr. (2000). Globalization: What's new? What's not? (and so what?). *Foreign Policy, 118,* 104–120.

Knox, P., & Agnew, J. (1998). *The geography of the world economy* (3rd ed.). London: Edward Arnold.

Kuhn, T. S. (1970). *The structure of scientific revolutions* (2nd ed.). Chicago: University of Chicago Press.

Kuhn, T. S. (1977a). *The essential tension: Selected studies in scientific tradition and change.* Chicago: University of Chicago Press.

Kuhn, T. S. (1977b). Second thoughts on paradigms. In F. Suppe (Ed.), *The structure of scientific theories* (2nd ed., pp. 459–482). Urbana: University of Illinois Press.

Lakatos, I. (1970). Falsification and the methodology of scientific research programmes. In I. Lakatos & A. Musgrave (Eds.), *Criticism and the growth of knowledge* (pp. 91–196). Cambridge, UK: Cambridge University Press.

Lapid, Y. (1989). The third debate: On the prospects of international theory in a post-positivist era. *International Studies Quarterly, 33,* 235–254.

McMichael, P. (2000). *Development and social change: A global perspective* (2nd ed.). Thousand Oaks, CA: Pine Forge Press.

Mittelman, J. H. (1997). Rethinking innovation in international studies: Global transformation at the turn of the millennium. In S. Gill & J. H. Mittelman (Eds.), *Innovation and transformation in international studies* (pp. 248–263). Cambridge, UK: Cambridge University Press.

Mittelman, J. H. (2000). *The globalization syndrome: Transformation and resistance.* Princeton, NJ: Princeton University Press.

Mittelman, J. H., & Norani Othman. (Eds.). (2001). *Capturing globalization.* London and New York: Routledge.

Norani Othman & Mandal, S. (Eds.). (2000). *Malaysia Menangani Globalisasi:*

Peserata atau Mangasi? [Malaysia responding to globalization: Participants or victims?]. Bangi, Malaysia: Penerbit Universiti Kebangsaan Malaysia [National University of Malaysia Press].

Olds, K. (2001). *Globalization and urban change: Capital, culture, and Pacific rim mega-projects.* Oxford, UK, and New York: Oxford University Press.

Olds, K., Dicken, P., Kelly, P., Kong, L., & Yeung, H. W. (Eds.). (1999). *Globalisation and the Asia-Pacific: Contested territories.* London: Routledge.

Panitch, L. (1996). Rethinking the role of the state. In J. H. Mittelman (Ed.), *Globalization: Critical reflections* (pp. 83–113). Boulder, CO: Lynne Rienner.

Peterson, S. (Ed.). (1992). *Gendered states: Feminist (Re)Visions of international relations.* Boulder, CO: Lynne Rienner.

Podestà, B., Gómez Galán, M., Jácome, F., & Grandi, J. (Eds.). (2000). *Ciudadanía y mundialización regional: La sociedad civil ante la integración regional.* Madrid: CIDEAL.

Puchala, D. J. (2000). Marking a Weberian moment: Our discipline looks ahead. *International Studies Perspectives, 1,* 133–144.

Robertson, R. (1992). *Globalization: Social theory and global culture.* Newbury Park, CA: Sage.

Rodrik, D. (1997). *Has globalization gone too far?* Washington, DC: Institute for International Economics.

Rosenau, J. N. (1997). *Along the domestic-foreign frontier: Exploring governance in a turbulent world.* Cambridge, UK: Cambridge University Press.

Ruggie, J. G. (1993). Territoriality and beyond: Problematizing modernity in international relations. *International Organization, 46,* 561–598.

Said, E. W. (1979). *Orientalism.* New York: Vintage Books.

Sassen, S. (1998). *Globalization and its discontents.* New York: New Press.

Scholte, J. A. (1999). Globalisation: Prospects for a paradigm shift. In M. Shaw (Ed.), *Politics and globalisation: Knowledge, ethics and agency* (pp. 9–22). London and New York: Routledge.

Scholte, J. A. (2000). Globalization: A Critical Introduction. London: Macmillan.

Seoane, J., & Taddei, E. (Eds.). (2001). Resistencias mundiales [De Seattle a Porto Alegre]. Buenos Aires, Argentina: Consejo Latinamericano de Ciencias Sociales.

Shaw, M. (1994). *Global society and internation relations: Sociological concepts and political perspectives.* Cambridge, UK: Polity Press.

Shaw, M. (Ed.). (1999). *Politics and globalisation: Knowledge, ethics and agency.* London and New York: Routledge.

Sorkin, M., (Ed.). (1992). *Variations on a theme park: The new American city and the end of public space.* New York: Hill & Wang.

Strange, S. (1996). *The retreat of the state: The diffusion of power in the world economy.* Cambridge, UK: Cambridge University Press.

Strange, S. (1998). *Mad money: When markets outgrow governments.* Ann Arbor: University of Michigan Press.

Sylvester, C. (1994). *Feminist theory and international relations theory in a postmodern era.* Cambridge, UK: Cambridge University Press.

Taylor, P. J. (1993). *Political geography: World-economy, nation-state, and locality* (3rd ed.). New York: Wiley.

Taylor, P. J., Johnson, R. J., & Watts, M. J. (1995). *Geographies of global change: Remapping the world in the late twentieth century.* Oxford, UK: Blackwell.

Thrift, N. (1996). *Spatial formations.* London: Sage.

Walker, R. B. J. (1993). *Inside/outside: International relations as political theory.* Cambridge, UK: Cambridge University Press.

Wallerstein, I. (2000). Globalization or the age of transition?: A long-term view of the trajectory of the world system. *International Sociology, 15,* 249–265.

Waltz, K. N. (1979). *Theory of international politics.* Reading, MA: Addison-Wesley.

Waltz, K. N. (1999). Globalization and governance. *PS: Political Science and Politics, 23,* 693–700.

Watson, J. L. (Ed.). (1997). *Golden arches east: McDonald's in East Asia.* Stanford, CA: Stanford University Press.

Weber, M. (1949). "Objectivity" in social science and social policy. In E. Shils & H. A. Finch (Eds., Trans.), *The methodology of the social sciences* (pp. 49–112). New York: Free Press.

Weiss, L. (1998). *State capacity: Governing the economy in a global era.* Cambridge, UK: Polity Press.

Wendt, A. (1999). *Social theory of international politics.* Cambridge, UK: Cambridge University Press.

3

Globalization and the Hyperdifferentiation of Space in the Less Developed World

RICHARD GRANT
JAN NIJMAN

In this chapter we are concerned with the spatial outcomes of globalization. In globalization debates the focus tends to be on economic or social trends and disparities, and "space" is rarely given explicit consideration. In our view, geography constitutes a vital and dynamic dimension in social processes and particularly in the processes of globalization. Geography is at once driving social change and reflective of it. We shall argue that explicit attention to space reveals trends and patterns that otherwise remain invisible.

Most observers believe that globalization tends to result not in homogenization across peoples and spaces, but in the increasing complexity of geographical patterns at all geographical scales. They suggest that the present phase of globalization entails an unprecedented restructuring of the world economy involving simultaneous integration and fragmentation, winners and losers, and increased volatility (Giddens, 2000; Rodrik, 1997; Smith, 1984, 1997; Grant & Short, 2002).

Current research has only just begun to reveal these complexities, and we are still far from reaching a consensus on the emerging patterns

(Sassen, 1991; Taylor, 1999; Marcuse & van Kempen, 2000; Wai-chung Yeung, 2002). In the words of Neil Smith, "The solidity of the geography of twentieth century capitalism at various scales has melted; habitual spatial assumptions about the world have evaporated. . . . Putting the jigsaw puzzle back together—in practice as well as theory—is a highly contested affair" (Smith, 1997, p. 171).

We think that these unresolved questions about the spatial outcomes of globalization explain, in part, the inconclusiveness of more general debates about the positives and negatives of globalization. Such inconclusiveness is especially salient in debates about development in less prosperous regions of the world. For example, in a recent International Monetary Fund (IMF) report, on one and the same page it is stated that "the distribution of income among countries has become more unequal" and that "the gaps may have narrowed." It is also stated that, despite "unparalleled growth," "far too many people are losing ground" (International Monetary Fund, 2002, p. 3).

In this chapter, we focus on the combined effects of liberalization and globalization on the economic geographies of Ghana and India. These are, obviously, very particular regions in the global economy and they are quite different in terms of size and cultural and regional settings. But for our purpose they share a number of important characteristics. Both countries are traditionally considered "underdeveloped," and both have undergone substantial policies of economic liberalization since the mid-1980s. The timing of these policy changes was essential since it coincided with an unprecedented mobility of global capital, ready to play its part in these newly available regional markets. Hence, Ghana and India represent comparable "laboratories" in which to examine the spatial impact of globalization.

One of the major difficulties of doing research on the spatial outcomes of globalization lies in the dominance of the state and the severe shortage of available data pertaining to alternative geographical units and scales. Indeed, one may well argue that the prevalence of the state in empirical research on economic globalization actually has the effect of hiding major transformations in the global space-economy at a variety of scales.

A good example of this can be found in a recent article on "globalization and its challenges" by the economist Stanley Fischer (2003; another example is Bhalla, 2002). Fischer examines whether the prosperity gap between rich and poor countries has increased during recent trends of globalization, a period he demarcates from 1980 to 2000. When using countries as a nondifferentiated unit of analysis (i.e., treating all countries the same), the data show a positive correlation between a country's gross national product (GNP/PC) in 1980 and its average economic growth rate from 1980 to 2000. This indicates a widening of the gap. But, says Fischer, if one differentiates countries on the basis of population size and weighs them accordingly, a different picture emerges. Mainly due to the behavior of China and India, the world's two most populous states, the data suggest that for most people in the world the correlation between a country's GNP/PC in 1980 and its average economic growth rate from 1980 to 2000 has been negative. This would suggest that poorer populations have grown faster, economically, than richer populations (both China and India had, of course low GNP/PC values in 1980 and have had relatively high growth rates since then).

While some may consider Fischer's approach innovative and telling (see *The Economist*, 2003), it illustrates a major problem and does not really resolve anything. The problem is that countries remain the spatial unit of analysis. In this case, this is particularly troublesome because of the enormous size of China and India. The suggestion that the Chinese and Indian populations as a whole have experienced economic growth during this time must sound absurd to anyone who is even slightly familiar with these countries' geographies or social structures. What is needed is a disaggregation of national data. One way of doing this is through geography.

In our research on Ghana and India, we have tried to mine the available and appropriate data pertaining to various geographical scales. The comparability of data over time and across spatial units is critical, but this also means that data availability to some degree has to steer the research and that it is difficult to go beyond more or less established administrative units (states, districts, etc.). Nevertheless, the findings are quite significant and point to what we believe is a hyperdifferentiation of space at various geographical scales inside Ghana and India. We postulate that this hyperdifferentiation of space occurs most visibly in peripheral zones of the world economy that have recently been drawn (more intensively) into the capitalist global system.

There are five parts to the rest of this chapter. First, we present a theoretical argument on the hyperdifferentiation of space. This is followed by a brief overview of the sweeping policy reversals in India and Ghana that opened the doors to economic globalization. Subsequently, we illustrate our argument with data at a variety of scales in Ghana and India before offering some conclusions.

A THEORY OF THE HYPERDIFFERENTIATION OF SPACE

We view economic globalization as a process that is essentially driven by a search for profits and that is first and foremost expressed in the increasing mobility of capital. The search for investment opportunities has been a constant factor since the inception of the capitalist world economy some five centuries ago. But the process of globalization, while progressive in the long run, has not been a linear one. It has been characterized by sudden accelerations, periods of stagnation, eras of steady advancement, and times of regression.

In general terms, the process has been conditioned by two additional factors aside from the ceaseless drive for capital accumulation. One relates to innovations in transport and communication technologies that enable accelerations of globalization. Examples include steamships and the telegraph in the latter part of the 19th century and mass air travel and telecommunications in the latter part of the 20th century. If capital accumulation is the primary driver of the process, technology is the primary facilitator. The other factor involves the potential mediator of the process: government policies. Governments can interfere in the market, impose regulations, impede the mobility of capital (and labor), or enable capital's free movement. Over time, state governments have altered their roles vis-à-vis the market, switching from spoilers to sponsors of the process of globalization. Even though there is notable variation among different states in the world, there have been general trends across the system. For example, in the late 19th and early 20th century liberalism prevailed across the advanced economies of the world. This came to an abrupt end—again across the whole system—with the onset of World War I.

The acceleration in the process of globalization since about 1980 is the combined result of these two factors: technological innovation (e.g., mass air travel, global communications, falling transport costs) and a powerful surge of transnational liberalism among an unprecedented number of countries around the world. Advances in communications technology (e.g., telephone, fax, email, video conferencing) have facilitated coordination of capital accumulation in diverse locations around the world. At least as significant as technological developments were the changes in government policies that took place in major national economies like the United States, the United Kingdom, and Germany as well in some smaller countries like the Netherlands and Sweden around and after 1980. By the latter part of the 1980s, the trend toward transnational liberalism had culminated in the "Washington Consensus."

The introduction of liberalization policies across the developing world ushered in the forces of globalization. Since then, academics and policymakers have debated the outcomes. There are two dominant views of globalization in the social sciences that we think are erroneous. First, globalization is generally, and incorrectly, understood as a historical process in which time increasingly prevails over space. In much thinking about globalization space is reduced to distance and relative distance is translated in time. For instance, the hypermobility of capital in the age of globalization is often thought to have diminished or even eradicated the meaning of distance. In what is clearly an erroneous leap in reasoning, some have taken this to presage the "end of geography" (O'Brien, 1991). We concur with Wai-chung Yeung (2002) and others (e.g., Smith, 1997) that geography has in fact become *more* important, not less so.

A second strand of thought, equally fallacious, views the world as increasingly homogeneous in economic and cultural terms (Ritzer, 2000). Proponents of this perspective emphasize the scaling up from the national to the global of the idea of "modernization," meaning that common global norms about markets, business practices, consumptions standards, and cultural practices are spreading everywhere.

We argue that globalization is in essence a geographical concept. We acknowledge the emergence of a new global economy and its geographical structure in the recent round of globalization. In our definition, *globalization* refers to a process of rescaling. It is the rescaling of economic relations so that these relations are increasingly conditioned at larger scales. This is not to imply that smaller scales matter less in an absolute sense. Rather, we recognize that economic processes must increasingly be understood at a variety of scales, from restricted local scales all the way up to the global scale. The latter represents the end, or rather culmination, of a scale spectrum that includes international regions, territorial states, substate regions, cities and towns, and neighborhoods. Once the process of globalization is conceived as one of rescaling, it follows that geography is as much in flux as history is. Geography is never static and, like history, its significance never expires. Thus geography does not become less relevant, but it does become more complex.

The increased mobility of capital is in part the result of market deregulation in many parts of the world. We think that the most dramatic accelerations of uneven development in this era of globalization occur in peripheral regions of the world economy that witnessed major shifts toward deregulation. These kinds of regions, we argue, have become characterized by a hyperdifferentiation of economic space, which forms a logical accompaniment to the hypermobility of capital or to the hypervolatility of financial markets. The reproduction of space has accelerated to unprecedented levels.

The notion of the hyperdifferentiation of space comprises more than established definitions of uneven development (Harvey, 1982; Smith, 1984). To be sure, in recent times, certain regions of the global economy have become more unevenly developed. But the hyperdifferentiation of space involves both quantitative change and qualitative transformations. There is increased economic divergence among certain places, but in addition there is an ongoing reconfiguration and redivision of the global space-economy at a variety of scales. In other words, not only is the developmental gap increasing among existing economic spaces, there is also a rapid creation and re-creation of entirely new spaces.

Increases in spatial differentiation are especially salient where the state retreats markedly from established dominance over the market, that is, where it substantially changes its policy from market interference to laissez-faire and does so at a time of high global mobility of capital. In such circumstances, spatial differentiation will be wrought at once by domestic and foreign capital. Hyperdifferentiation takes shape simultaneously at various scales and is expressed in the economic fracturing and redivision of conventional spatial entities, such as national states, regions, provinces, and cities.

In the rest of this chapter, we substantiate our argument with evidence from India and Ghana. The common characteristic of these countries (critical for our purpose) is the profound change in their central government policies around the middle of the 1980s, which moved away from a pronounced regulatory regime to a free market. As such, both countries exposed themselves rapidly to global market forces, and they were, differentially, swept up in the tide of economic globalization. Ghana and India are similar, then, in their histories of the recent international political economy: they both moved from colonialism, to "nationalism," to globalism. The two countries are very different in terms of size, physical geography, culture, main economic activities, and regional context. They are a good pair, therefore, with which to examine the pervasive effects of global capitalism.[1]

LIBERALIZATION AND GLOBALIZATION IN GHANA AND INDIA

After independence (for India in 1947, for Ghana in 1957), national development policies and ideological perspectives in the two countries were quite similar. Kwame Nkrumah and Pandit Nehru, respectively, articulated a nonaligned position in world affairs and promoted national economic policies of self-reliance. An elaboration of the nature of these policies and of their effects on the economic geographies of the two countries is beyond the scope of this chapter, but a few general points provide some historical context for our main argument.

Ghana's and India's economic policies during their "national" phase implied severe curtailments of the free market. Concretely, this was achieved through the imposition of high trade tariffs, the nationalization of industries, widespread subsidies for selected consumption and production, tight regulation of foreign investments, and so forth. At a basic level, the purpose was to transform an unbalanced, fragmented, and dependent colonial economy into a diversified, integrated, and sustainable national economy.

From a geographical point of view, these policies resulted in considerable constraints on the mobility of capital (especially foreign or global capital), and these, in turn, impeded the penchant for uneven development inherent to the free market. While there continued to be differences among various regions (particularly inside the huge and diverse areas that make up the Indian economy), it is hard to find strong evidence for increasing spatial differentiation in this historical period. Indeed, geographically even development was often an integral goal of Ghanaian and Indian development strategies.

It is quite clear that the policy environments in Ghana and India changed drastically after the mid-1980s. In both countries reforms entailed both short-term stabilization measures and long-term economic growth measures. The global engagement of the two economies increased, and all the indices of economic globalization showed increasing global integration at the aggregate level.

Ghana has been acclaimed as the "star pupil of structural adjustment" (Leechor, 1994, p. 194) because of the government's successes in implementing liberalization policies without backtracking. The growth of foreign investment, trade, and foreign corporate collaborations and the internationalization of the stock markets have been extensive. Some liberal commentators have deplored the allegedly slow pace of reform in India, especially when compared to the more liberal policy environments in Southeast Asia. But India has come a long way in dismantling one of the most densely regulated economies in the world.

While there is no question that the external sectors of the Ghanaian and Indian economies (i.e., the globalization of these national economies) are now much more significant than before the 1980s, it remains unclear how globalization has affected the overall economic development of these countries. In both, reform is widely believed to have resulted in significant economic growth at the aggregate national level (though some dispute the causality). Indian GDP grew by 4.1% per year during the 1990s. This compares to an average growth rate of 1.8% per year from 1965 to 1988, which used to be derisively referred to as the "Hindu rate of growth." The growth of the Ghanaian economy accelerated at a similar pace. Ghana experienced a consistent growth in GPD at an average of 5–6% between 1984 and 1991 and of 4% since 1992, compared with negative growth rates from 1974 to 1983.

But opinions vary widely about the effects of liberalization and globalization for development. Table 3.1 and 3.2 contain a collection of views from, respectively, Ghana and India.

The main questions in the development debate revolve around the relationships between growth, equality, and poverty. Considerable research has been done in Ghana and India, and large amounts of data have been collected. Some argue that growth has contributed to the alleviation of poverty. Others argue that growth has caused poverty to rise. And yet still others maintain that growth has led to more inequality.

Liberalization and globalization affect various groups in various places differentially, which makes it hard to generalize about their overall effects. It appears that reforms often either have been associated with or have been unable to prevent growing inequalities. The differential effects of liberalization and globalization are not confined to people according to socioeconomic characteristics such as class—the effects also vary across space. It is important not only *who* you are but *where* you are.

We argue that one important reason for the inconclusiveness of these debates in Ghana and India lies in the spatial differentiation in Ghana and India of globalization and economic development. Nationally aggregated statistics, then, have lost much of their meaning. The question is to what extent the transition from a highly regulated, insulated, and socialist-oriented regime toward a liberalized and globalized market environment has resulted in a faster differentiation of economic space.

The uneven spatial development of India and Ghana since the inception of market liberalization is best observed at a variety of scales distinguishable in the main geographical "theatres of accumulation."[2] In the next two sections, we present evidence for trends of spatial differentiation in the two countries, with particular attention to the roles of

---

TABLE 3.1. Inconclusive Debates in Ghana: Selected Statements

"The balanced sheet of liberalization is thus far favorable in Ghana" (Bourguignon & Morrison, 1992, p. 89).

"The adjustment program in Ghana has had a positive impact on the poor" (World Bank, 1996, p. 105).

"There were significant pockets [of poverty] of deterioration" (United Nations Development Programme, 1997, p. 61).

"Overall, SAPs [structural adjustment policies] have not been universally effective in Ghana" (Konadu-Agyemang, 2000, p. 481).

"SAPs may have made matters worse, and re-widened the gaps" (Konadu-Agyemang, 2000, p. 475).

TABLE 3.2. Inconclusive Debates in India: Selected Statements

"All the people who once said market forces would worsen India's poverty have gone quiet" (The Economist, 2000, p. 18).

" . . . poverty is unlikely to have increased in the recent past. . . . The incidence of poverty might in fact have declined" (World Bank, 1995, p. xxiii).

"While there was a marked decline in both rural and urban poverty rates between 1973–1985, there is no sign of anything comparable since"p. 3516).

"India's growth process during the last two decades does not seem to have been a virtuous one—it has polarized the economy. It has been a period of growth with inequality" (Nagaraj, 2000, p. 2837).

Greater Mumbai and Accra as the main regional engines of uneven development.

THE FRAGMENTATION OF INDIA
Interregional Differentiation

Debates in India about economic growth, equality, and development have a long history of intensity and contentiousness. But recent research findings are unequivocal about trends of economic divergence among the states in the Indian federation (see, e.g., Shaw, 1999; Kurian, 2000; Nagaraj, 2000; and Chakravorty, 2000). The increasing regional disparity is illustrated with an increasing coefficient of variation among states' gross domestic product (NSDP) from 31.10 in 1980 to 39.98 in 1995 (adapted from Shand & Bhide, 2000, p. 3749). The GINI coefficient[3] for incomes in all of rural India increased from 30.10 in 1983 to 30.56 in 1997 and rose further to 37.8 in 2002 (United Nations Development Programme, 2003). For urban India, the value increased from 34.08 to 36.54 over the same time period (Datt, 1999, p. 3517).[4] World Bank research (e.g., Milanovic, 2002) has also demonstrated that openness made income distribution worse in India.[5]

Within the Indian federation, Maharashtra was already one of the most "advanced" states of India before the mid-1980s. Since liberalization it has extended its lead (see Figure 3.1). From 1980 to 1995 Maharashtra had the highest economic growth rate of all Indian states at 85%—the average was 52% (Shand & Bhide, 2000, p. 3749). Rapid economic growth in Maharashtra coincided with substantial increases in foreign investment. For all of India, incoming foreign direct investment (FDI) increased fivefold between 1990 and 1999, with the largest share, 15.6%, going to Maharashtra. Karnataka and Tamil Nadu shared second place at 6.8% (Indian Investment Centre, 2000).

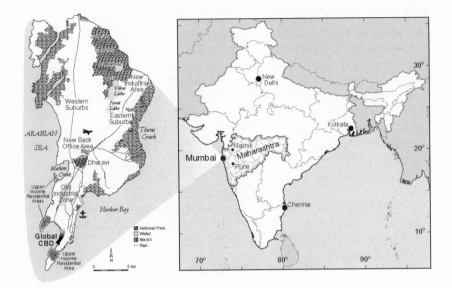

FIGURE 3.1. Scales of uneven development in India.

Maharashtra's agricultural share of the state economy is the smallest of all Indian states, and its share of the tertiary sector is the biggest. Between 1990 and 1995 the state had the fastest growth of the share of the service sector and by far the steepest growth of the finance sector (Shand & Bhide, 2000, pp. 3750, 3755). Maharashtra is also unequalled in the number of per capita Internet connections: it has about three times as many as Karnataka, which ranks second, and about 68 times as many as Orissa, which ranks last (United Nations Development Programme, 2001, p. 41).

Intraregional Differentiation

If India is increasingly too diverse to be represented in a meaningful way with singular development statistics, the same is true for Maharashtra. It is the most urbanized of all Indian states. In 1991 almost 40% of the population was classified as urban, as compared to an average of 25.7% for all of India. But most parts of the state are still rural, and some interior parts are very remote and isolated. The state consists of 31 districts and 41,336 towns and villages, most of which are a long way from Mumbai, in space as well as in time.

Of all approved foreign collaborations in Maharashtra in 1997,

82% of the Indian partners were headquartered in Mumbai. In 1989 and 1993 Mumbai's share was, respectively, 83% and 86% (Indian Investment Centre, various years). Elsewhere in the state, foreign investments were increasingly concentrated in the so-called Mumbai–Pune corridor and in the city of Pune itself. Pune's share of foreign collaborations went from 9% (1989) to 11% (1993) to 13% (1997). Indeed, data from a variety of sources show that Pune has drawn considerable manufacturing activity from Mumbai, while the latter increasingly specializes in finance and producer services—yet another form of spatial differentiation. From 1989 to 1997, in the rest of the state of Maharashtra, taken together, there was a relative decline in foreign collaborations.

In terms of development measures, Maharashtra is an interesting case of intrastate disparities. The state enjoys a very high level of per capita income (surpassed only slightly by Punjab): in 1997 it was almost 60% higher than the national average (Kurian, 2000, p. 541). But in the same year rural poverty there was the second highest in the country, after Bihar (Dev & Ranade, 1999, p. 51). A recent comparison of agroclimatic regions within India's states revealed that Maharashtra has become the most disparate of all. The difference in the incidence of poverty between Maharastra's central inlands and the Bombay region is not matched in any other state.[6]

The contrasts have been growing since the introduction of liberalization policies and the onset of globalization. GINI coefficients for rural and urban areas in 1983 and 1994 show Maharashtra as the only state in India where rural inequality has increased (the coefficient rose from 28.82 to 30.65). Urban inequality in Maharashtra increased as well, from 34.25 to 35.67, and is now the highest of all states (World Bank, 1997, pp. 53–67).

Intraurban Differentiation

The spatial differentiation of the impact of globalization is evident in Greater Mumbai, even to the casual observer. Foreign companies are highly concentrated in the "island city," the southern part of the peninsula. And within the island city, a distinct global CBD (central business district) has emerged that is separate from the traditional bazaar as well as from the former colonial center of the city that was largely "Indianized" after 1947.

This global CBD is centered in Nariman Point. What London's City is to Great Britain, Nariman Point is to India: a remarkable cluster of global linkages in finance and producer servers—banks, investment firms, consulting agencies, real estate companies, accounting firms, and so on. Yet, about 8 miles to the north, is Dharavi, known as "the biggest slum on earth" and as "Mumbai's sweatshop" (Seabrook, 1996). There, economic activities are of an entirely different order, ranging from tanneries (which clean and cut hides and make leather) to soap making and garbage picking (Sharma, 2000).

Real estate values in the global CBD skyrocketed after liberalization, in part as a result of the entry of foreign-controlled multinationals (Nijman, 2000). In 1995 and 1996 this business area was the most expensive in the world. Residential real estate prices spiraled up as well, in part in response to the influx of money from nonresident Indians. In the mid-1990s apartments in Malabar Hill changed hands for the equivalency of U.S. $2.5 million. In Dharavi, in contrast, the going price for a lot big enough for a tent was about U.S. $3,000—for many the investment of a lifetime (Patel, 1996; see also Eyre, 1990).

Charles Correa, one of India's most famous architects and noted public intellectuals, raging about Mumbai's poverty and disparity, called the city

> pathological—a terrifying morass of filth and decay. Dharavi sprawls in stinking mockery of Bombay's pretensions to sophistication and prosperity. The pavements heave with hawkers, forcing pedestrians onto crumbling streets where traffic is stuck in a gridlock that never stops. Overlooking this madness are tower-block apartments worth a million dollars; the greatest juxtaposition on earth of rich and poor. (quoted in Thomas, 1997, p. 26)

THE FRAGMENTATION OF GHANA

As in India, uneven development in Ghana since the introduction of liberalization policies is also best observed at a variety of scales within "theatres of accumulation." Figure 3.2 represents the different scales of analysis in Ghana.

Interregional Differentiation

Recent research findings are explicit about trends of economic divergence among the regions of Ghana (see, e.g., Konadu-Ageymang, 1998, 2000; Yeboah, 2000; Ofori, 2002; and Owusu, 2001). The increasing differentiation is illustrated by an increase in the GINI coefficient from 35.50 in 1960 to 39.80 in 2000 (Bhalla, 2002, p. 220), which remained at this level in 2002 (United Nations Development Programme, 2003). World Bank research on globalization, openness, and foreign direct investment also confirms that liberalization has made the income distribu-

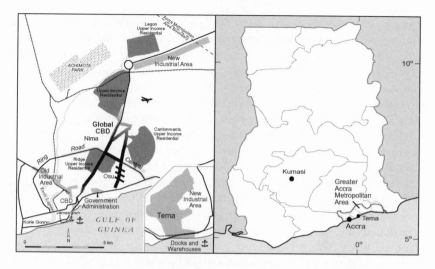

FIGURE 3.2. Scales of uneven development in Ghana.

tion worse in Ghana (Milanovic, 2002). Apart from the Greater Accra Metropolitan Area (GAMA) and the Ashanti Region, the country remains rural, and there is arguably an overconcentration of population in these two urban economic centers (Ghana Statistical Service, 2002).

GAMA has emerged as the gateway and primate region in the country. It is the most diversified in the country, contributing more than 20–30% of GDP (Ministry of Finance, 2001, p. 20). Spatial disparities in the regional geography of FDI and the headquartering of foreign companies have accelerated since the introduction of liberalization policies. Areas outside of GAMA account for a declining proportion of FDI. For instance, the share of FDI of Ghana's second most prosperous economic region (the Ashanti Region) declined from 18% in the 1970s to 9% in 1999 (Grant, 2001). The Ghana Investment Promotion Centre (GIPC) (2001) estimates that Accra accounts for 86% of Ghana's FDI. Moreover, 84% of all companies (both domestic and foreign) in Ghana are located in GAMA (Grant, 2002). The growth of GAMA has been in part facilitated by the availability of land for commercial development because of the deterioration of the traditional landholding system. In contrast, land outside of the region remains outside of the free market because the traditional landholding system is still in place and serves as a barrier to capitalist development.

There is unevenness to the geography of well-being and access to basic services in Ghana. Researchers have underscored the hardships and poverty that have emerged in the liberalization era, particularly in rural areas (Owusu, 2001). Konadu-Agyemang's (2000) research shows that while there were some modest improvements in education, health care, and access to basic services (electricity, water, etc.), interregional variations were essentially unchanged nationally, and that intraregional variation may be higher now than it was 15 years ago. For instance, only 6% of rural residents have access to electricity compared to 81% of Accra residents having access (Konadu-Agyemang, 2000, p. 477). Even Ghana's Ministry of Finance (2001) acknowledges that the incidence of poverty (measured in terms of income) in the Upper West and Upper East Regions now approaches a staggering 90% of the population.

The effects of globalization on spatial development appear to lead to hyperdifferentiation at the national level. Konadu-Agyemang (2000, p. 475) notes that liberalization policies "have made matters worse, and rewidened the gaps." There appears to be increasing differentiation between GAMA as a command, gateway, and service center and other regions in Ghana on the margins of global capitalist and national development.

Intraregional Differentiation

Within GAMA we can differentiate the Accra District, the Tema District, and the Ga District. Apart from the Accra District, only the Port of Tema (in the Tema District but in close proximity to Accra) and its surrounding area could be considered as economically developed. Most of the remainder of the Tema and Ga districts reveal periurban development, where informal economic activities such as home-based economic enterprises predominate.

The gaps between Accra and Tema and other periurban districts are apparent even to the casual observer. Economic development has been concentrated in the Accra–Tema corridor. The rise of this commercial hub has also been facilitated by the growth and expansion of Tema port, Kotoka international airport, and expenditures on developing the road network in and around the center of Accra. Earmarking the largest portion of infrastructural expenditures for a "gateway project" in the city of Accra has widened the gap between it and the rest of the country.

Intraurban Differentiation

Within the Accra district there is a sharp contrast between the formation of a "new global CBD" in the Osu–Cantonments area containing domestic and foreign companies (Grant & Nijman, 2002) and the low-income housing areas where the majority of people live and work. The global CBD is a remarkable cluster of finance and producer service companies as well as upscale retail establishments and restaurants. The upper end of the housing market is close to this area (e.g., East Legon). Here private developers are building four-bedroom houses that can cost up to U.S. $390,000. This is in sharp contrast to lower income, overcrowded, deteriorating rental accommodations where 60% of Accra's residents live and work without basic amenities (proper roads, drainage, water and waste disposal systems). There, economic activities are of an entirely different order, mostly microenterprises in production and services, particularly located in homes and along the major thoroughfares, in road reservations, and in areas reserved for other purposes within the built environment (Yankson, 2000). Typically, home-based enterprises have been characteristic of residential communities on the urban fringe, but they are now representative of all low-income areas in Accra (Grant & Yankson, 2003).

The United Nations Development Programme notes (1997, p. 61) that "against a general trend of economic recovery and growth, the decline in living standards in Accra is a major source of concern." Despite the globalization of aspects of the Accra economy, poverty levels in the city may have tripled during the liberalization era (Konadu-Agyemang, 1998, p. 134), and a noticeable poverty gap has emerged between the poor and an emerging middle class (Accra Study Team, 1998, p. 14; Ghana Statistical Service, 2002). The GINI coefficient of income distribution in Accra increased from 38.30 in 1985 to 43.00 in 1994 (Aryeetey & Aryeetey, 1996).

The increasing number of individuals seeking refuge in the informal economy illustrates the fragmentation of the economy. The ratio of workers in the informal sector to those in the formal sector in Accra increased from 2:1 in 1980 to 5:1 by 1990 (Aryeetey & Aryeetey, 1996). The retrenchment of 70,000 government workers since the introduction of liberalization policies, many of whom continued to stay in Accra, makes the employment situation even more difficult. In addition, low-income residents are spending a greater share of their incomes on feeding themselves (up to 60% in some instances) (Accra Study Team, 1998). Vulnerable groups are borrowing money for food and/or relying on someone outside the household for food security. Household indebtedness is also on the increase (monthly interest rates on debts can be as high as 50% in low-income areas [Ghana Statistical Service, 1988, 1995]). Increases in malnutrition among children under age 5, deteriorating environmental standards, growing informalization of jobs (especially hawking), relative depravation, decreasing access to health care, social exclusion (Pellow, 2002), and declining nutritional diets have all been documented for particular Accra groups (Benneh et al., 1993). As a result, more and more individuals rent single-room accommodations.

Residential areas like Cantonments, Airport RE, the Ridge, Legon, and most recently Osu have prospered from the liberal economy. New gated housing developments with resort amenities are being built in East Legon to cater to the elite. East Legon is out of touch with most Ghanaians' reality, illustrated in part by private developers' offers of preconstruction house models with names like "The Monarch," "The King," and "The Emperor." This is in sharp contrast to the low-income areas where most people have their "family home" (particularly the Ga—the original settlers) in overcrowded and dirt-ridden places, without proper roads, sanitation, and green space.

CONCLUSIONS

If the spatial outcomes of globalization are highly uneven, it is all the more important to acknowledge the geographical coordinates of empirical research on globalization. India and Ghana represent "peripheral" zones in the world economy that have recently increased their exposure to global capital. The effects of globalization there, as everywhere, must be understood in a geographical–historical context.

Even though the empirical evidence remains sketchy due to the difficulty of obtaining relevant data, it seems quite clear that the spatial economies of Ghana and India are rapidly becoming much more uneven at various scales. We measured this hyperdifferentiation of space in terms of capital accumulation (mainly foreign investments) and prosperity (mainly growth, equality, and poverty). To be sure, national statistics on economic development in these two countries appear to conceal more than they reveal.

The hyperdifferentiation of space is a corollary of the hyperspeed of capital movement in the current global era: the historically unprecedented velocity of capital shifts in the current information era cannot but bring about corresponding spatial reconfigurations. The shift in foreign investment flows from manufacturing to services is another factor that implies that much capital is becoming ever less embedded in particular locales. The most recent illustrations of these fast developments are the rapid emergence since 2001 of international 1-800 call centers in Bangalore (in value, overtaking the importance of software production) and of back-office facilities in Accra that processed New York City's traffic citations. How long such facilities will remain in place is entirely uncertain: by 2003, the back-office work on traffic citations had already returned from Accra to New York.

Current trends of hyperdifferentiation of space are different from older notions of uneven development in that the process has become much more intense, particularly in regions such as India and Ghana, since the mid-1980s. Historically, there is no precedent for a region undergoing such important policy changes toward a free market at a time of such highly mobile global capital. It is this combination of contextual factors that has been at the basis of the hyperdifferentiation of space in India and Ghana.

It is, of course, not easy to *prove* that the hyperdifferentiation of space in both countries is caused by globalization per se. One of the main challenges of globalization studies is to distinguish global dynamics from domestics (or local) processes. To be sure, the key to the changes in India and Ghana lies in the implementation of liberalization policies. This dramatic shift toward deregulation should itself be understood as part of the globalization of free-market ideologies. Liberalization, in turn, has caused enormous changes in the domestic economy as well as in the external economic sector. It is hard to say to what extent global capital has been exclusively responsible for the unprecedented fragmentation of Ghana and India, but it seems clear that it played a significant part, as is evident in the rapid increases in foreign direct investment, foreign corporate collaborations, and foreign corporate control. Thus, if Accra and Mumbai function as the main "theatres of accumulation," this is to a substantial degree due to global capital investments.

We do not think that the spatial differences in Ghana and India will be evened out in the future. Simply put, there is not enough time or opportunity for equalization as one might expect in a traditional neoclassical market environment. Given the frantic circulation of capital, we do not anticipate a great deal of stability of *any* sort in these regions as long as the mobility of capital is not constrained. Thus, there is not likely to be a movement toward equilibrium at any of the examined scales, but current uneven configurations also may not hold for a very long time. This is the endless division and redivision of space in the global free-market economy, and it is especially vigorous and visible in regions such as Ghana and India.

As of yet, little empirical research has been done on the spatial effects of globalization in the less developed world, even if this seems critical for the evaluation of present policies at many levels. In the current neoliberal climate, many governments have put their hopes on laissez-faire development strategies that strengthen regional clusters of economic activity. The notion of "regional push" connotes the role of regional engines of growth in the wider economy (Scott, 2002). In many countries, the geographic agglomeration of economic activities is now primarily based in the large urban centers that serve as drivers of the national economy. Accra and Mumbai are good examples.

The accompanying spatial dislocations, however, may be greater and more consequential than is currently realized.[7] Before too long the accelerated spatial differentiation of national economies will pose economic and political challenges to central and regional governments in countries such as India and Ghana. This may well result in renewed government policies aimed at correcting such spatial imbalances for the sake of more even, stable, and sustainable development.

ACKNOWLEDGMENTS

This research is part of a larger project on spatial polarization in Accra and Mumbai that is supported by the National Science Foundation, Grant No. BCS 0213648. We thank Chris Hanson for the cartographic productions in this chapter. An earlier version of this chapter was published in *Tijdschrift voor Economische en Sociale Geografie* (2004).

NOTES

1. For a more elaborate argument about the choice of Ghana and India as comparable case studies, see Grant and Nijman (2002).
2. We borrow the term "theatres of accumulation" from Armstrong and McGee (1985).
3. The GINI coefficient is the most widely used measure of inequality. It measures the extent to which the distribution of income or consumption among individuals or households within a country deviates from a perfectly equal distribution (0 signifies perfect equality, 100 means that one person holds all the income).
4. The GINI coefficient for all of India changed slightly from 32.2 in 1980 to 32.7 in 2000 (Bhalla, 2002, p. 218), but such general and spatially aggregated figures seem to have little meaning because they conceal vast regional differences.
5. Milanovic (2002) used data from household budget surveys and examined the impact of openness and foreign direct investment on relative income shares at each end of the income spectrum.
6. Squared Poverty Gap Index figures for Maharashtra's coast area and for the central inlands were, respectively, 0.67 and 7.85 (adapted from World Bank, 1998, p. 54).
7. There are a few critical voices in Ghana and India calling for more attention to the geographical dimension of development strategies. See Ofori (2002) and Banerjee Guha (2002).

REFERENCES

Accra Study Team. (1998). *Promoting urban food and nutrition security for the vulnerable in the Greater Accra Metropolitan Area.* Accra, Ghana: World Health Organization.

Armstrong, W., & McGee, T. (1985). *Theatres of accumulations: Studies in Asian and Latin American urbanization*. New York: Methuen.

Aryeetey, E., & Aryeetey, E. (1996). *An urban perspective on poverty in Ghana: A case study of Accra* (Report of a Study Commissioned by the Canadian International Development Agency). Accra, Ghana: Institute of Statistical Social and Economic Research.

Banerjee Guha, S. (2002, November 2–9). Critical geographical praxis: Globalization and socio–spatial disorder. *Economic and Political Weekly*. Retrieved from *http://www.epw.org.in*

Benneh, G., Songsore, J., Nabila, J. S., Amuzu, A. T., Tutu, K., & Yangyuoru, Y. (1993). *Environmental problems and the urban household in the Greater Accra Metropolitan Area (GAMA)—Ghana*. Stockholm: Stockholm Environment Institute.

Bhalla, S. (2002). *Imagine there's no country: Poverty, inequality, and growth in the era of globalization*. Washington, DC: Institute for International Economics.

Bourguignon, F., & Morrisson, C. (1992). *Adjustment and equity in developing countries*. Paris: Organization for Economic Co-operation and Development.

Chakravorty, S. (2000). How does structural reform affect regional development?: Resolving contradictory theory with evidence from India. *Economic Geography, 76*, 367–394.

Datt, G. (1999, March 5). Has poverty declined since economic reforms? *Economic and Political Weekly*, 3516–3518.

Dev, S., & Ranade, A. (1999). Persisting poverty and social insecurity: A selective assessment. In K. S. Parikh (Ed.), *India development report, 1999–2000* (pp. 49–67). Mumbai, India: Indira Ghandi Institute for Development Research.

The Economist. (2000, March 4). Waiting for the new India, p. 18.

The Economist. (2001, September 29). Globalization and its critics. A survey of globalization. pp. 1–30.

The Economist. (2003, August 23). Economics focus: Catching up, p. 62.

Eyre, L. A. (1990). The shanty towns of central Bombay. *Urban Geography, 11*, 130–152.

Fischer, S. (2003). Globalization and its challenges. *American Economic Review, 93*, 1–30.

Ghana Investment Promotion Centre. (2001). *Statistics on registered projects*. GIPC, Ghana: Author.

Ghana Statistical Service. (1988). *Ghana living standards survey report*. Accra, Ghana: Author.

Ghana Statistical Service. (1995). *Ghana living standards survey: Report of the third round GLSS 3*. Accra, Ghana: Author.

Ghana Statistical Service. (2002). *2000 population and housing census: Special report on urban localities*. Accra, Ghana: Author.

Giddens, A. (2000). *Runaway world: How globalization is reshaping our lives*. New York: Routledge.

Grant, R. (2001). Liberalization and foreign companies in Accra, Ghana. *Environment and Planning A, 33*, 997–1014.

Grant, R. (2002). Foreign companies and globalizations: Evidence from Accra, Ghana. In R. Grant & J. R. Short (Eds.), *Globalization and the margins* (pp. 130–149). London: Palgrave/Macmillan.

Grant R., & Nijman, J. (2002). Globalization and the corporate geography of cities in the less-developed world. *Annals of the Association of American Geographers, 92,* 320–340.

Grant, R., & Short, J. R. (Eds.). (2002). *Globalization and the margins.* London: Palgrave/Macmillan.

Grant, R., & Yanskon, P. (2003). City profile: Accra. *Cities, 20,* 65–74.

Harvey, D. (1982). *The limits to capital.* Chicago: University of Chicago Press.

Indian Investment Centre. (various years). *List of approved foreign collaborations.* New Delhi, India: Author.

International Monetary Fund. (2002). Globalization: Threat or opportunity? *IMF Issues Briefs.* Washington, DC: Author. April 12, 2000, corrected January 2002. Available *http://www.imf.org/external/np/exr/ib/2000/041200.htm#III.*

Konadu-Agyemang, K. (1998). Structural adjustment programs and the perpetuating of poverty and underdevelopment in Africa: Ghana's experience revisited. *Scandinavian Journal of Development Alternatives and Area Studies, 17,* 127–144.

Konadu-Agyemang, K. (2000). The best of times and the worst of times: Structural adjustment programs and uneven development in Africa: The case of Ghana. *Professsional Geographer, 52,* 469–483.

Kurian, N. J. (2000, February 12). Widening regional disparities in India: Some indicators. *Economic and Political Weekly,* pp. 538–550.

Leechor, C. (1994). Ghana: Frontrunner in adjustment. In I. Husain & R. Faruqee (Eds.), *Adjustment in Africa: Lessons from country studies* (pp. 169–180). Washington, DC: World Bank.

Marcuse, P., & van Kempen, R. (Eds.). (2000). *Globalizing cities: A new spatial order?* Oxford, UK: Blackwell.

Milanovic, B. (2002). *Can we discern the effects of globalization on income distribution?: Evidence from household budget surveys* (World Bank Policy Research Paper 2876). Washington, DC: World Bank.

Ministry of Finance. (2001). *Interim poverty reduction strategy, 2000–2002.* Accra, Ghana: Author.

Nagaraj, R. (2000, August 5). Indian economy since 1980: Virtuous growth or polarization? *Economic and Political Weekly,* pp. 2831–2838.

Nijman, J. (2000, February 12). Mumbai's real estate market in the 1990s: De-regulation, global money, and casino capitalism. *Economic and Political Weekly,* pp. 575–582.

O'Brien, R. (1991). *Global financial integration: The end of geography.* London: Pinter.

Ofori, S. (2002). *Regional policy and regional planning in Ghana.* Burlington, VT: Ashgate.

Owusu, H. (2001). Spatial integration, adjustment and structural transformation in sub-Saharan Africa: Some linkage pattern changes from Ghana. *Professional Geographer, 53,* 230–247.

Patel, S. (1996). Bombay's urban predicament. In S. Patel & A. Thorner (Eds.), *Bombay: Metaphor for modern India* (pp. xi–xxxiii). Bombay, India: Oxford University Press.

Pellow, D. (2002). Sabon Zongo: On the margins of the margins. In R. Grant & J. R. Short (Eds.), *Globalization and the margins* (pp. 111–129). London: Palgrave/Macmillan.

Ritzer, G. (2000). *The McDonaldization of society.* Thousand Oaks, CA: Pine Forge Press.

Rodrik, D. (1997). *Has globalization gone too far?* Washington, DC: Institute for International Economics.

Sassen, S. (1991). *The global city: New York, London, Tokyo.* Princeton, NJ: Princeton University Press.

Scott, A. (2002). Regional push: Towards a geography of development and growth in low- and middle-income countries. *Third World Quarterly, 23,* 137–161.

Seabrook, J. (1996). Bombay in the nineties. In J. Seabrook (Ed.), *In the cities of the South: Scenes from a developing world* (pp. 42–73). London: Verso.

Shand, R., & Bhide, S. (2000, October 14). Sources of economic growth: Regional dimensions of reforms. *Economic and Political Weekly,* pp. 3747–3757.

Sharma, K. (2000). *Rediscovering Dharavi: Stories from Asia's largest slum.* New Delhi, India: Penguin Books India.

Shaw, A. (1999, April 17). Emerging patterns of urban growth in India. *Economic and Political Weekly,* pp. 2390–2397.

Smith, N. (1984). *Uneven development: Nature, capital and the production of space.* Cambridge, UK: Blackwell.

Smith, N. (1997). The satanic geographies of globalization: Uneven development in the 1990s. *Public Culture, 10,* 169–189.

Taylor, P. J. (1999). Places, spaces and Macy's: Place–space tensions in the political geography of modernities. *Progress in Human Geography, 23,* 7–26.

Thomas, C. (1997, January 25). Ghetto blaster. *The Times Magazine* (London), pp. 26–29.

United Nations Development Programme. (1997). *Ghana human development report.* Accra, Ghana: Author.

United Nations Development Programme. (2001). *Human development report 2001: Making new technologies work for human development.* New York: Oxford University Press.

United Nations Development Programme. (2003). *Human development indicators 2003.* Available online at *http://www.undp.org/hdr2003/indicator/indic_126_1_1.html.*

Wai-chung Yeung, H. (2002). The limits of globalization theory: A geographic perspective on global economic change. *Economic Geography, 68,* 285–303.

World Bank (1996). *World development report: Attacking poverty.* New York: Oxford University Press.

World Bank. (1997). *India: Achievements and challenges in reducing poverty.* Washington, DC: Author.

World Bank. (1998). *Reducing poverty in India: Options for more effective public services.* Washington, DC: Author.

Yankson, P. (2000). Houses and residential neighbourhoods as work places in urban areas: The case of selected low income residential areas in Greater Accra Metropolitan Area (GAMA) Ghana. *Singapore Journal of Tropical Geography*, *21*, 200–214.

Yeboah, I. (2000). Structural adjustment and emerging urban form in Accra, Ghana. *Africa Today*, *7*, 61–89.

4

The Global Apparel Value Chain
What Prospects for Upgrading
by Developing Countries?

GARY GEREFFI
OLGA MEMEDOVIC

Although it is generally accepted that the clothing industry played a leading role in East Asia's early export growth, the degree to which international trade in this industrial sector can be the basis of sustained economic growth for developing countries has been questioned. Under what conditions can trade-based growth be a vehicle for genuine industrial upgrading, given the frequent criticisms of low-wage, low-skill, assembly-oriented export activities? Do Asia's accomplishments in trade-led industrialization contain significant lessons for other regions of the world?

This chapter looks at these and related questions, using a global value chain framework. A *value chain* is the range of activities involved in the design, production, and marketing of a product, although there is a critical distinction between buyer-driven and producer-driven value chains. Japan in the 1950s and 1960s, the East Asian newly industrializing economies (NIEs) in the 1970s and 1980s, and China in the 1990s became world-class exporters primarily by mastering the dynamics of buyer-driven value chains.

The key to East Asia's success was the move from mere assembly of imported inputs (traditionally associated with export processing zones, or EPZs) to a more domestically integrated and higher value-added form of exporting known as full-package supply, or OEM (original equipment manufacturing) production. (Throughout this chapter, "OEM production," "specification contracting," and "full-package supply" are used as broadly synonymous terms. In addition, "assembly," "production sharing," and "outward processing" refer to similar processes, even though a specific term may be favored in a particular region.) Japanese companies and some firms in the East Asian NIEs moved on from OEM export to original brand-name manufacturing (OBM), supplementing their production expertise with the design and then the sale of their own branded merchandise at home and abroad. The OEM model at the international level is a form of commercial subcontracting in which the buyer–seller linkage between overseas buyers and domestic manufacturers allows for a greater degree of local learning about the upstream and downstream segments of the apparel chain. (See Table 4.1 for further descriptions of these production systems.)

East Asia's ability to establish links with a wide range of lead firms in buyer-driven chains enabled it to make the transition from assembly to full-package supply. Lead firms are the primary sources of material inputs, technology transfer, and knowledge. In the apparel value chain, different types of lead firms use different networks and source from different parts of the world. Retailers and marketers in developed countries tend to rely on full-package sourcing networks, buying ready-made apparel primarily from Asia, where manufacturers in Hong Kong (now with the official name Hong Kong Special Administrative Region [SAR] of China), Taiwan Province of China, and the Republic of Korea historically specialized in this type of production. But as wages have risen, multilayered sourcing networks have been developed; low-wage assembly can be done in other parts of Asia, Africa, or Latin America while the NIE manufacturers coordinate the full-package production process. Branded manufacturers, by contrast, tend to create production networks that focus on apparel assembly using imported inputs. Full-package sourcing networks are generally global, while the production networks of branded manufacturers are predominantly regional. Manufacturers in the United States use Mexico and the Caribbean Basin, European Union (EU) firms look to North Africa and Eastern Europe, and Japan and the East Asian NIEs to lower-wage regions within Asia.

First, the global value chain framework is outlined, with emphasis on the structure and dynamics of buyer-driven chains. Second, the role of each of the big buyers (retailers, marketers, and manufacturers) in forging global sourcing networks in the apparel value chain is examined.

TABLE 4.1. International Production Systems

Assembly is a form of industrial subcontracting in which garment sewing plants are provided with imported inputs for assembly, most commonly in export processing zones (EPZs).

Original equipment manufacturing (OEM) is a form of commercial subcontracting. The supplying firm makes a product according to a design specified by the buyer; the product is sold under the buyer's brand name; the supplier and buyer are separate firms; and the supplier lacks control over distribution.

Original brand name manufacturing (OBM) is the upgrading by manufacturers from the production expertise of OEM to first the design and then the sale of their own branded products.

Third, the evolution and upgrading of apparel sourcing networks in Asia are considered. Industrial upgrading in the Asian context is examined through the process of building, extending, coordinating, and completing international production and trade networks. Fourth, the implications of the Asian experience for apparel sourcing in North America and Europe are assessed. Both regions are moving beyond assembly production and establishing full-package or OEM models in order to promote regionally integrated apparel value chains. The Japanese pattern of apparel sourcing, which is highly concentrated on a few suppliers, is contrasted with U.S. and European patterns, and the differences are traced to trade policy. The final section of the chapter offers conclusions regarding upgrading options within the global apparel industry.

GLOBAL VALUE CHAINS

In global capitalism, economic activity is international in scope and global in organization. "Internationalization" refers to the geographical spread of economic activities across national boundaries. As such, it is not a new phenomenon. It has been a prominent feature of the world economy since at least the 17th century when colonial powers began to carve up the world in search of raw materials and new markets. "Globalization" is more recent, implying functional integration between internationally dispersed activities.

Industrial and commercial firms have both promoted globalization, establishing two types of international economic networks. One is "producer-driven" and the other is "buyer-driven" (Gereffi, 1994, 1999). In *producer-driven value chains*, large, usually transnational, manufacturers play the central roles in coordinating production networks (including their backward and forward linkages). This is typical of capital and technology-intensive industries such as automobiles, aircraft, computers, semiconductors, and heavy machinery. *Buyer-driven value chains* are those in which large retailers, marketers, and branded manufacturers play the pivotal roles in setting up decentralized production networks in a variety of exporting countries, typically located in developing regions. This pattern of trade-led industrialization has become common in labor-intensive consumer goods industries such as garments, footwear, toys, handicrafts, and consumer electronics. Tiered networks of third world contractors that make finished goods for foreign buyers carry out production. Large retailers or marketers that order the goods supply the specifications.

Firms that fit the buyer-driven model, including retailers like Wal-Mart, Sears, and J. C. Penney; athletic footwear companies like Nike and Reebok; and fashion-oriented apparel companies like Liz Claiborne, The Gap, and The Limited, generally design and/or market—but do not make—the branded products they order. They are "manufacturers without factories," with the physical production of goods separated from their design and marketing. Unlike producer-driven chains, where profits come from scale, volume, and technological advances, in buyer-driven chains profits come from combinations of high-value research, design, sales, marketing, and financial services that allow the retailers, designers, and marketers to act as strategic brokers in linking overseas factories and traders with product niches in their main consumer markets (Gereffi, 1994). Profitability is greatest in the concentrated parts of global value chains that have high entry barriers for new firms.

In producer-driven chains, manufacturers of advanced products like aircraft, automobiles, and computers are the key economic agents both in terms of their earnings and their ability to exert control over backward linkages with raw material and component suppliers, and forward linkages into distribution and retailing. The lead firms in producer-driven chains usually belong to international oligopolies. Buyer-driven value chains, by contrast, are characterized by highly competitive and globally decentralized factory systems with low entry barriers. The companies that develop and sell brand-name products have considerable control over how, when, and where manufacturing will take place, and how much profit accrues at each stage. Thus large manufacturers control the producer-driven value chains at the point of production, while marketers and merchandisers in buyer-driven value chains exercise the main leverage at the design and retail stages.

Apparel is an ideal industry for examining the dynamics of buyer-driven value chains. The relative ease of setting up clothing companies, coupled with the prevalence of developed country protectionism in this sector, has led to an unparalleled diversity of garment exporters in the third world. Furthermore, the backward and forward linkages are extensive, and help to account for the large number of jobs associated with the industry (see Appelbaum, Smith, & Christerson, 1994). The apparel value chain is organized around five main parts: (1) raw material supply, including natural and synthetic fibers; (2) provision of components, such as the yarns and fabrics manufactured by textile companies; (3) production networks made up of garment factories, including their domestic and overseas subcontractors; (4) export channels established by trade intermediaries; and (5) marketing networks at the retail level (see Figure 4.1).

There are differences between these parts, such as geographical location, labor skills, labor conditions, technology, and the scale and type of enterprises, which also affect market power and distribution of profits among the main firms in the chain. Entry barriers are low for most garment factories, although they become progressively higher when moving upstream to textiles and fibers; brand names and stores are alternative competitive assets that firms can use to generate significant economic rents. The lavish advertising budgets and promotional campaigns needed to create and sustain global brands, and the sophisticated and costly information technology employed by megaretailers to develop "quick response" programs that increase revenues and lower risks by getting suppliers to manage inventories, have allowed retailers and marketers to displace traditional manufacturers as the leaders in many consumer goods industries. In apparel, the split between manufacturing and marketing that prompted the emergence of "lean retailing" (i.e., the model of frequent shipments by suppliers to fill ongoing replenishment orders by retailers, based on real-time sales information collected at the retailer's stores on a daily basis) was caused by the development of several key information technologies. These included bar coding and point-of-sale scanning used to provide immediate and accurate information on product sales; electronic data interchange (EDI) used by the retailer to restock; and automated distribution centers to handle small restocking orders, rather than the traditional warehouse system used for large bulk shipments (Abernathy, Dunlop, Hammond, & Weil, 1999).

A major hypothesis of the global value chains approach is that national development requires linking up with the most significant lead firms in an industry. These lead firms are not necessarily the traditional vertically integrated manufacturers, nor are they necessarily involved in making finished products. Lead firms, such as fashion designers or private-label retailers, can be located upstream or downstream from manufacturing, or they can be involved in the supply of critical components (e.g., microprocessor companies like Intel or software firms like Micro-

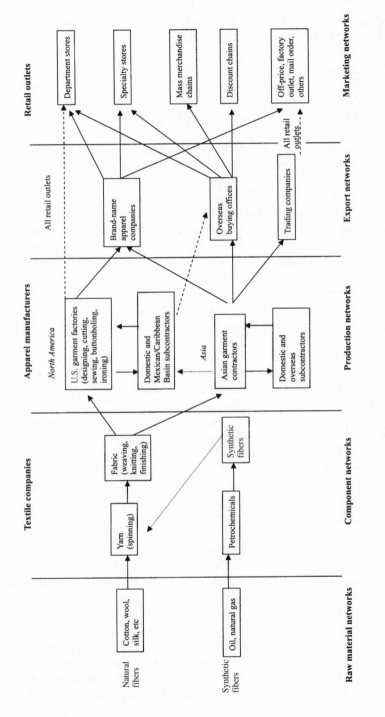

FIGURE 4.1. The apparel commodity chain. From Appelbaum and Gereffi (1994, p. 46). Copyright 1994 by Temple University Press. Reprinted by permission.

soft in the computer industry). What distinguishes lead firms from nonlead firms is that they control access to major resources (such as product design, new technologies, brand names, and consumer demand) that generate the most profitable returns.

BIG BUYERS AND GLOBAL SOURCING

The retail sector in the United States and other developed economies is undergoing a major restructuring. Global retailing is dominated by large organizations that are developing greater specialization by product (the rise of specialized stores selling only one item, such as clothes, shoes, or office supplies) and price (the growth of high-volume, low-cost discount chains). Furthermore, keeping the distribution pipeline filled means that these retailers are developing strong ties with global suppliers, particularly in low-cost countries (Management Horizons, 1993). Nowhere are these changes more visible than in apparel. Between 1987 and 1991 the five largest soft goods chains in the United States increased their share of the national apparel market from 35% to 45% (Dickerson, 1995, p. 452). By 1995, the five largest retailers—Wal-Mart, Sears, Kmart, Dayton Hudson Corporation, and J. C. Penney—accounted for 68% of all apparel sales. The next top 24 retailers, all billion-dollar corporations, represented an additional 30% of these sales (Finnie, 1996, p. 22).[1] Thus, the 29 biggest retailers made up 98% of all U.S. apparel sales. The two top discount giants, Wal-Mart and Kmart, control one-quarter of all apparel (by unit volume, not value) sold in the United States.

Although the degree of market power that is concentrated in large U.S. retailers may be extreme, a similar shift from manufacturers to retailers and marketers appears to be under way in other developed countries. Retailing across the EU has been marked by substantial concentration in the 1990s. In Germany, the largest clothing retailers (C&A, Quelle, Metro/Kaufhof, Karstadt, Otto, and increasingly H&M and Zara) in 1992 accounted for 28% of its economy, and the United Kingdom's two top clothing retailers (Marks & Spencer and the Burton Group) controlled over 25% of the market in 1994 (L'Observatoire Européen du Textile et de l'Habillement, 1995, pp. 11–13). Marks & Spencer, the United Kingdom's largest and most successful retailing firm, with 134 franchise stores in 25 countries in 2001, has adopted a new sourcing strategy that significantly shifts buying from the United Kingdom to low-labor-cost regions. While the company traditionally prided itself on the fact that at least 90% the goods sold in its U.K. stores were made in the United Kingdom, this "Buy British" focus began to erode in the 1990s. Marks & Spencer, which had an 11% share of the U.K. clothing market in 2001, planned to source more than 70% its apparel from lower cost countries by 2002 (Tait, 2000; Davies, 2002). In both France and Italy, the role of independent retailers has declined since the mid-1980s, while the share of specialized chains, franchise networks, and hypermarkets is rising rapidly. In Japan, cost-conscious consumers have contributed to a decline in the leading role played by high-fashion department stores such as Seibu and Isetan. New specialty apparel retailers offering lower prices have proliferated, and many now offer Chinese apparel, which accounted for over 60% of apparel imports into Japan in 2000 (Onozuka, 2001).

For buyer-driven value chains, the major significance of growing retailer concentration is the resulting expansion of global sourcing. Whereas in 1992 about 49% of all retail apparel sold in the United States was made in that country, by 1999 the proportion of domestically made U.S. retail apparel dropped to just 12% (Rabon, 2001, p. 55). As each type of buyer in the apparel value chain has become more involved in offshore sourcing, the competition between retailers, marketers, and manufacturers has intensified, leading to a blurring of traditional boundaries and a realignment of interests.

Retailers used to be garment manufacturers' main customers, but they have now become their competitors in certain manufacturing functions. With consumers demanding better value, retailers have turned to imports. In 1975, only 12% of apparel sold by U.S. retailers was imported; by 1984, this had doubled (American Apparel Manufacturers Association, 1984). By the mid-1990s, retailers accounted for approximately one-half of all apparel imported into the United States and Europe (Jones, 1995, pp. 24–26; Scheffer, 1994, pp. 11–12). These trends mark the rise in what is known as "vertical retailing," whereby a diverse array of national department stores (e.g., J. C. Penney and Sears), discount chains (e.g., Wal-Mart and Kmart), and specialty retailers (e.g., The Gap, The Limited Inc., and Benetton) have taken on manufacturing responsibilities to produce private-label or store-brand lines. Today, retailers' overseas offices go well beyond their original buying functions, and they are actively engaged in product design, fabric selection and procurement, and monitoring contracted sewing as well as other production functions handled by offshore manufacturers (Dickerson, 1999, pp. 464–466; Speer, 2001). Private-label goods, which are estimated to cover 15–25% of the U.S. apparel market during the 1990s (Dickerson, 1995, p. 460; Abend, 2000, p. 36), can disrupt the business of both manufacturers and well-known designer lines.

A notable feature of buyer-driven chains has been the creation since the mid-1970s of prominent marketers with well-known brands but that carry out no production. They include companies like Liz Claiborne,

Nike, and Reebok, which were "born global" since their sourcing has always been overseas. As pioneers in global sourcing, branded marketers were instrumental in providing overseas suppliers with knowledge that subsequently allowed them to upgrade their position in the apparel chain.

In order to deal with this competition, branded marketers have adopted several new strategies that will alter the content and scope of their global sourcing networks: reassigning certain support functions (such as pattern grading, marker making, and sample making) to contractors; reducing their purchase and redistribution activities, by handing them over to contractors, as well as their supply chains; using fewer but more capable manufacturers; adopting more stringent vendor certification systems to improve performance; and shifting their sourcing from Asia to the western hemisphere. In essence, marketers have recognized that overseas contractors can manage the whole production process, restricting their competitive edge to design and brandings.

With foreign producers providing similar quantity, quality, and service as domestic producers, but at lower prices, apparel manufacturers in developed countries have been caught in a squeeze. They are responding in different ways. In the United States and Europe, an "if you can't beat them, join them" attitude has evolved among many smaller and middle-sized firms. They feel they cannot compete with the low cost of foreign goods and are defecting to the ranks of importers or to relocating their manufacturing plants abroad.

For many larger manufacturers the decision is no longer whether to engage in foreign production, but how to organize and manage it. These firms supply intermediate inputs (cut fabric, thread, buttons and other trim) to extensive networks of offshore suppliers, typically located in neighboring low-cost countries with reciprocal trade agreements that allow goods assembled offshore to be reimported with a tariff charged only on the value added by foreign labor. This international subcontracting system exists worldwide. In the United States it is called the "807/9802 program" or "production sharing" (U.S. International Trade Commission, 1997), with sourcing networks predominantly located in Mexico, Central America, and the Caribbean. In Europe it is known as "outward-processing trade" (OPT), and the principal suppliers are in North Africa and Eastern Europe (L'Observatoire Européen du Textile et de l'Habillement, 1995); and in Asia, manufacturers from relatively high-wage economies like Hong Kong SAR have outward-processing arrangements (OPAs) with China and other low-wage countries (Birnbaum, 1993).

A significant countertrend is emerging among established apparel manufacturers, however. They are reducing their production activities and building up the marketing side of their operations by capitalizing on both brand names and retail outlets. Sara Lee Corporation, one of the largest apparel producers in the United States—whose stable of famous brand names includes L'Eggs hosiery, Hanes, Playtex, Wonderbras, Bali, and Coach leather products—has "deverticalized" its consumer products divisions, a fundamental reshaping that moved it out of making the brand-name goods it sells (Miller, 1997). Other well-known manufacturers such Phillips-Van Heusen and Levi Strauss & Co. are also building global brands, frequently through acquisitions of related product lines, while many of their production facilities are being closed or sold to off-shore contractors.

GLOBAL SOURCING IN APPAREL

The world textile and apparel industry has undergone several production migrations since the 1950s, all involving Asia. The first was from North America and Western Europe to Japan in the 1950s and early 1960s, when Western textile and clothing production was displaced by a sharp rise in imports from Japan. The second shift was from Japan to Hong Kong, Taiwan Province of China, and the Republic of Korea, which dominated global textile and clothing exports in the 1970s and early 1980s. In the late 1980s and the 1990s there was a third migration, from the Asian "Big Three" (Hong Kong SAR, Taiwan Province of China, and the Republic of Korea) to other developing economies. In the 1980s, production moved principally to Mainland China, but also to several Southeast Asian countries (Indonesia, Thailand, Malaysia, and the Philippines) and Sri Lanka. In the 1990s, new suppliers included South Asian and Latin American apparel exporters (Khanna, 1993; Gereffi, 1998).

This most recent shift is seen most clearly in changes in apparel imports to the United States, the world's largest market, from 1983 to 2001. In 1983, the Asian Big Three, plus China, were responsible for two-thirds of these imports; by 2001 this share had dropped to 27%. There are two main trends: first, a shift within Asia with the Big Three's share being reduced, first by China, then by Southeast Asia and South Asia; and second, a growth in non-Asian imports, particularly from Central America and the Caribbean, which nearly doubled its contribution from 8% in 1990 to 15% in 2001, and, most notably, Mexico, which multiplied its share nearly fivefold from 3% to 15%.

Why did these shifts occur? Neoclassical economics has the simplest explanation: the most labor-intensive segments of the apparel value chain will be based in countries with the lowest wages. This view is supported by the sequential relocation of textile and apparel production from the United States and Western Europe to Japan, then to the Asian Big Three and China, when each new tier of entrants had significantly lower wage rates than its predecessor. The cheap labor argument does not hold up as well, however, in the case of new Asian and Caribbean suppliers, whose market share expanded even though their wage rates are often considerably higher than China's. Furthermore, although the share of imports represented by Hong Kong SAR, the Republic of Korea, and Taiwan Province of China declined in the 1990s, these NIEs still ranked among Asia's top apparel exporters to the United States in 2001, despite having the highest apparel labor costs in the region, excluding Japan (International Labour Organization, 1995, pp. 35–36).

The dynamic perspective to comparative advantage, which argues that government policies will may play a major role in shaping the location of apparel export activities, helps to explain these discrepancies. A critical factor in the sharp decline of Taiwan Province of China's and the Republic of Korea's apparel exports in the late 1980s was not only their rising wage rates, but the sharp appreciation of their currencies vis-à-vis the dollar after the Plaza Agreement was signed in 1985. Between 1985 and 1987, the Japanese yen was revalued by nearly 40% and the New Taiwan dollar by 28%; from 1986 to 1988 the Korean won appreciated by 17%.

The most important influence on U.S. apparel imports, however, are quotas and preferential tariffs. Quotas on apparel and textiles items were regulated by the Multi-Fiber Arrangement (MFA) of the early 1970s. It has been used by the United States, Canada, and some European countries to impose quantitative limits on imports in a wide variety of products. Although these were designed to protect firms in developed countries from a flood of low-cost imports that threatened to disrupt major domestic industries, the result was the opposite: protectionism increased the competitive capabilities of developing countries' manufacturers, who learned to make more sophisticated and therefore more profitable products. Protectionism also increased the competition from overseas suppliers to the United States and Europe, as an ever-widening circle of exporters was needed to meet booming North American and European demand. The creation of the EU and the North American Free Trade Agreement (NAFTA) has led to the imposition of preferential tariffs within regional markets, which has generated a major shift in global sourcing dynamics.

The ability of the East Asian NIEs to sustain their export success over several decades and to develop a multilayered sourcing hierarchy within Asia is only partially related to wage rates and national policies. From a value chain perspective, East Asia must be seen as part of an interrelated regional economy (Gereffi, 1998). The apparel export boom in the less-developed southern tier of Asia has been driven to a significant extent by the industrial restructuring of the northern-tier East Asian NIEs. As Northeast Asian firms began moving their production offshore, they found ways to coordinate and control their sourcing networks, ultimately focusing on the more profitable design and marketing areas to sustain their competitive edge. This transformation can be conceptualized as a process of industrial upgrading, based in large measure on building economic and social networks between buyers and sellers.

The East Asian NIEs are generally taken as the archetype for industrial upgrading in developing countries. They made a rapid transition from the initial assembly phase of export growth (typically utilizing EPZs located near major ports) to a more generalized system of incentives that applied to all export-oriented factories in their economies. The next stage for Taiwan Province of China, the Republic of Korea, Hong Kong SAR, and Singapore was OEM production. East Asian firms soon became full-range package suppliers for foreign buyers, and developed an innovative entrepreneurial capability that involved the coordination of complex production, trade, and financial networks (Gereffi, 1995).

The OEM export role has many advantages. It helps local entrepreneurs to learn foreign buyers' preferences, including international standards for price, quality, and delivery. It also generates substantial backward linkages in the domestic economy, as OEM contractors are expected to develop reliable sources of supply. Moreover, OEM production expertise increases over time and spreads across different activities. Suppliers learn about the downstream and upstream segments of the apparel value chain from the buyer and this can become a powerful competitive weapon.

Countries such as the East Asian NIEs thus retain an enduring competitive edge in export-oriented development. However, East Asian producers face intense competition from lower cost exporters in other parts of the third world, and the price of their exports to Western countries has been increased by sharp currency appreciations since the Plaza Agreement. They therefore need to establish forward linkages to developed country markets, where the biggest profits are made in buyer-driven value chains. Some firms in the East Asian NIEs are pushing beyond OEM to the OBM role by integrating their manufacturing expertise with the design and sale of their own branded goods.

The Republic of Korea is the most advanced of the East Asian NIEs in OBM production, with its own brands, including automobiles (Hyundai), electronic products (Samsung), and household appliances (Samsung and Goldstar), being sold in North America, Europe, and Japan. Taiwanese companies have pursued OBM in computers, bicycles, sporting equipment, and shoes, but not apparel. In Hong Kong SAR, clothing companies have been the most successful in making the shift from OEM to OBM. Well-known local retailers include the women's clothing chain Episode, which is controlled by Hong Kong SAR's Fang Brothers Group, one of the foremost OEM suppliers for Liz Claiborne since the 1970s; Giordano, Hong Kong's most famous clothing brand; and Hang Ten, a less expensive line that in the late 1990s was the largest foreign clothing franchise in Taiwan Province of China (Granitsas, 1998).

An important mechanism facilitating the move to higher value-added activities for mature export industries like apparel in East Asia is triangle manufacturing (Gereffi, 1999). The essence of triangle manufacturing, which was initiated by the East Asian NIEs in the 1970s and 1980s, is that U.S. (or other overseas) buyers place their orders with the NIE manufacturers they have previously sourced from, who in turn shift some or all of the requested production to affiliated offshore factories in low-wage countries (e.g., China, Indonesia, or Guatemala). These factories can be wholly owned subsidiaries of the NIE manufacturers, joint-venture partners, or simply independent overseas contractors. The triangle is completed when the finished goods are shipped directly to the overseas buyer under the U.S. or European import quotas issued to the exporting country. Triangle manufacturing thus changes the status of NIE manufacturers from being established suppliers for U.S. retailers and designers to being middlemen in buyer-driven value chains that can include as many as 50–60 exporting countries.

In each of the East Asian NIEs, a combination of domestic supply-side constraints (labor shortages, high wages, and high land prices) and external pressures (currency revaluation, tariffs, and quotas) led to the internationalization of the textile and apparel network by the late 1980s and early 1990s. Typically, the internationalization of production was sparked by quotas, but the process was accelerated as supply-side factors became unfavorable. Quotas determined when the outward shift of production began, while preferential access to overseas markets and social networks determined where firms went. In this division of labor, skill-intensive activities, which provided relatively high gross margins, such as product design, sample making, quality control, packing, warehousing, transport, quota transactions, and local financing in the apparel industry, stayed in East Asia and labor-intensive activities were relocated.

In Hong Kong SAR internationalization was triggered by textile import restrictions imposed by the United Kingdom in 1964, which led manufacturers to shift production to Singapore, Taiwan Province of China, and Macao SAR of China, where the Chinese population had cultural and linguistic affinities with Hong Kong SAR investors. Macao SAR also benefited from its proximity to Hong Kong SAR, and Singapore qualified for Commonwealth preferences for imports into the United Kingdom. In the early 1970s Hong Kong apparel firms targeted Malaysia, the Philippines, and Mauritius. This second round of outward investment again was prompted by quota restrictions, coupled with specific host country inducements. For example, Mauritius established an export-processing zone in an effort to lure in Hong Kong SAR investors, particularly knitwear manufacturers who directed their exports to European markets that offered preferential access in terms of low tariffs.

The greatest spur to the internationalization of Hong Kong's textile and apparel companies was the opening up of the Chinese economy in 1978. At first, production was subcontracted to state-owned factories, but eventually an elaborate outward-processing arrangement was set up that relied on an assortment of manufacturing, financial, and commercial joint ventures. The relocation of industry to the Chinese mainland led to the dismantling and relocation of Hong Kong's manufacturing sector during the late 1980s and early 1990s. In 1991, 47,000 factories employed 680,000 workers, 25% less than the peak of 907,000 recorded in 1980 (Khanna, 1993, p. 19). The decline was particularly severe in textiles and apparel. Employment in the textile industry fell from 67,000 in 1984 to 36,000 in 1994, a drop of 46%. Meanwhile, clothing jobs plummeted by 54% in a single decade, from 300,000 in 1984 to 137,000 in 1994 (De Coster, 1996a, p. 65). In 1995 Hong Kong entrepreneurs operated more than 20,000 factories employing an estimated 4.5–5 million workers in the Pearl River Delta alone in the neighboring Chinese province of Guangdong (De Coster, 1996b, p. 96). Considering that total employment in Hong Kong industry had shrunk to 386,000 in 1995, or just over 15% of the workforce (Berger & Lester, 1997, p. 9), Hong Kong manufacturers in effect increased their domestic labor force well over 10-fold through their outward processing arrangement with China.

This extreme reliance of Hong Kong SAR apparel manufacturers on low-cost Chinese labor could make them vulnerable (Berger & Lester, 1997, pp. 158–162). First, although Guangdong Province has low wages and an abundant workforce, both wages and land costs have risen rapidly. As costs in Guangdong go up, Hong Kong SAR manufacturers who wish to retain a Chinese-based production system will have to move their facilities further into China, where they will once again encounter bad roads, inadequate water and power systems, and lack of a commercial infrastructure. Second, as production moves inland, it will be increasingly difficult to attract enough Hong Kong SAR managers. Rather than trying to replicate the Pearl River Delta pattern on a large scale further inland, it might be better to try to upgrade operations at the Guangdong plants. Third, new low-cost apparel-exporting Asian countries are emerging—India, Indonesia, Myanmar, Sri Lanka, Vietnam, and others—while Mexico and the Caribbean Basin countries loom as cheap production sites closer to the U.S. market. Hong Kong SAR has no special advantages in many of these places, which suggests that it should avoid being locked into low-wage offshore manufacturing networks and instead take fuller advantage of the global trend toward service-enhanced manufacturing, where it retains a strong competitive edge.

The internationalization of Korean and Taiwanese apparel producers also began as a response to quota restrictions. Korean garment firms lacking sufficient export quotas set up offshore production in quota-free locations like Saipan, a U.S. territory in the Mariana Islands. More recent waves of internationalization were the result of rising wages and worker shortages at home. Latin America and Southeast and South Asia have attracted the largest numbers of Korean companies. Latin America (the Dominican Republic, Guatemala, Honduras, etc.) is attractive because of its proximity to the United States and easy quota access, while the pull of Asian countries such as Indonesia, Sri Lanka, and Bangladesh is their wage rates, which are among the lowest in the world.

When Taiwanese firms moved offshore in the early 1980s, they also confronted binding quotas. Although wages in the late 1970s and early 1980s were still relatively low, quota rents were high. Firms had to buy quotas (whose value in secondary markets fluctuated widely) in order to expand their exports, thereby causing a fall in profitability for firms without sufficient quota (Appelbaum & Gereffi, 1994). This led to a growing emphasis on nonquota markets by textile and apparel exporters. Quota markets (the United States, the EU, and Canada) accounted for over 50% of Taiwan Province of China's textile and apparel exports in the mid-1980s, but this declined to 43% in 1988 and to 35% in 1991. By 2000, Taiwan Province of China's textile and apparel exports to the United States, Europe, and Canada remained at about one-third of the total of $13.8 billion. However, China and Hong Kong SAR alone accounted for 53% of textile exports of $10.4 billion, and several Southeast Asian nations (Thailand, the Philippines, Indonesia, and Malaysia) received another 12.8%, while the United States had 68% of Taiwan Province of China's $3.4 billion in apparel exports.[2] The fact that textiles represented three-quarters of Taiwan Province of China's total textile and apparel trade, and that most of these textile exports were going to low-wage countries in Asia, reinforces the importance of triangular manufacturing in the region, with Taiwan Province of China providing a growing proportion of textile inputs for many of Asia's leading apparel exporters.

The Asian financial crisis of 1997–1998 did not have a major effect on the region's textile and apparel exports because the latter were concentrated in industries that relied heavily on labor-intensive technologies, with relatively little reliance on costly foreign inputs or high levels of external debt. Most of the region's apparel exports are financed by letters of credit from U.S. and European buyers, rather than by local financial resources. In some respects, textile and apparel exports in Asia may have received a short-term boost from the region's financial crisis because these exports generated vital sources of foreign exchange, leading textile and apparel firms to expand overseas sales while more capital- and technology-intensive export industries were struggling to regain their financial stability.

APPAREL SOURCING IN NORTH AMERICA

The analysis of the apparel value chain in Asia has significant implications for the future of the textile and apparel sector in North America. Since Asian supply to the United States has primarily been directed to filling the OEM orders of retailers and branded marketers, apparel manufacturers in North America will need to develop the capability to carry out full-package supply. Previously this has only been done by the East Asia NIEs for the U.S. mass market, or by the fashion centers of Europe for high couture.

Between 1990 and 2002, U.S. apparel imports rose from $25.0 billion to $63.7 billion. Figure 4.2 helps to identify trade shifts among the main suppliers. Those countries in the innermost circle each account for 10% or more of the total value of clothing imports in 2002, while each of those in the outer ring makes up only 1.0–1.9% of total imports. In other words, the relative importance of national apparel exporters decreases between the inner rings and the outer ones.

Several key aspects of the direction and magnitude of change in the U.S. apparel trade are revealed in Figure 4.2. First, there are striking regional differences in the pattern of imports. The NIEs in Northeast Asia are becoming less important, South and Southeast Asia are growing slowly or not at all, and imports from China, Mexico, and to a lesser degree the Caribbean Basin are booming. Second, despite considerable mobility during the 1990s, there is a strong core–periphery pattern that dominates the geography of export activity. Only four economies (Hong Kong SAR, the Republic of Korea, China, and Mexico) were core suppliers in the past decade, and only China and Mexico held that distinction in 2002. There are 21 suppliers in the outer two rings (indicating 1–4% shares of the market), none in the middle ring, and just three countries in the inner two rings (6% or more of U.S. apparel imports). Third, for most countries the degree of change from 1990 to 2002 was relatively

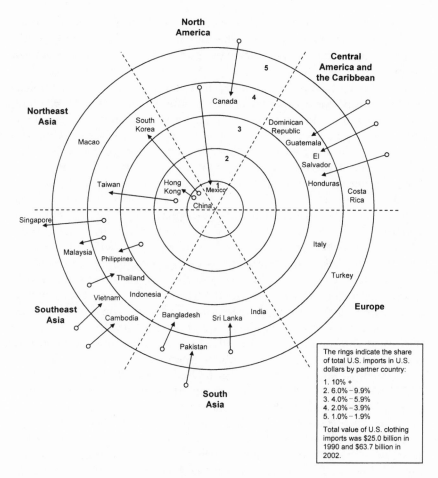

FIGURE 4.2. Shifts in the regional structure of U.S. apparel imports from 1990 to 2002. The 2002 position corresponds to the ring where the country's name is located; the 1990 position, if different, is indicated by a small circle. The arrows represent the magnitude and direction of change over time. Compiled from official statistics of the U.S. Department of Commerce, U.S. imports for consumption, customs value.

modest (changing their position by one ring or not at all), but Mexico improved its position substantially, moving from the fourth circle (2–3.9% of U.S. apparel imports) in 1990 to the core (over 10%) in a decade. Nonetheless, inward shifts of even one ring may be significant for smaller economies, given the substantial overall growth of U.S. apparel imports in the 1990s.

However, two important features of U.S. apparel sourcing are not revealed. First, there are two contending production systems: export-processing assembly (production sharing) and full-package supply (OEM production). The countries that have penetrated the U.S. market most deeply either have been experts at OEM supply (Hong Kong SAR, Taiwan Province of China, and the Republic of Korea) or they are trying to develop full-package capabilities (China and Mexico). The other countries shown in the figure carry out simple assembly only. Second, different kinds of networks are involved and these networks link the countries in the figure in different ways. Triangle manufacturing in East Asia has already been discussed, but the networks relevant to the North American sourcing mix still need to be considered.

If the complete apparel value chain is seen as including raw materials, natural and synthetic fibers, textiles, apparel, and the distribution of apparel to retailers, the Mexican and U.S. value chains are quite distinct. Mexico has several large, reasonably successful synthetic fiber companies, a multitude of export-oriented assembly firms that send products to the United States using U.S. inputs, and an emergent retail sector that is fashioning a number of strategic alliances with its large U.S. counterparts. The weakest link in the Mexican production chain has been textiles. Most of its textile companies are undercapitalized, technologically backward and inefficient, and produce poor-quality goods. By contrast, the United States is strong in synthetic fibers, textiles, and retailing, but its garment production capability is limited, especially for women's and children's clothing. The Mexican chain thus appears to be strongest where the U.S. chain is weak: garment production.[3]

The picture is more complex if North America is expanded to include Central America and the Caribbean. Production sharing in Latin America is centered on Mexico and the Caribbean Basin because of the region's low wages and proximity to the U.S. market, where over 90% of its exports go. Virtually all the production is low value-added, which is a direct result of U.S. policy. Under the tariff schedule provision HTS 9802.00.80 (formerly clause 807), companies engaged in production sharing have an incentive to minimize locally purchased inputs as only components made in the United States are exempt from import duties when the finished product is shipped back there. This is a major impediment to increasing integration between export activity and the local economy, and it limits the usefulness of production sharing as a stepping-stone to higher stages of industrialization.

From a regional perspective, Mexico competes most directly with the Caribbean Basin Initiative (CBI) countries for the U.S. market. By the early 1990s, EPZs had become a leading source of exports and manufacturing employment in a number of Caribbean countries, of which the Dominican Republic is a prime example. In the mid-1990s, there were

430 companies employing 164,000 workers in 30 free-trade zones; three-quarters of the firms were involved in textiles and apparel (Burns, 1995, p. 39). By 2000, apparel exports were still a primary source of employment for many countries in the Caribbean Basin, with 145,000 apparel employees in the Dominican Republic, 110,000 in Honduras, 77,000 in Guatemala, 60,000 in El Salvador, nearly 40,000 in Costa Rica, and 20,000 in Nicaragua. Although Mexico had almost 560,000 apparel workers, they tended to be in much smaller plants than the large export factories established in Central America and the Caribbean (Bair & Gereffi, 2002, p. 33; also see Mortimore, 2002; and Matthews, 2002).

The rivalry among neighboring EPZs to offer transnational companies the lowest wages fosters a perverse strategy of competitive devaluation, whereby currency depreciations are viewed as a means to increase international competitiveness (Kaplinsky, 1993). Export growth in the Dominican Republic's EPZs skyrocketed after a sharp depreciation of its currency against the dollar in 1985; similarly, Mexico's export expansion was facilitated by recurrent devaluations of the Mexican peso, most notably in 1994–1995. Devaluations heightened already substantial wage differences in the region. Hourly compensation rates for apparel workers in the early 1990s were $1.08 in Mexico, $0.88 in Costa Rica, $0.64 in the Dominican Republic, and $0.48 in Honduras, compared with $8.13 in the United States (International Labour Organization, 1995, pp. 35–36). Although devaluations may attract users of unskilled labor to production sites, the advantages evaporate when other countries simultaneously engage in wage-depressing devaluations, which lower local standards of living while doing nothing to improve productivity.

Three models of competition stand out when examining the North American apparel sector and its prospects for change: the East Asian, the Mexican, and the Caribbean Basin. It would be misleading, however, to think of these as inherently national or regional patterns. Rather, the success and limitations of East Asian, Mexican, and Caribbean Basin apparel producers are determined by two factors: their location (not nationality) and the transnational networks of which they are part. Ultimately, to be successful in the global economy requires an understanding of how to use organizational networks to penetrate international markets. The three models of competition identified here use networks and markets quite differently.

The *East Asian model* is based on highly successful textile and apparel exporters from Hong Kong SAR, Taiwan Province of China, and the Republic of Korea, which have moved through a sequence of roles from assembly to OEM to OBM. The East Asian NIEs developed and refined their OEM capabilities in the 1960s and 1970s by establishing close ties with U.S. retailers and marketers, and then "learning by watching" in order to build their export competence (Gereffi, 1997). The performance trust built up through successful business transactions with U.S. buyers enabled suppliers in the East Asian NIEs to use their OEM expertise internationally via triangle manufacturing, that is, the East Asian manufacturers became intermediaries between U.S. buyers and apparel factories in Asia and other developing regions in order to take advantage of lower labor costs and favorable quotas. The creation of these global sourcing networks helped East Asian NIEs to sustain their international competitiveness when domestic economic conditions and quota constraints threatened the original, bilateral OEM relationships. The East Asian NIEs have gone beyond OEM by shifting to higher value upstream products (e.g., exports of textiles and fibers rather than apparel), moving downstream from OEM to OBM in apparel, and switching to new value chains where their export success in apparel can be replicated.

When the phase-out of MFA tariffs is implemented in 2005, in accordance with the World Trade Organization (WTO) Agreement on Textiles and Clothing, a considerable consolidation of apparel exports from the largest low-wage suppliers can be expected. China (including Hong Kong SAR) is likely to become even more dominant as the world's export leader after 2005, with Indonesia, Vietnam, India, Mexico, and Turkey moving into the second tier at the global level, although Mexico and Turkey are primarily regional suppliers for the U.S. and EU markets, respectively. The Republic of Korea and Taiwan Province of China will continue to exploit their niche as suppliers of textile inputs to the major Asian apparel exporters, and they are likely to retain smaller but still significant exports of relatively high-value apparel items in which quality, product development, timely delivery, and related services are at a premium.

The emerging *Mexican model* involves a transition from assembly to OEM (or full-package) production. The key factor here has been NAFTA. The passage of NAFTA in 1994 began to remove the trade restrictions that had locked Mexico into an assembly role. The *maquiladora* system required suppliers in Mexico to use U.S. inputs in order to gain duty-free access to the U.S. market. The progressive 10-year phase-in period for NAFTA shows, step by step, how more and more of the apparel supply chain (such as cutting, washing, and textile production) is relocating to Mexico as specific tariff restrictions on each of these stages are eliminated. The East Asian NIEs did not employ the production-sharing provisions established by the 807/9802 U.S. trade regime because the distances involved made U.S. textile inputs impractical. In addition, U.S. textile mills had neither the manufacturing capability nor the desire because of their mass-production orientation to supply the wide range of fabrics required for women's clothing and fashion-oriented apparel, which became the specialty of East Asian exporters. This created an OEM niche that East Asian apparel companies were quick to exploit.

However, NAFTA alone does not guarantee Mexico's success. While the massive peso devaluations of 1994–1995 made it attractive to U.S. apparel manufacturers with international subcontracting operations, Mexico has traditionally lacked the necessary infrastructure for full-package production of garments. From a value-chain perspective, the solution to how to complete the transition to full-package supply and develop new production and marketing niches is to forge linkages to lead firms that can supply the needed resources and tutelage. In other words, Mexico needs to develop new and better networks in order to compete with East Asian suppliers for the United States full-package market.

U.S. firms have shown a strong interest in transferring missing pieces of the North American apparel supply chain to Mexico, but there is still a problem with who controls critical nodes of the chain and how to manage the dependency relationships this implies. Thus far, U.S. firms are in control of the design and marketing segments, while Mexican companies are in a good position to maintain and coordinate the production networks. For the foreseeable future, Mexico is likely to retain a mix of assembly plants linked to U.S. branded manufacturers and a new set of full-package producers linked to private-label retailers and marketers. As more of the critical apparel inputs become available in Mexico, U.S. inputs will decline and traditional Mexican assembly plants will be replaced by vertically integrated manufacturers or by clusters of related firms that compete through localized networks, such as the jeans producers in Torreón (Bair & Gereffi, 2001).

The *Caribbean Basin model* is almost exclusively limited to EPZ assembly using the 807/9802 trade regime. The CBI countries did not receive NAFTA parity until October 2000, and therefore they encountered quota restrictions, higher tariffs, and more limited possibilities for vertical integration than Mexico. Nonetheless, they have had considerable success with export assembly. They are expanding their position in the U.S. market, primarily through large assembly plants linked to the production-sharing operations of U.S. transnationals (Gereffi, 2000, Table 7.3; Mortimore, 2002). However, CBI exporters are losing ground to Mexican firms that can export similar goods to the United States more cheaply and quickly. They need to develop new networks with U.S. retailers and marketers if they are to acquire the skills and resources needed to move into the more diversified activities associated with full-package production.

Sustained competitiveness in the international apparel industry involves continual changes in economic roles and capabilities. New exporters constantly enter the global supply chain, which is pushing existing firms to cut costs, upgrade, or exit the market. There is a need to run faster to stay in the same place. To facilitate both adjustment and survival in a volatile, export-oriented sector such as apparel, industrial upgrading typically requires organizational linkages to buyers and suppliers in developed countries' markets. Mexico is using networks with U.S. firms to try to occupy niches that have been the stronghold of East Asian suppliers, and the CBI countries are trying to keep up with Mexico. If Mexico is to take over the North American apparel market, it must learn from U.S. lead firms in the chain, and also seize control of those opportunities that would allow it to expand its domestic and regional capabilities and options.

WORLD MARKET TRENDS

A closer look at leading apparel exporters in the 1980s and 1990s reveals both a broadening and a deepening of global sourcing networks. If apparel exports worth $1 billion are taken as a threshold for major players in the global industry, Table 4.2 shows a striking stair-step pattern of market entry.[4] In 1980, only Hong Kong SAR, the Republic of Korea, Taiwan Province of China, China, and the United States were major exporters. By 1990, Indonesia, Thailand and Malaysia in Southeast Asia had joined them, as had India and Pakistan in South Asia and Tunisia in North Africa. The largest newcomer in 1990 was Turkey, whose total of $3.4 billion in clothing exports placed it fifth in world rankings, behind the four Northeast Asian powerhouses. In 2001, new members of the billion-dollar club included the Philippines and Vietnam in Southeast Asia, Bangladesh and Sri Lanka in South Asia, Morocco in Africa, and three East European countries. Mexico had a meteoric rise, with clothing exports soaring from $0.1 billion in 1990 to $8.5 billion in 2001. The top four developing economy apparel exporters in 2001 were China ($41.2 billion), Hong Kong SAR ($24.2 billion), Mexico ($8.5 billion), and Turkey ($7.1 billion).

Notwithstanding these high absolute levels of apparel shipments, the world's leading suppliers vary widely in the importance of apparel as an export item. The countries most reliant on apparel exports are Bangladesh (79%), Mauritius (57%), Sri Lanka (53%), and the Dominican Republic (48%), while in Tunisia and Morocco apparel represents 40% and 33% of total exports, respectively, and in Romania, Pakistan, and Turkey 20–25% (see Table 4.2).

TABLE 4.2. Patterns of Entry to World Market and Apparel Concentration Ratios for World's non-EU Top Apparel Exporters, 1980–2001

Region/ country	Population (millions) 2001[a]	GNI (U.S.$ billions) 2001[a]	GNI/ capita (U.S.$) 2001[a]	Total national exports (U.S.$ billion)			Apparel exports to the world market (U.S.$ billion)			Apparel as % of total national exports			Hourly apparel labor costs (wages and fringe benefits; U.S.$) 1998[d]
				1980[b]	1990[b]	2001[b]	1980	1990	2001	1980	1990	2001	
Northeast Asia													
China	1,272	1,131	890	19.3	64.9	299	1.7	10.2	41.2	8.6%	15.7%	13.8%	0.43
Hong Kong	6.7	170	25,330	20.8	83.8	196	5.3	15.7	24.2	25.4%	18.7%	12.3%	5.20
South Korea	47	448	9,460	18.5	66.5	163	3.1	8.3	4.7	17.0%	12.4%	2.9%	2.69
Taiwan	22[c]	287[c]	12,876[c]	21.1	71.3	152	2.6	4.2	3.0	12.3%	5.8%	2.0%	4.68
Southeast Asia													
Indonesia	209	145	690	23.6	28.1	61.3	0.6	2.9	5.1	2.4%	10.3%	8.3%	0.16
Thailand	61	118	1,940	6.9	23.8	70.8	0.3	2.9	4.0	4.2%	12.2%	5.6%	0.78
Philippines	78	81	1,030	6.1	8.4	35.3	0.3	0.7	2.7	4.9%	8.4%	7.6%	0.76
Malaysia	24	79	3,330	13.8	30.5	96.6	0.2	1.4	2.3	1.2%	4.5%	2.4%	1.30
Vietnam	80	33	410	0.2	1.4	14.9	0.0	0.1	1.6	7.3%	5.0%	10.6%	0.22
South Asia													
India	1,032	477	460	8.1	18.5	47.8	0.6	2.6	5.6	7.9%	14.2%	11.7%	0.39
Bangladesh	133	49	360	0.8	1.5	6.6	0.0	0.6	5.2	0.2%	42.0%	78.8%	0.30
Sri Lanka	19	16	880	1.1	2.0	5.1	0.1	0.7	2.7	10.5%	33.9%	52.5%	0.44
Pakistan	141	60	420	2.6	5.7	10.0	0.1	1.1	2.4	4.2%	18.5%	23.9%	0.24
Central and Eastern Europe													
Turkey	69	167	2,530	2.8	13.0	33.0	0.1	3.4	7.1	4.6%	25.9%	21.4%	1.84
Romania	22	39	1,720	11.9	5.7	12.0	0.4	0.4	3.0	3.1%	7.6%	24.6%	1.04
Poland	39	164	4,230	15.0	12.4	37.5	0.6	0.4	2.0	4.2%	3.0%	5.4%	2.77
Hungary	10	49	4,830	8.6	9.7	32.3	0.3	0.4	1.4	4.0%	3.8%	4.4%	2.12
Africa													
Tunisia	10	20	2,070	2.1	3.5	7.0	0.3	1.1	2.8	15.7%	32.9%	39.5%	0.98[e]
Morocco	29	35	1,190	2.4	4.3	7.5	0.1	0.7	2.5	4.6%	17.2%	33.0%	1.36
Mauritius	1.2	5	3,830	0.5	1.3	1.6	0.1	0.6	0.9	17.2%	50.9%	56.7%	1.03
Caribbean Basin													
Dominican Rp	8.5	19	2,230	0.8	2.1	4.9	0.0	0.8	2.4	0.0%	35.7%	48.1%	1.48
Costa Rica	3.9	16	4,060	1.1	1.6	4.9	0.0	0.1	0.4	1.9%	3.7%	8.2%	2.52
North America													
Mexico	99	550	5,530	16.4	29.2	168	0.1	0.1	8.5	0.3%	0.4%	5.0%	1.51
USA	285	9,781	34,280	240	418	778	1.3	2.7	7.5	0.1%	0.6%	1.0%	10.12
World Totals	6,130	31,400	5,120	2,014	3,471	6,482	39.6	110.6	217.0	2.0%	3.2%	3.3%	na

[a] World Bank (2003). Gross National Income (GNI) is now used by the World Bank instead of the closely related measure, Gross National Product.

[b] World Trade Analyzer, based on United Nations trade data. Apparel is defined as SITC 84.

[c] Data from Taiwan Government Information Office website: *http://www.gio.gov.tw/taiwan-website/5-gp/yearbook/chpt09.htm*

[d] Werner International, Inc., "Hourly Labor Cost in the Apparel Industry."

[e] Data for 1996.

Table 4.3 provides information on whether apparel has risen or fallen in rank among the leading export items (measured at the two-digit SITC level) of the world's biggest non-EU apparel exporters. In Northeast and Southeast Asia, it has declined in importance, except in China where it remains the top export item, and in Indonesia and Vietnam where apparel has climbed to third place. However, in much of South Asia, Africa, the Caribbean Basin, and Central and Eastern Europe, apparel is the leading export, and frequently has been for a decade or more. Sub-Saharan Africa lags behind the other developing regions in apparel sourcing, largely because of poor transportation and communication infrastructure in many countries, its shortage of concentrated pools of low-wage labor, and a difficult political and cultural environment for foreign investors.

CONCLUSION

This chapter uses the global value chains framework to explain the transformations in production, trade, and corporate strategies that have altered the global apparel industry and changed the prospects for developing countries in entering and moving up these chains. The apparel industry is identified as a buyer-driven value chain that contains three types of lead firms: retailers, marketers, and branded manufacturers. As apparel production has become global and competition has intensified, each type of lead firm has developed extensive global sourcing capabilities. While deverticalizing out of production, they are building up their activities in the high-value-added design and marketing segments of the apparel chain, leading to a blurring of boundaries and a realignment of interests and opportunities within the chain.

Industrial upgrading in apparel is primarily associated with the shift from assembly to full-package production. Compared with the mere assembly of imported inputs, full-package production fundamentally changes the relationship between buyer and supplier in a direction that gives far greater autonomy and learning potential for industrial upgrading to the supplying firm. Full-package production is needed because the retailers and marketers that order the garments do not know how to make them. The East Asian NIEs of Hong Kong SAR, Taiwan Province of China, the Republic of Korea, and China have used the full-package role to create an enduring edge in export-oriented development. However, NAFTA and a decline in the importance of East Asian apparel exports to the United States have created favorable conditions for the extension of full-package production to North America. Prominent apparel suppliers to Europe, such as Turkey and several East European

TABLE 4.3. Position of Apparel Among Leading Export Items, 1980–2001

Exporting country	2001 SITC[a]	Top export item, Description	Top export item, U.S.$ billions	% of total exports	Apparel rank[b] 1980	1990	2001	Trends in apparel ranking 1980–1990	1990–2001
Northeast Asia									
China	84	Apparel	41.2	13.8	4	1	1	Up	Same
Hong Kong	77	Electrical machinery	31.3	16.0	1	1	2	Same	Down
South Korea	77	Electrical machinery	23.0	14.1	1	1	11	Same	Down
Taiwan	77	Electrical machinery	34.7	22.8	1	5	13	Down	Down
Southeast Asia									
Indonesia	33	Petroleum	7.5	12.3	6	4	3	Up	Up
Thailand	77	Electrical machinery	10.1	14.2	8	1	4	Up	Down
Philippines	77	Electrical machinery	15.8	44.7	6	1	3	Up	Down
Malaysia	77	Electrical machinery	23.1	23.9	9	6	7	Up	Down
Vietnam	33	Petroleum	3.2	21.6	4	5	3	Down	Up
South Asia									
India	66	Nonmetallic mineral mfg.	6.2	13.0	4	2	3	Up	Same
Bangladesh	84	Apparel	5.2	78.8	13	1	1	Up	Same
Sri Lanka	84	Apparel	2.7	52.5	4	1	1	Up	Same
Pakistan	65	Textile yarn and fabrics	4.9	49.1	4	2	2	Up	Same
Central and Eastern Europe									
Turkey	84	Apparel	7.1	21.4	6	1	1	Up	Same
Romania	84	Apparel	3.0	24.6	3	4	1	Down	Up
Poland	78	Road vehicles	3.4	9.0	6	10	5	Down	Up
Hungary	76	Telecommunications	4.0	12.5	8	9	6	Down	Up
Africa									
Tunisia	84	Apparel	2.8	39.5	2	1	1	Up	Same
Morocco	84	Apparel	2.5	33.0	7	1	1	Up	Same
Mauritius	84	Apparel	0.9	56.7	2	1	1	Up	Same
Caribbean Basin									
Dominican Rp	84	Apparel	2.4	48.1	34	1	1	Up	Same
Costa Rica	05	Vegetables and fruit	1.0	20.3	9	7	4	Up	Up
North America									
Mexico	78	Road vehicles	29.6	17.6	27	35	6	Down	Up
United States	77	Electrical machinery	90.5	11.6	38	37	25	Up	Up
World Totals	78	Road vehicles	580.3	9.0	15	9	8	Up	Up

Note: From World Trade Analyzer, based on United Nations trade data.

countries (e.g., Romania, Poland, and Hungary), also appear to be adopting the full-package model.

Three models of competition are evident in the North American market: the East Asian, Mexican, and Caribbean Basin. Each model presents different perspectives and challenges for industrial upgrading. The United States continues to define the terms of change, and U.S. firms lead the process toward mass customization and agile manufacturing. Mexico needs to develop new and better networks in order to compete with East Asian suppliers for the U.S. full-package market. The Caribbean Basin model, almost exclusively limited to assembly, would have to develop networks with U.S. retailers and marketers if companies are to acquire the skills and resources needed to move into the more diversified activities associated with full-package production. The Mexican and Caribbean experiences can be generalized as full-package and assembly models applicable to other regional contexts.

There has been a dramatic consolidation of the retailer segment of buyer-driven value chains in the United States, and a growth in the strength of retailers as opposed to apparel manufacturers in the EU and Japan. While retailing and marketing are becoming more concentrated, manufacturing is splintering. To a certain degree, this trend is propelled by the information revolution giving retailers better day-to-day market information about consumer purchasing decisions, allowing them to demand more from their suppliers in terms of inventory management, quick response, more frequent deliveries, and the like. As retailers develop their own private-label collections, they also change the competitive dynamics of the textile and apparel supply chain, since they become competitors (rather than customers) of traditional apparel manufacturers and designers. Finally, retailers are pushing globalization in a direct way as importers, and by demanding lower prices from manufacturers which in turn forces them to go overseas. Because they themselves do not have production experience, however, the retailers in buyer-driven chains are dependent upon the suppliers in their global sourcing networks. In Asia, a number of these manufacturers are integrating forward from specification contracting (the OEM, or full-package role) to developing and selling their own brands (the OBM role). In North America, textile companies are forming production clusters with local apparel firms in Mexico to assure themselves of a customer base. Thus a growing concentration at the retail end of the value chain is generating networks of collaborators as well as competitors in the upstream segments of the chain.

How does the control structure of value chains affect industrial upgrading in developing countries? First, the comparison of apparel imports to the United States and the EU reveals distinct regional patterns of sourcing. While both the United States and the EU source heavily out of Asia, they each have nearby sourcing bases as well: Mexico, Central America, and the Caribbean Basin for the United States, and Eastern–Central Europe and North Africa for the EU. More importantly, these different regional supply bases for apparel are organized in different kinds of networks. The Asian sourcing is done on the basis of direct imports and specification contracting, while the Caribbean and Mediterranean Basin sourcing patterns use forms of international subcontracting in which U.S. and EU textiles are sent to nearby low-wage countries for assembly into garments. The controlling agents in these two networks are different: they are retailers and designers in the Asian trade, and textile and apparel manufacturers for outward processing trade.

The possibilities for integrated local industrial development are greater in the OEM model where Asian manufacturers have developed an important form of social capital in the guise of the multifaceted and dense networks utilized to offer full-package supply. In the outward-processing or production-sharing pattern, the production networks are much thinner in the supplying countries. One of the most interesting emerging responses is the effort by textile firms, apparel companies, and retailers in the United States and Mexico to emulate the OEM model of the East Asian NIEs by constructing similar kinds of full-package networks in the North American context. This requires supportive policies at the macroeconomic level (participation in NAFTA), as well as capable Mexican firms that are able to anchor global production and sourcing networks within the full-package model. However, the downturn in U.S. apparel sales in the early 2000s, the profitability problems, bankruptcies, and mergers faced by major textile firms in the United States (such as Burlington Industries, Cone Mills, and Guilford Mills), and the likelihood of greatly increased import competition from China and other low-cost competitors after the phase-out of apparel quotas by the WTO in 2005, make sustained success in the global apparel industry a challenging and still very elusive goal.

ACKNOWLEDGMENTS

This chapter was originally prepared by Gary Gereffi, Department of Sociology, Duke University, in cooperation with United Nations Industrial Development Organization (UNIDO) staff member Olga Memedovic, from the Strategic Research and Economy Branch.

The views expressed therein, the designations employed, as well as the presentation of material in this publication do not imply the expressions of any opinion whatsoever on the part of the Secretariat of the UNIDO concerning the legal status of any country, territory, city, or area or of its authorities, or concerning the delimitation of its frontiers or boundaries. Designations such as "industrialized," "developed," and "developing" countries are intended for statistical convenience and do not necessarily express a judgment about the stage reached by a particular country or area in the development process. Mention of firm names or commercial products does not imply endorsement by UNIDO. Material in this chapter may be freely quoted but acknowledgment is requested, together with a copy of the publication containing the quotation or reprint.

The editors have made modest cuts in the original article in order to better fit the parameters of this book.

NOTES

1. These figures refer to the retail market comprised of companies with publicly held stock.
2. These statistics are derived from World Trade Analyzer, a database developed by Statistics Canada using United Nations trade statistics.
3. Empirical support for this argument is provided in Office of Technology Assessment (1992, Ch. 9) and Gereffi, Spener, and Bair (2002).
4. Intra-EU apparel exports are excluded from the total for European countries in this table.

REFERENCES

American Apparel Manufacturers Association. (1984). *Apparel manufacturing strategies*. Arlington, VA: Author.

Abend, J. (2000). Battle of the brands. *Bobbin, 42*(2), 36–40.

Abernathy, F. H., Dunlop, J. T., Hammond, J. H., & Weil, D. (1999). *A stitch in time: Lean retailing and the transformation of manufacturing—Lessons from the apparel and textile industries*. New York: Oxford University Press,

Appelbaum, R. P., & Gereffi, G. (1994). Power and profits in the apparel commodity chain. In E. Bonacich, L. Cheng, N. Chinchilla, N. Hamilton, & P. Ong (Eds.), *Global production: The apparel industry in the Pacific Rim* (pp. 42–64). Philadelphia: Temple University Press.

Appelbaum, R. P., Smith, D., & Christerson, B. (1994). Commodity chains and industrial restructuring in the Pacific Rim: Garment trade and manufacturing. In G. Gereffi & M. Korzeniewicz (Eds.), *Commodity chains and global capitalism* (pp. 187–204). Westport, CT: Praeger.

Bair, J., & Gereffi, G. (2001). Local clusters in global chains: The causes and consequences of export dynamism in Torreon's blue jeans industry. *World Development, 29*(11), 1885–1903.

Bair, J., & Gereffi, G. (2002). NAFTA and the apparel commodity chain: corporate strategies, interfirm networks, and industrial upgrading. In G. Gereffi, D. Spener & J. Bair (Eds.), *Free trade and uneven development: The North American apparel industry after NAFTA* (pp. 23–50). Philadelphia: Temple University Press.

Berger, S., & Lester, R. K. (1997). *Made by Hong Kong.* New York: Oxford University Press.

Birnbaum, D. (1993). *Importing garments through Hong Kong.* Hong Kong: Third Horizon Press.

Burns, J. G. (1995, September). Free-trade zones: Global overview and future prospects. *Industry, Trade, and Technology Review,* pp. 35–47.

Davies, M. (2002, April 10). Marks & Spencer sales rise delights the City. Press Association Limited news release (retrieved from LexisNexis Academic online search).

De Coster, J. (1996a). Hong Kong and China: The joining of two giants in textiles and clothing. *Textile Outlook International, 68,* 63–79.

De Coster, J. (1996b). Productivity: A key strategy of the Hong Kong textile and clothing industry. *Textile Outlook International, 68,* 80–97.

Dickerson, K. G. (1995). *Textiles and apparel in the global economy* (2nd ed.). Englewood Cliffs, NJ: Prentice-Hall.

Dickerson, K. G. (1999). *Textiles and apparel in the global economy* (3rd ed.). Englewood Cliffs, NJ: Prentice-Hall.

Finnie, T. A. (1996). Profile of Levi Strauss. *Textile Outlook International, 67,* 10–37.

Gereffi, G. (1994). The organization of buyer-driven global commodity chains: How United States retailers shape overseas production networks. In G. Gereffi & M. Korzeniewicz (Eds.), *Commodity chains and global capitalism* (pp. 95–122). Westport, CT: Praeger.

Gereffi, G. (1995). Global production systems and third world development. In B. Stallings (Ed.), *Global change, regional response: The new international context of development* (pp. 100–142). New York: Cambridge University Press.

Gereffi, G. (1997). Global shifts, regional response: Can North America meet the full-package challenge? *Bobbin, 39*(3), 16–31.

Gereffi, G. (1998). Commodity chains and regional divisions of labour in East Asia. In E. M. Kim (Ed.), *The four Asian tigers: Economic development and the global political economy* (pp. 93–124). San Diego, CA: Academic Press.

Gereffi, G. (1999). International trade and industrial upgrading in the apparel commodity chain. *Journal of International Economics, 48*(1), 37–70.

Gereffi, G. (2000). The transformation of the North American apparel industry: Is NAFTA a curse or a blessing? *Integration and Trade, 4*(11), 47–95.

Gereffi, G., & Korzeniewicz, M. (Eds.). (1994). *Commodity chains and global capitalism.* Westport, CT: Praeger.

Gereffi, G., Spener, D., & Bair, J. (Eds.). (2002). *Free trade and uneven develop-*

ment: The North American apparel industry after NAFTA. Philadelphia: Temple University Press.

Granitsas, A. (1998, May). Back in fashion: Hong Kong's leading garment makers are going global—learning to add value and high technology. *Far Eastern Economic Review,* pp. 52–54.

International Labour Organization. (1995). *Recent developments in the clothing industry* (Report No. 1). Geneva: Author.

Jones, J. (1995, March). Forces behind restructuring in United States apparel retailing and its effect on the United States apparel industry. *Industry, Trade, and Technology Review,* pp. 23–27.

Kaplinsky, R. (1993). Export processing zones in the Dominican Republic: Transforming manufactures into commodities. *World Development, 21*(11), 1855–1856.

Khanna, S. R. (1993). Structural changes in Asian textiles and clothing industries: The second migration of production. *Textile Outlook International, 49,* 11–32.

L'Observatoire Européen du Textile et de l'Habillement. (1995). *The EU textile and clothing industry 1993/94.* Brussels: Author.

L'Observatoire Européen du Textile et de l'Habillement. (1996). *The EU textile and clothing industry 1995.* Brussels: Author.

Management Horizons. (1993). *Global retailing 2000.* Columbus, OH: Author.

Mathews, D. T. (2002). Can the Dominican Republic's export-processing zones survive NAFTA? In G. Gereffi, D. Spener, & J. Bair (Eds.), *Free trade and uneven development: The North American apparel industry after NAFTA* (pp. 308–323). Philadelphia: Temple University Press.

Miller, J. P. (1997, September 15). Sara Lee plans "fundamental reshaping." *Wall Street Journal,* pp. A3, A10.

Mortimore, M. (2002). When does apparel become a peril?: On the nature of industrialization in the Caribbean Basin. In G. Gereffi, D. Spener, & J. Bair (Eds.), *Free trade and uneven development: The North American apparel industry after NAFTA* (pp. 287–307). Philadelphia: Temple University Press.

Office of Technology Assessment, U.S. Congress. (1992). *US–Mexico Trade: Pulling together or pulling apart?* (ITE-545). Washington, DC: United States Government Printing Office.

Onozuka, A. (2001). Japan faces dilemma over Chinese imports. *Bobbin, 43*(1), 60–65.

Rabon, L. (2001). Technology outlook 2001: U. S. suppliers weather winds of change. *Bobbin, 42*(5), 54–60.

Scheffer, M. (1994). *The changing map of European textiles: Production and sourcing strategies of textile and clothing firms.* Brussels: L'Observatoire Européen du Textile et de l'Habillement.

Speer, J. K. (2001). The Bobbin top 50: Apparel retailers take on a new face. *Bobbin, 42*(10), 45–66.

Tait, N. (2000). Marks & Spencer dramatically shifts supplier base. *Bobbin, 41*(6), 20–25.

Taiwan Statistical Data Book (1999). Council for Economic Planning and Development, Republic of China.

U.S. International Trade Commission. (1997). *Production sharing: Use of United States components and materials in foreign assembly operations, 1992–1995* (USITC Publication No. 3032). Washington, DC: Author.

World Bank. (2003). *World development indicators online.* Washington, DC: Author.

5

A System on the Brink
*Pitfalls in International Trade Rules
on the Road to Globalization*

Keith E. Maskus

Ten years ago the antiglobalization movement was unfocused, incoherent, and shrill. Significant concerns were raised about the potential implications of such trade agreements as the North American Free Trade Agreement (NAFTA) and the World Trading Organization (WTO), both of which were in their final stages of approval at that time. However, the claims made by globalization critics generally lacked serious intellectual foundations, were based more on selective anecdotes than on systematic evidence, and at times seemed willfully misleading (Wallach & Sforza, 1999). It was, in any event, not difficult for politicians to dismiss such arguments.

Moreover, in the early 1990s the world was awash in a kind of globalization euphoria in which it seemed that the economic collapse of the Soviet system had put to rest the notion that central planning and rigorous trade protection could build a competitive economy. Rather, persuaded (and at times compelled) by the so-called Washington Consensus of privatization, deregulation, and liberalization of goods and financial markets, many developing and transition economies rapidly opened their borders and reduced the role of government in their citizens' lives (Stiglitz, 2002). Openness to external competition was hailed as a significant, and perhaps necessary, contributor to sound economic growth (Frankel & Romer, 1998; Dollar & Kraay, 2001). Moreover, liberalization advocates relied on the almost self-evident proposition that opening up to trade would present citizens of poor countries more opportunities to engage in the modern economic system and rapidly lift them out of poverty (World Bank, 1993). Partly because of this optimism, great numbers of developing countries joined the WTO in the late 1990s, culminating in the entry of China in 2001.

In the early years of the 21st century, however, attitudes are changing rapidly about the virtues and costs of globalization through adherence to the international trading system. Many critics argue that the promises of freer trade—more rapid growth, greater poverty reduction through expanded opportunities, faster convergence in incomes through technology transfer—simply have not materialized for large numbers of impoverished people. There is no clear correlation in recent policy episodes among trade liberalization, growth, and poverty alleviation. Indeed, in many countries more openness has been accompanied by rising inequality, at least in the medium term (McCulloch, Winters, & Cirera, 2001). Moreover, the perception is increasingly common that reducing trade and investment barriers induces excessive exploitation of environmental resources or worsens the conditions of work.[1]

Perhaps most significantly, criticisms of globalization have become deeper and intellectually more rigorous over time.[2] In no small part this reflects the fact that many serious trade analysts, who recognize the underlying promise of greater international integration, are nonetheless vitally concerned about the forms in which this integration is taking place (Stiglitz, 2002; Rodrik, 1997). For international trade economists, including myself, this dissatisfaction stems from three fundamental and interrelated factors. First, the world is far more complex than either the earlier critics or the advocates of globalization could describe with their basic messages. To illustrate, both of the competing claims that free trade either harms or improves the physical environment are true, false, and absurd at the same time. They are true because one can find meaningful examples of both claims. They are false because one can find counterexamples. And they are absurd because an indirect policy instrument—tariff cuts—aimed at other objectives could never be expected to have simple and unidirectional impacts on environmental use (Antweiler, Copeland, & Taylor, 2001).

Second, the WTO agreements contain a far more comprehensive set of disciplines, defined as rules limiting the ability of governments to interfere with trade, than did its predecessor, the General Agreement on Tariffs and Trade (GATT). The former system encouraged governments to reduce border restraints on trade and attempted to restrain policy-makers from violating their commitments on such reductions, albeit with no ability to compel such discipline. In contrast, the WTO moves far beyond simple disciplines on border restraints into limiting the ability of governments to adopt policies that traditionally have been the preserve of domestic sovereignty (Hoekman & Kostecki, 2001). Examples include intellectual property rights (IPRs), limitations on entry into services provision, the regulation of food safety, and investment policy. Even the ability to define and sustain national health and environmental policies may be restricted by WTO rules under certain circumstances. These limits to domestic policy flexibility arise putatively because domestic regulations are "trade-related," and therefore may interfere with the efficient flow of goods and investment (Maskus, 2002). However, even to economists it is not a self-evident proposition that external restraints on trade and regulatory policies provide enough economic gains to outweigh the potential harm done to social policies.[3]

Third, the rules that were negotiated in founding the WTO inevitably reflect the economic interests of powerful lobbyists in different nations (Hoekman & Kostecki, 2001). The world is obviously characterized by major economic and political imbalances of power among countries, with the United States, the European Union (EU), and a few other major states (or aggregations of states) having significant power to set the rules of the game. Seen in this context, it is no surprise that influential protectionists were able largely to exempt agriculture and labor-intensive manufactures (such as apparel) from liberalization commitments in earlier trade negotiations. Nor is it surprising that the United States and the EU pushed so strongly for disciplines in IPRs, investment, and services, which are clearly items of net export advantages for their firms (Maskus, 2000). However, while developing countries may gain in the long run from more rational regulations of this kind, it is also possible they will be made worse off and, in any event, there are few domestic constituencies that favor them (World Bank, 2001). Thus the trading system is increasingly viewed as unbalanced and driven by special interests (Oxfam, 2002).

An ironic feature of this situation is that the WTO system can fairly be characterized as simultaneously being too limited and being too comprehensive. It is "limited" in that it has failed to achieve significant liberalization in rich-country agricultural trade policies or to discipline the use of antidumping tariffs. Progress in these areas would greatly enhance the benefits of the system for many poor countries but face withering opposition in the developed countries (World Bank, 2001). It is "comprehensive" in its insistence on minimum standards in IPRs, services deregulation, product standards, and other areas. As noted, these rules reach into the domestic policy realm of national governments and are not necessarily welfare-enhancing. For this reason, the system attracts criticism both from trade economists who strongly favor a multilateral, but minimalist, set of rules (Srinivasan, 2000) and from those who think it questionable even to consider a global set of rules (Rodrik, 2001). Or, to put it more simply, the antiglobalists have a case.

In this chapter I explore some of these issues from the perspective of an international trade economist who has been closely involved in this debate, particularly as it concerns IPRs. In the next section I argue that economic theory does not support a simple view of trade liberalization, implying that issues are complex and require empirical study. In the following section I discuss the reasons why a system of trade rules is important but question whether the WTO structure is sensible from the standpoint of economic development. In the final section I offer thoughts on how this system can be improved, both to make it more sustainable and more friendly to development.

THE TYRANNY OF THE SECOND BEST

On its face it is a simple concept: free trade raises the well-being of all countries by inducing them to specialize their resources in those goods they produce relatively most efficiently. Specialization makes people, land, and capital more productive, while the opportunity to trade permits countries to import a wider range of goods at lower costs (Markusen, Melvin, Kaempfer, & Maskus, 1995). Further, it is theoretically and empirically plausible that competition through trade raises a country's long-term growth rate by expanding access to global technologies and promoting innovation (Grossman & Helpman, 1991; Coe, Helpman, & Hoffmaister, 1997). These simple precepts have had a powerful impact on national and global trade policymaking, especially in recent decades.

Reality is far more complicated, of course. First, the argument for aggregate income gains from openness to trade and investment is conditioned by a number of strong assumptions, including the obviously false one that liberalization occurs against the backdrop of an economy that is not otherwise distorted (Markusen et al., 1995). The *theory of the second best* demonstrates that reducing or removing one market distortion (e.g., removing restrictions on trade) in the presence of other market distortions (e.g., international monopoly power or inadequate domestic regulation of the environment) cannot be guaranteed to raise total national income. The potential gains from a more liberal trade environment can be overcome by the potential losses arising from exacerbating another market problem.

To take a simple example, an economy with weakly regulated and nontransparent financial institutions and distorted real estate markets could be made worse off by completely opening its markets to international financial flows (Rodrik & Velasco, 2000; Stiglitz, 2002). Massive capital inflows to finance speculative construction lending and capital flight (the departure of locally owned capital) to escape domestic risk can decidedly raise the volatility of interest rates and exchange rates, with a negative effect on output growth (Stiglitz, 2002). Another illustration is that economies with a comparative advantage in goods made with technologies that pollute the air and water can be made worse off from trade liberalization as such industries expand in the absence of appropriate environmental taxes. Put simply, the productivity gains from reallocation of production resources can be outweighed by the costs of worsened pollution (Bovenberg & van der Ploeg, 1994).

Note carefully that the logic of second-best theory does not imply that openness to globalization processes necessarily reduces welfare. Rather, the claim is that the issues are complex, dependent on circumstances, and must be resolved through careful empirical study. It is quite possible, for example, for trade liberalization to generate large income gains, which are widely dispersed through the economy, by destroying inefficient and high-cost domestic service monopolies (World Bank, 2001). Income gains surely outweigh costs to the natural environment in some cases, and themselves provide further opportunities for individuals and households. For example, after Vietnam phased out the quotas restricting its exports of rice, the real net incomes and consumption standards of households and villages in rice-growing areas rose rapidly, permitting families to shift children from the workplace to education (Edmonds & Pavcnik, 2002).

One does not need to appeal to specific distortions to argue that globalization has a powerful ability to redistribute income and wealth within an economy. Standard trade theory shows that abundant factors (e.g., skilled workers in the United States) will gain from openness to trade and investment while scarce factors (e.g., unskilled workers) will be made worse off (Markusen et al., 1995). However, this is a long-run, full-employment result that again fails to account for such important market rigidities as the inability of workers to get retrained or to relocate, the fact that wealth effects from owning assets with declining values can dominate income gains, and that job-reducing recessions following liberalization can be long-lasting. Moreover, the impacts of openness on poverty depend at least as much on price effects and consumption patterns as they do on job status and employment location (McCulloch et al., 2001).

International trade economists have long appreciated the logic that trade liberalization is not necessarily a good thing when cast in a light broader than basic income effects. The fact that most continue to argue for openness reflects an empirical judgment, perhaps more imagined than real, that the competition and opportunities it affords people generate substantial net gains, almost independent of underlying economic and political circumstances. It is almost self-evident, for example, that the commerce clause of the U.S. Constitution, which forbids individual states to restrict cross-border trade, has contributed to enormous gains in U.S. productivity and incomes by encouraging interstate trade in capital, labor, and goods. Advocates of open trade also point to the experience of East Asia's fast-growing economies (Japan, Korea, Hong Kong, Singapore, Malaysia, and now China) as supportive of the gains that can emerge from strong integration with global markets (World Bank, 1993).

The reality of East Asia is far more complicated, of course, as many analysts have pointed out (Campos & Root, 1996; Lardy, 1998; Posen, 1998). While these economies, far more than others, successively aligned their domestic relative prices with international prices—a key component of international integration—they retained a significant role for governments to intervene strategically in the domestic and international marketplaces. Japan and Korea, for example, followed a policy of reasonably rigorous competition among domestic manufacturing firms while restraining the ability of foreign firms to enter through trade and blocking their entry through foreign direct investment (FDI) (Lawrence, 1993; Young, 1994). At the same time the combination of subsidies to large exporters and the insistence of those firms on competitive exchange rates helped build the large manufacturing concerns that dominate their export structures. Whether this policy actually improved welfare in Japan and Korea remains an open question, for perhaps consumers would have been better served by more open markets and less industrial concentration. Nonetheless, East Asian history demonstrates that complete liberalization and deregulation is not the sole path to rapid growth.

In this context, it is highly doubtful that uniform application of the so-called Washington Consensus of rapid trade liberalization, deregulation of capital markets, privatization, and market-based determination of exchange rates makes sense for *all* developing and transition economies at *all* times. The adherence (often exaggerated and misunderstood by globalization critics) of the International Monetary Fund (IMF), the World Bank, and the WTO to this credo seems puzzling. Why not permit countries to run their own development policies and render their decisions on trade liberalization subservient to the development goal? This is Rodrik's (2001) essential message.

Perhaps the answer is simply that the IMF, the World Bank, and the

WTO are captured by the politically and economically powerful in the rich countries, which do not have essential interests in promoting development of the poor nations (Stiglitz, 2001). There is both logic and evidence favoring this view, and I will expand on this point later. Ultimately, however, trade economists find this view simplistic, for it suggests that the globalization policy architecture is solely a cynical application of power relations. However, an alternative view is that (at least) the WTO represents a serious effort to establish rules of the game under which countries must restrain themselves from indulging their instincts to harm themselves and their trading partners (Hoekman & Kostecki, 2001). As discussed next, this is the essential justification for a world trading system based on rules.

EXTERNALITIES, COORDINATION, AND ALL THAT

Antiglobalization interests often advocate elimination of the WTO, usually on the grounds that its rules embody fundamental restraints on the exercise of national sovereignty (Wallach & Sforza, 1999). This simple view fails to appreciate that a multilateral set of rules that restrict the exercise of mutually harmful trade policies is central for the efficient operation of international trade and the acquisition of mutual gains from exchange (Hoekman & Kostecki, 2001).

Why Do We Need a WTO?

The fundamental problem is that open international trade is a global public good in that it would be underprovided if individual nations were left to their own devices (Kaul, Conceicao, Le Goulven, & Mendoza, 2003). Countries would not ordinarily be expected to unilaterally adopt a liberal trade regime for two reasons. First, the classic "optimal tariff" argument from trade theory is that some nations are large enough to affect international prices in their favor through imposing import tariffs or export taxes (Markusen et al., 1995). This ability is vested in large countries and, absent the nondiscrimination requirements of the WTO, could be used to set prices to the disadvantage of small exporters. Second, there is a powerful asymmetry among domestic economic interests within a country in terms of lobbying for trade policies. Specifically, producers and their employees that compete with imports find it relatively easy to organize themselves in order to achieve high import barriers (Magee, Brock, & Young, 1989; Grossman & Helpman, 2002). In contrast, consumers are diffuse and difficult to organize, while exporters have little standing to lobby for reduced tariffs in foreign export markets. Put differently, there is an inherent bias in virtually all countries toward a policy of high import restraints, despite the evident costs in reduced efficiency and growth.

The WTO may be conceptualized as an auction house and a coordinating device that overcomes these basic problems (Hoekman & Kostecki, 2001). It is an auction house within which member governments may trade off access to their own markets in return for access to foreign markets. In this regard, it mobilizes export interests to lobby their own governments to engage in mutual trade liberalization, bypassing the nearly impossible task of organizing consumers for this purpose. The fact that the WTO (and before it GATT), through its periodic rounds of trade negotiations, is able to convert essentially mercantilist policy tendencies into a bias toward globally liberal trade regimes is its particular genius. Under this system, average global tariffs on manufacturing goods have diminished greatly since the 1950s, while more recently countries have given up substantial recourse to the use of nontariff barriers (NTBs).[4] No other international organization shares a similar purpose, nor can it point to a similar degree of success.[5]

Being an auction house is not sufficient to ensure success, however, because individual governments always have a private incentive to cheat on mutual trade agreements. Put differently, the tariff reductions agreed to in the WTO are *time-inconsistent policies* because governments would be expected to raise tariffs in later periods due to domestic pressures (Bagwell & Staiger, 2003). Thus the WTO needs to have mechanisms in place that coordinate the mutual interests of nations in open trade. Necessarily this means constraining governments from reneging on their agreed-upon obligations. One primary method for doing this is that tariffs are bound from above (and usually at the negotiated rates), meaning that countries cannot unilaterally raise them (Hoekman & Kostecki, 2001). Similar language applies to NTBs. Still another is the pair of nondiscrimination obligations within the WTO. Member governments must provide equal treatment to all other members under the most-favored nation (MFN) clause. And under the national treatment obligation governments may not treat foreign firms in a way that disadvantages their commercial opportunities in comparison with domestic firms.

Backing up this coordination function is the WTO dispute resolution system, under which a violation by one government of an agreed-upon obligation may be challenged by other governments that believe their commercial interests have been harmed. The system was given considerably more power in the transition from the GATT to the WTO (Hoekman & Kotecki, 2001). Specifically, defendant governments may no longer refuse to have a panel report issued, panels have stronger adjudicatory powers, and an offending nation that fails to reform its policy violation must pay for the privilege through a system of fines and retaliatory tariffs. Note that defendant governments may refuse to change their policies. Good examples are the EU's banana import-control regime and its ban on imports of beef from cattle raised with growth hormones, and the U.S. export subsidies under the Foreign Sales Corporation Act. In so refusing, however, they are subject to monetary or trade damages that may alter their domestic political calculus.

Indeed, the WTO is really a form of "soft" federalism (Krueger, 1998). It is federalism in the sense that it limits the rights of governments to act unilaterally in the trade policy realm, rather like (although far more weakly) the commerce clause of the U.S. Constitution. It restricts itself to disciplining violations made only by member governments and not to regulating the actions of private firms in international trade (Maskus, 2000). It is soft in that, while the available penalties are sufficient to make a difference in policymaking, countries retain the right to act unilaterally, at a cost, in those cases where domestic political sensitivities preclude policy reforms.

In principle, these basic features of the WTO—nondiscrimination, mutual access guarantees, no backsliding, and effective dispute resolution—make the system highly advantageous for small and developing countries (Martin & Winters, 1996; Hoekman & Kostecki, 2001). In a world characterized by strictly unilateral trade policies, such countries would have little protection from, or recourse against, the politically determined restraints imposed by large and rich nations. Perhaps more fundamentally, most developing countries themselves are heavily subject to domestic politics and would be unlikely to move toward unilateral liberalization.[6] Over time, the bound nature of trade concessions reached at the WTO has the effect of committing governments not to raise their barriers later. Indeed, this requirement to commit to openness in the face of domestic pressure seems to be a central reason that virtually all developing nations have joined the WTO or are in the process of negotiating accession.

But Why So Many Rules?

The description I have just provided is unobjectionable to most trade economists and might be considered the "core competence" of the WTO. However, the WTO was formed in a way that vastly expanded its purview beyond restrictions against trade in manufactures and primary commodities. This extension underlies much of the international controversy about the trading system (Hoekman & Kostecki, 2001; Bagwell & Staiger, 2003).

There are, in fact, three component agreements that make up the WTO. First, the prior GATT, with its rules on tariff bindings, NTBs, subsidies, and antidumping, supplemented by the extended dispute resolution system, is the WTO's foundation. Notable among the new rules on trade in goods is a commitment to phase out by the year 2005 the bilateral quotas on trade in textiles and apparel under the Multi-Fiber Arrangement (Martin & Winters, 1996). Attempts to achieve reductions in trade barriers in agriculture were frustrated. A significant clarification and strengthening of procedures governing the permissibility of trade restrictions to limit trade in goods that might be considered unsafe was set out in the Agreement on Sanitary and Phytosanitary Measures (SPS) and the Agreement on Technical Barriers to Trade (TBT) (Maskus & Wilson, 2001).

Second, different liberalization principles have been applied to trade in services via the new General Agreement on Trade in Services (GATS). Under this approach, governments may reserve particular forms of service provision (e.g., public health or education) from liberalization commitments, while the forms of liberalization refer less to taxes on provision and more to foreign rights of ownership, local establishment, and movement of persons (World Bank, 2001). Third, the Agreement on Trade-Related Aspects of Intellectual Property Rights (TRIPS) obligates member countries to adopt and enforce minimum standards of protection for patents, copyrights, trademarks, and related devices in their own legislation (Maskus, 2000). A related effort to establish an Agreement on Trade-Related Investment Measures (TRIMS) in order to reduce global regulation of direct foreign investment achieved minor results.

A fascinating question is whether this considerable extension of trading rules into the so-called new areas of product standards, services, and intellectual property rights (IPRs) may be justified in terms of the need for global policy intervention. Again, the essential rationale for supranational regulation is that, in its absence, private actors and individual governments will not provide an appropriate quantity of global public goods and nations will find themselves in a "prisoner's dilemma" of costly unilateralism. This question has been addressed elsewhere (Maskus, 2002; Hoekman & Kostecki, 2001; Kaul et al., 2003) and cannot be discussed fully here. However, some observations are important for setting up the ensuing commentary on the trading system and development.

Stated purely in terms of political economy, the interjection of these new regulatory requirements into the trading regime was necessary to achieve a broad agreement in the Uruguay Round, which established the WTO (Krueger, 1998). The United States and the EU remain the main players in setting the rules of the WTO game, and the executive branches of both needed to align important export interests in favor of a comprehensive pact in order to move their domestic policymaking bodies in that direction. Financial and business services are an area of comparative advantage for both regions, while the United States and several European countries are significant net exporters of goods and technologies protected by IPRs (World Bank, 2001). Without arranging these lobbying interests, it would have been difficult to overcome domestic political opposition to opening apparel markets, cutting tariffs, and strengthening dispute resolution. Even with them, it proved impossible to achieve any effective liberalization of agricultural markets in the rich nations (Martin & Winters, 1996). As for product safety standards, an effective set of WTO agreements was required to mediate between green concerns and business interests in the industrial countries.

That commercial interests drive the WTO agenda is neither new nor surprising. For example, negotiations on the SPS Agreement largely ignored the interests of developing-country food exporters and the SPS Agreement now raises the specter of standards-based, protectionist restraints on their trade (Maskus & Wilson, 2001). However, the influence of private actors in the area of intellectual property rights was particularly remarkable (Maskus, 2000; Sell, 2003). In essence, the corporate leaders in five major industries—pharmaceuticals, biotechnology, agricultural technologies, software, and recorded entertainment—were able to elevate IPRs to the top of U.S. and EU commercial policy priorities, and thence to global obligations in the TRIPS Agreement. That these interests stand to gain considerably from the new regime is not in doubt. Recent econometric evidence suggests that the stronger patent rights required of member states could transfer as much as US$20 billion per year in patent royalties and license fees to firms headquartered in the United States (World Bank, 2001). Even if this estimate exaggerates potential transfers, there should be similar processes at work in the areas of copyright sectors, trademarks, and plant variety protection. In brief, the TRIPS Agreement may be characterized fairly as the greatest piece of strategic trade and industrial policy ever managed on behalf of technology exporters in rich countries (World Bank, 2001; Commission on Intellectual Property Rights, 2002).

Some analysts might find that claim overly cynical, and, indeed, it is possible to assess the need for global rules in the new areas on grounds other than commercial politics. There are, in essence, four criteria that economists consider in judging the suitability of incorporating regulatory standards into the ambit of a multilateral trading agreement (Maskus, 2002). First, does the existence of differing standards distort international trade (i.e., are they "trade-related")? Of course, virtually any significant business regulation framework affects costs and trade flows, so the answer to this question is generally "yes." For example, the efficiency of such producer services as telecommunications, finance, and shipping bears a direct and sharp relationship to international competitiveness in trade (World Bank, 2001). Variations in patent rights across countries have been shown to affect trade and investment flows considerably (Maskus, 2000). The fact that some countries, such as those in the EU and Japan, have extremely rigorous inspection and certification requirements in food products clearly restricts trade opportunities as well (Maskus & Wilson, 2001). Based on this criterion, nearly any set of regulations could be the subject of WTO rules.

Second, are there policy coordination failures that exist because countries acting individually fail to recognize the mutual gains that could come from joint rule setting? As noted above, this is the essential justification for the WTO itself in the area of border trade restrictions. By cutting and binding tariffs, governments jointly reduce inefficient taxes on trade in goods and share the resulting efficiency gains (Bagwell & Staiger, 2003).

However, the issue is more complicated in regulatory areas. Consider opening up to short-term financial flows via banking liberalization. The resulting allocation of capital can be efficient and raise global productivity. However, the volume of such flows and the speed at which they enter and depart markets can overwhelm thin and poorly regulated local banking systems (Rodrik & Velasco, 2000).

With respect to IPRs, it is possible to argue that countries individually tend to select overly weak rights because they fail to recognize the benefits their standards provide to foreign consumers (Grossman & Lai, 2002; Scotchmer, 2003). If countries do not account for this externality in setting policy, a patchwork system of unilateral regulations will fail to generate a globally optimal extent of innovation. However, even if a global requirement for minimum (but strong) national standards could increase national or global welfare in the aggregate, it would need to be accompanied with side payments in the short term to ensure that all nations actually would gain. The reason is that countries that are significant net importers of intellectual property surely would suffer significant short-term costs from stronger standards through higher prices and license fees and administrative charges (World Bank, 2001). There are no such side payments on offer within the WTO.

Similarly, tighter product standards may help improve global food safety but might also be employed at protectionist levels that do little to improve health but substantially restrict trade (Maskus & Wilson, 2001). In this context, a global agreement to restrict excessive zeal can be welfare improving and the WTO has a role to play. Thus the argument for including services and standards is sensible, if dependent on circumstances, while the case for incorporating IPRs into the WTO is tenuous and conditional on so-far-absent transfer payments.

A third criterion is that if governments fail to meet their agreed-upon obligations, there is reasonable scope for using the WTO dispute resolution processes to offset the harm done. The WTO rules focus on commercial damages and give no standing to complaints about weak social protection that is difficult to value in economic terms. In this context, assessing damages from rules violations in services, product standards, and IPRs, while far from straightforward, is conceptually possible in most instances (Maskus, 2002). Thus arbitration and judicial panels presumably can manage much of the tensions inherent in trade-related regulations in these areas.

Finally, an important concern is whether the interjection of these difficult regulatory areas into the WTO threatens to sink the trading system itself (Barfield, 2001). The WTO has a small secretariat that cannot provide sufficient capacity to analyze the intricacies of fiduciary regulation or IPRs. It is therefore heavily dependent on external expertise, including that from interested private parties, in interpreting its rules and issuing panel reports, while the volume and complexity of such cases can overwhelm available resources.

More broadly, the fact that liberalization in services, harmonization of IPRs, and acceptance of rigorous product standards all tend to favor the interests of richer nations with significant regulatory capacities has diminished the legitimacy of the WTO in the eyes of many developing countries (Oxfam, 2002). When the GATT was a loose confederation of best-efforts commitments on tariff cuts, it faced no significant doubts about its scope or legitimacy. Now that the WTO is a quasi-regulatory multilateral body with rules that extend deeply into domestic policy-making, many countries wonder if its organization and mandates are designed more to protect entrenched interests than to liberalize global trade (Srinivasan, 2000; Rodrik, 2001). This skepticism resides most obviously in the area of IPRs but exists regarding services and food standards also.

Based on these criteria, it is fair to say that, as a matter of economic principles, it was not unreasonable to incorporate these new functional areas into the WTO. However, they are so complex and controversial that they raise doubts about the potential for the WTO to improve development prospects, not to mention concerns expressed about such social issues as environmental protection and worker rights. In its next round of negotiations, a failure to rebalance at least the development concerns could discourage many developing countries from continued adherence to the system (World Bank, 2001; Oxfam, 2002). In what follows I set out some of the major concerns that have been expressed about WTO rules and how they affect development prospects and the perceived fairness of the trading system.

INTEGRATING TRADE AND DEVELOPMENT OBJECTIVES

There are many objectives one might wish the global trading system to achieve. Recent history suggests that the primary goal is to enhance the efficiency of international trade in ways that are advantageous to the economic interests of exporters and large multinational firms, typically headquartered in the wealthy countries (Srinivasan, 2000; Maskus, 2000). This efficiency should, it is argued, raise growth and productivity while inducing capital and technology to flow to the locations in which they can reap the highest rewards (Dollar & Kraay, 2001). In principle, these gains would be shared across all countries engaged in the system.

However, some observers argue that the system should promote certain social goals, most prominently international income equity, social development, and environmental protection (Rodrik, 2001; Wallach & Sforza, 1999). A liberal trade regime also is considered by many political scientists to be a means of reducing the likelihood of armed conflict (Mansfield, 1994). To many economists it is inappropriate to larder the WTO system with extensive expectations and obligations as regards social objectives (Srinivasan, 2000). It is reasonable, however, to argue that international trade opportunities are beneficial largely to the extent that they expand development opportunities and do so in a broad-based way across countries. Put differently, international trade rules should be designed in a way that enhances growth prospects, making foreign commerce a vehicle for economic development, rather than an end in itself (World Bank, 2001; Rodrik, 2001; McCulloch et al., 2001; Commission on Intellectual Property Rights, 2002).

Achieving a sensible and systemic integration of global trade policy with the needs of developing and transitional economies, while recognizing the importance of sustainability, will take years of analysis and negotiation. Predicting the outcome of such a process is impossible, especially in light of the extensive political opposition that would emerge within certain industry groups and nongovernmental organizations (NGOs). However, the general outline of such a change can be summarized here, stated in terms of certain basic principles.

First, negotiators need to avoid the "one-size-fits-all" approach that has characterized recent agreements on IPRs, food safety standards, customs procedures, and other regulatory arrangements (Rodrik, 2001). It is axiomatic that democratic countries tend to adopt stronger regulatory regimes and social protection schemes as they get richer and technologically more energetic (Maskus, 2000). Requiring rapid implementation of developed-country minimum standards without safeguards or opt-out procedures is to invite developing countries not to adhere to their obligations. This problem is already evident in the area of IPRs, where poor countries are struggling to meet their patent obligations in pharmaceuticals and cannot reasonably enforce restrictions against piracy. Similarly, harmonized customs inspection practices raise considerable costs for authorities in poor countries with little promise of net benefits (Finger & Schuler, 1999). More fundamentally, globalized regulatory regimes may not fit well with domestic social objectives. Health, sanitation, and education policies, for example, tend to be quite different in developing countries from those in rich countries (Commission on Macroeconomics and Health, 2001). To layer strong intellectual property standards onto underfunded health systems and educational processes can be inimical to economic development.

Second, negotiators need to be more cognizant of the costs they are imposing on member countries. According to World Bank (2001) estimates, the annual fixed costs for a typical poor economy to meet its administrative obligations in customs enforcement, IPRs, and product standards can exceed its annual development budget. It is both unreasonable and unrealistic to expect this kind of investment in regulatory systems that mainly benefit foreign firms, at least in the medium term. Thus the depth of regulatory obligations demanded by the rich nations of the poor countries should be closely conditioned on the amount of financial and technical assistance the former are willing to provide, according to well-designed timetables for phasing in new requirements. Failing this, permitting the developing countries some forms of temporary and declining derogations through offsetting trade policies may be in order. The WTO so far has fallen far short of this simple principle, and this failure lies at the core of increasing discontent.

Third, and most fundamentally, WTO members need to recognize the basic asymmetry that now underlies the entire negotiating endeavor. When countries were exchanging trade "concessions" through tariff cuts, they were employing mercantilist terminology to justify welfare-enhancing market opening. In an environment where negotiations turn increasingly to disciplines on domestic regulations, a "concession" by one member may not be in its direct interest. Again, for example, stronger patent rights in medicines or exclusive private rights in new seed varieties can be costly for poor countries. Similarly, a more liberal environment for inward investment in services can generate both benefits and costs, with the net balance not necessarily being positive. And a requirement that food safety standards in poor countries meet the rigorous requirements of such standards in rich countries can be costly to the former without improving local health status.

In this context, for the developed nations to have insisted on such changes in regulatory systems, including more to come in competition policy and environmental protection, without offering significant market access in their own economies in return is the core problem facing the WTO today (Maskus, 2002). Most obviously, the EU, Japan, and the United States are playing a dangerous game by failing to open their markets wider to imports of agricultural goods. The potential gains to developing-country food exporters from even modest liberalization of such trade restrictions are immense, easily outdistancing any possible scope for foreign assistance (World Bank, 2001). There is much evidence that the demands of rich countries for strong international standards in food safety are yet another means of keeping out imports (Maskus & Wilson, 2001).

Similar comments apply to the protectionist antidumping laws in the United States and the EU. These laws are increasingly deployed in ways that ruin small exporting firms in developing countries and impose exceedingly high economic costs (Blonigen & Prusa, 2003). Unfortunately, these governments may make considerable use of such restraints when the trade quotas on apparel and textiles are phased out in 2005. In such a case, many developing-country exporters would be worse off than before and could hardly expect to gain from their adherence to the trading system.

Finally, it is important that the rules of the trading game pay more attention to the service-export interests of developing nations. There is substantial scope for gains to poor countries from an international agreement permitting larger flows of temporary workers to developed countries, where demand for low-skilled workers remains high (McCulloch et al., 2001). This labor exchange would seem to be an appropriate counterpart to the demand by rich governments that their firms be permitted unimpeded and protected access to foreign markets for capital and technology.

CONCLUDING REMARKS

To summarize, the international trading system is complex and becoming more complicated through its requirements on such regulatory circumstances as intellectual property rights, services, and standards. This change in obligations, ushered in by the establishment of the WTO itself, was performed in a highly asymmetrical way. Developed countries provided little in the way of effective market access and, indeed, their markets remain highly protected as regards agriculture, apparel, and labor services—the very items of export interest for developing and transition economies. Redressing this asymmetry in future negotiations is central to the sustainability of the system.

The market liberalization demanded of developing countries by WTO membership should, in the long run, generate large efficiency and growth benefits. However, that same liberalization imposes significant administrative and adjustment costs on these countries. Viewed strictly in terms of domestic political economy, it will be difficult for these nations to discharge these obligations effectively without the potential for phase-ins and temporary derogations, along with other safeguards.

Thus for its own survival the system needs to be reoriented toward integrating trade policy obligations with development needs. In this chapter I have discussed the general outline of how such integration might proceed. This task will be difficult and contentious and will require more patience than rich country authorities have displayed in recent years. Nevertheless, the hard work of moving in that direction must be undertaken with determination if the trading regime is to be brought back safely from the brink on which it is now poised.

NOTES

1. Maskus (2004) surveys recent literature.
2. For example, Oxfam (2002) recognizes potential gains from an open trade regime but cogently questions other aspects of the trading system.
3. See the chapters in Bhagwati and Hudec (1996).
4. See several chapters in Martin and Winters (1996).
5. See several chapters in Krueger (1998).
6. Many countries did so in the 1980s and 1990s, of course. The most likely explanation is that benefits available under the IMF and the World Bank often have been strongly conditioned on freer trade regimes.

REFERENCES

Antweiler, W., Copeland, B. R., & Taylor, M. S. (2001). Is free trade good for the environment? *American Economic Review, 91,* 877–908.

Bagwell, K., & Staiger, R. W. (2003). *The economics of the world trading system.* Cambridge, MA: MIT Press.

Barfield, C. E. (2001). *Free trade, sovereignty, democracy: The future of the World Trade Organization.* Washington, DC: AEI Press.

Bhagwati, J., & Hudec, R. E. (Eds.). (1996). *Fair trade and harmonization: Prerequisites for free trade?: Vol. 1. Economic analysis.* Cambridge, MA: MIT Press.

Blonigen, B., & Prusa, T. J. (2003). Antidumping. In E. Kwan Choi & J. Harrigan (Eds.), *Handbook of international economics* (pp. 251–284). Oxford, UK: Blackwell.

Bovenberg, A. L., & van der Ploeg, F. (1994). Environmental policy, public finance and the labor market in a second-best world. *Journal of Public Economics, 55,* 349–390.

Campos, J. E., & Root, H. L. (1996). *The key to the Asian miracle.* Washington, DC: Brookings Institution Press.

Coe, D. T., Helpman, E., & Hoffmaister, A. W. (1997). North–South R&D spillovers. *Economic Journal, 107,* 134–149.

Commission on Intellectual Property Rights. (2002). *Integrating intellectual property rights and development policy.* London: Commission on Intellectual Property Rights.

Commission on Macroeconomics and Health. (2001). *Investing in health for development.* Geneva: World Health Organization.

Dollar, D., & Kraay, A. (2001). *Trade, growth, and poverty.* Unpublished manuscript, World Bank, Washington, DC.

Edmonds, E., & Pavcnik, N. (2002). *Does globalization increase child labor?: Evidence from Vietnam* (NBER Working Paper No. 8760). Boston: National Bureau of Economic Research.

Finger, J. M., & Schuler, P. (1999). *Implementation of Uruguay Round commitments: The development challenge* (Policy Research Working Paper No. 2215). Washington, DC: World Bank.

Frankel, J. A., & Romer, D. (1999). Does trade cause growth? *American Economic Review, 89,* 379–399.

Grossman, G. M., & Helpman, E. (1991). *Innovation and growth in the global economy.* Cambridge, MA: MIT Press.

Grossman, G. M., & Lai, E. L. C. (2002). *International protection of intellectual property* (Working Paper No. 790). Munich, Germany: Center for Economic Studies and IFO Institute for Economic Research.

Grossman, G. M., & Helpman, E. (2002). *Interest groups and trade policy.* Princeton, NJ: Princeton University Press.

Hoekman, B. M., & Kostecki, M. M. (2001). *The political economy of the world trading system: The WTO and beyond* (2nd ed.). Oxford, UK: Oxford University Press.

Kaul, I., Conceicao, P., Le Goulven, K., & Mendoza, R. U. (Eds.). (2003). *Providing global public goods: Managing globalization.* Oxford, UK: Oxford University Press.

Krueger, A. O. (Ed.). (1998). *The WTO as an international organization.* Chicago: University of Chicago Press.

Lardy, N. R. (1998). *China's unfinished economic revolution.* Washington, DC: Brookings Institution Press.

Lawrence, R. Z. (1993). Japan's different trade regime: An analysis with particular reference to the keiretsu. *Journal of Economic Perspectives 7,* 3–20.

Magee, S. P., Brock, W. A., & Young, L. (1989). *Black hole tariffs and endogenous policy theory: Political economy in general equilibrium.* London: Cambridge University Press.

Mansfield, E. D. (1994). *Power, trade, and war.* Princeton, NJ: Princeton University Press.

Markusen, J. R., Melvin, J. R., Kaempfer, W. H., & Maskus, K. E. (1995). *International trade: Theory and evidence.* New York: McGraw-Hill.

Martin, W., & Winters, L. A. (1996). *The Uruguay Round and the developing countries.* New York: Cambridge University Press.

Maskus, K. E. (2000). *Intellectual property rights in the global economy.* Washington, DC: Institute for International Economics.

Maskus, K. E. (2002). Regulatory standards in the WTO. *World Trade Review, 1,* 135–152.

Maskus, K. E. (2004). Trade and competitiveness aspects of environmental and labor standards in East Asia. In K. Krumm & H. Kharas (Eds.), *East Asia integrates: A trade policy agenda for shared growth* (pp. 115–134). Washington, DC: World Bank.

Maskus, K. E., & Wilson, J. S. (Eds.). (2001). *Quantifying the impact of technical barriers to trade: Can it be done?* Ann Arbor: University of Michigan Press.

McCulloch, N. L., Winters, A., & Cirera, X. (2001). *Trade liberalization and poverty: A handbook.* London: Centre for Economic Policy Research.

Oxfam. (2002). *Rigged rules and double standards: Trade, globalisation and the fight against poverty.* London: Author.

Posen, A. S. (1998). *Restoring Japan's economic growth.* Washington, DC: Institute for International Economics.

Rodrik, D. (1997). *Has globalization gone too far?* Washington, DC: Institute for International Economics.

Rodrik, D. (2001). *The global governance of trade as if development really mattered.* Geneva: United Nations Development Program.

Rodrik, D., & Velasco, A. (2000). Short-term capital flows. In *Annual World Bank Conference on Development Economics 1999* (pp. 59–90). Washington, DC: World Bank.

Scotchmer, S. (2003). *The political economy of intellectual property treaties.* Unpublished manuscript, University of California at Berkeley.

Sell, S. K. (2003). *Private power, public law.* Cambridge, UK: Cambridge University Press.

Srinivasan, T. N. (2000). *Developing countries and the world trading system: Emerging issues.* Tokyo: Asian Development Bank Institute.

Stiglitz, J. E. (2002). *Globalization and its discontents.* New York: Norton.

Wallach, L., & Sforza, M. (1999). *Whose trade organization? Corporate globalization and the erosion of democracy.* Washington, DC: Public Citizen.

World Bank. (1993). *The East Asian miracle.* New York: Oxford University Press.

World Bank. (2001). *Global economic prospects and the developing countries: Making trade work for the world's poor.* Oxford, UK: Oxford University Press.

Young, A. (1994). Lessons from the East Asian NICs: A contrarian view. *European Economic Review, 38,* 964–973.

6

Caught Behind the Eight Ball
Impeding and Facilitating Technology Diffusion to Developing Countries

MICHAEL W NICHOLSON

International technology transfer carries immense importance in a world with such stark current inequalities of wealth. The richest countries, as defined by traditional measures such as gross domestic product (GDP), are host to a majority of the world's investments in research and development (R&D), an indicator of dynamic economic activity. This concentration of industrial research may be an efficient outcome of actions by economic agents actively pursuing market-based incentives for the innovation and dissipation of technological advances, or it could be a significant means by which poor countries stay poor. The salient feature of these inequalities of R&D for globalization and development concerns the manner in which new technology gets transferred from the rich to the poor.

This chapter approaches the broad subject of technological development and diffusion from the perspective of international impacts of intellectual property rights (IPRs) and the role of the multinational enterprise. Multinational firms engage in foreign direct investment (FDI) or licensing in order to leverage technological innovations into new markets. The economic literature regarding technological innovation and the incentives for concentration of industrial R&D entails a role for multinationals to transfer technology from an innovative core to the rest of the world. These transfers develop the technological capabilities of the countries that host these activities. The issues surrounding the economics of technology are particularly important regarding recent multilateral actions by the World Trade Organization (WTO) countries to set minimal international standards for IPRs, the Trade-Related Aspects of Intellectual Property Rights (TRIPs) agreement.[1]

In the next section I discuss the technological divide between the North and the South. Then I offer a description of theories that use free agents who behave according to economic incentives. In the following section I focus on the empirical evidence of technology transfer, analyzing the various channels in which these activities occur. Such transfer carries different results depending on whether it occurs by intent of the firm or includes a shift in control of proprietary knowledge. In the final section I offer some conclusions.

THE TECHNOLOGY DIVIDE

Contemporary global technology flows occur within the historical context of a deep technological divide. This divide has a loose geographical demarcation that gives rise to the popular distinctions between a wealthy "North" and a less-wealthy "South," terminology adopted for convenience in the present chapter. The economic models analyzed below assume the simple existence of this demarcation, while economic evidence indicates the line may be a consequence of geopolitical forces. Gallup, Sachs, and Mellinger (1998) suggest that location and climate bring large effects on income levels and income growth through transportation costs, human health, agricultural productivity, and access to natural resources. Coastal, temperate, northern hemisphere economies enjoy a density of wealth, as measured by GDP per square kilometer, and 28 of the 30 countries with the largest GDP per capita in 1995 were north of the Tropic of Cancer.[3] The core economic regions of the world account for 3% of the world's inhabited land area, 13% of the world's population, and 32% of the world's income.

In a historical picture of North–South wealth differences, Diamond (1997) describes environmental factors that may be fundamental to the economic dominance of Europe and her former colonies. He argues that historical technological diffusion of plant species and domesticated animals worked most effectively *within* ecological zones, favoring the East–West axis of Eurasia over the North–South axis of Africa and the Americas. These historical advantages of Europe and Asia, combined with the high transportation costs for the landlocked regions of South America,

Africa, and Central Asia, contribute to the contemporary dichotomy of wealth and technology.

Historical advantages of ecology serve as an analogy to contemporary economic advantages of technology, in which static differences strengthen productivity gaps that lead to dynamic advantages of wealth. The existing North–South technology divide demonstrates both a static and a dynamic differentiation, and the relationship of the regions across this gap entails both the development and the diffusion of technology. The United Nation's (1999) *Human Development Report* provides a host of statistics on the current stratification of technology. The fifth of the population living in the wealthiest countries of the world in 1998 produced 86% of global GDP, exported 82% of world trade, and undertook 68% of all foreign direct investment. They also enjoyed 74% of the world's telephone lines. The fifth in the poorest countries did not muster 1% of any of these indicators of wealth.

Investment in research and development (R&D) is similarly concentrated among industrialized countries. In 1993, 10 countries hosted 85% of global research and development expenditures. Between 1977 and 1996, these 10 countries controlled 95% of U.S. patents, but developing countries managed less than 2%. In addition, residents of industrialized countries controlled more than 80% of the patents granted to developing countries (United Nations, 1999). The share in world R&D expenditures by countries in the South was only 4% in 1990, a decline from 6% in 1980, despite well-documented gains in research activity in Southeast Asia over that time period (Correa, 2000). In 1998, 29 OECD countries spent $250 billion in R&D, more than the GDP of the 30 poorest countries combined, and accounted for 91% of the 347,000 new patents granted in the world (United Nations, 2001). Research and development is also highly concentrated within industrialized countries, as the United States conducted 50.4% of all R&D in the OECD in 1985, and three countries accounted for over 80% (Grossman & Helpman, 1991).

However, this concentration in technological development does not necessarily imply a similar concentration in its international availability. Through diffusion, the technological developments from this concentrated innovative core have tremendous impacts on the rest of the world. Eaton and Kortum (1996) show that about half of all growth in the industrialized countries, and over 90% of growth for countries other than the five leading research economies, can be traced to innovations originating outside their barriers. The United Nations (2001) cites technological breakthroughs as fundamental in lowering mortality rates in Asia, Africa, and Latin America throughout the 20th century, in particular for reducing undernutrition in Southeast Asia between 1970 and 1997.

One explanation for the concentration of innovation and the diffusion of technology across national borders involves economic incentives. With this line of reasoning, the concentration of industrial R&D among the most technologically advanced countries of the world occurs because resources, in general, are allocated to regions where they earn the highest return. Countries on the technology frontier enjoy the lowest costs of basic research, as well as the ability to transform innovations into marketable commodities. Since technology enjoys increasing returns, and industrial R&D generally earns higher returns in regions with existing technology, the efficient outcome is a concentration of industrial R&D in the North.

These economic forces are supported by an international legal structure that encourages this tendency toward concentrated R&D. Recently, WTO member countries signed a multilateral agreement to strengthen intellectual property rights (IPRs), which often enhance the value of existing technology and may exacerbate the incentives for concentration. Such agreements are politically volatile, since the repercussions may involve substantial transfers of wealth from the South to the North and may significantly impede access to technology by developing countries. The next section investigates these arguments with a discussion of economic considerations on the international protection of intellectual property and its role in technological development.

INCENTIVES FOR INNOVATION
Intellectual Property Rights

The key issue regarding the economic incentives of innovation is the contrast between the static and the dynamic effects of technology. At any point in time, a fixed stock of technology exists in the world. Technology is a non-rival good, which by definition means that its use by one does not preclude its use by another. Consider the alphabet as an example of non-rival technology. If one person uses the alphabet, then another person can use the alphabet without detriment to the first person's use. In fact, the value of literacy may rise with the size of the literate population, which suggests that the benefits of the first user increase due to the presence of the second user. (This is not true of, say, a loaf of bread, since the first user would have less to eat.) The optimal static solution would be for everyone to have access to that technology, since everyone benefits from its use without negatively impacting its use by another. Correspondingly, the most efficient policy would ensure the greatest dissemination of existing technology.

On the other hand, the dynamic effect recognizes that technological innovations are costly to develop, and industrial R&D will only occur if expected returns to the innovation will exceed that cost. Legal restrictions on the use of technology, such as IPRs, exclude the use of technology by someone other than the holder of that property right. These rights yield temporary economic profits, known as "monopoly rents," to their owners. The optimal dynamic policy would ensure that these rents exactly cover the cost of developing the innovation.

The contrast between the optimal static and dynamic results leads to complex legal and economic wrangling about the proper level of intellectual property protection. A welfare-based conception of IPRs balances the desire for access to technology as well as its exclusion. This follows the tradition of English law, which considers the allocation of property rights as a fundamental element of smoothly functioning markets. From its inception, the stated goal for U.S. patent policy has been to promote the public welfare, recognizing the fundamental trade-off that trade secrets are granted to the public domain in exchange for a limited exclusivity (Ryan, 1998). TRIPs adopted this language of welfare-based protection of intellectual property, with Article 7 focusing on "technological innovation" and the "transfer and dissemination" of technology (Watal, 2000).

TRIPs entered initial negotiations at the Uruguay Round of the General Agreement on Tariffs and Trade (GATT), in large part due to unilateral actions by the United States, by far the largest producer and seller of technology goods. In the early 1980s multinational firms in the United States mobilized to improve the protection of IPRs internationally. A divergence existed at that time between the North and the South, especially in Latin America and Southeast Asia, for product and process patents. This held especially true in food, agriculture, chemical, and pharmaceutical patents. As of 1988, the World Intellectual Property Organization identified 31 developing countries that excluded pharmaceuticals products, eight that excluded pharmaceuticals processes, and 12 that excluded chemical products. Finland, Greece, Iceland, Monaco, Portugal, and Spain also excluded product patents for pharmaceuticals at that time (Watal, 2000).

Correa (2000) suggests that numerous factors converged to motivate the activity on intellectual property protection, including the growth of technology in importance to international competition, high externalities in the production of knowledge with limited appropriation, the increasing elimination of trade barriers, and a perception in the United States that its technological supremacy was being challenged by Japan and the newly industrializing countries. Pfizer and IBM, leading a consortium of firms in pharmaceuticals, electronics, and entertainment, initiated a proposal at the GATT for multilateral actions to strengthen IPRs (Ryan, 1998). In a 1984 amendment to Section 301 of the 1974

Trade Act, the U.S. Congress gave the U.S. Trade Representative the legal right to tariff impositions on intellectual property (Watal, 2000). With a mandate for "adequate and effective protection" of IPRs, the U.S. Trade Representative began successful bilateral actions against Brazil and the Republic of Korea. After the United States established a credible threat, developing countries acceded to the Treaty of Marrakech by 1994, giving life to TRIPs (Ryan, 1998). Compliance with the agreement was staggered according to relative levels of national wealth, with developed countries required to comply by 1996 and developing countries by 2000 (Maskus, 2000). Least developed WTO member nations have until 2006 to comply with the major provisions, and until 2016 to comply with some pharmaceutical patent protection.

Negotiations about the language and intent of the TRIPs agreement are ongoing. Debates about ambiguities in the treatment of some IPRs such as geographical indicators were contested at the most recent WTO ministerial meeting in Doha, Qatar. The hot issue remains the Doha Declaration on public health, which grants provisions to countries to take measures to induce compulsory licensing of necessary pharmaceutical products in cases of national emergency.

One static implication of TRIPs may be to enhance the value of existing technology (specifically, in many cases, pharmaceutical innovations). IPRs protect technology by extending the length of the patent over time, broadening the scope of the patent across product space, and widening enforcement across multiple sectors. A strengthening of IPRs thus benefits the current holders of patents. By raising the price of technology, they imply a welfare transfer—through the terms of trade—from technology consumers to technology producers. This would naturally benefit the innovative core mentioned above.

Moreover, the dynamic effect may be to exacerbate the concentration of industrial R&D in the North. Since technology is a non-rival good, those who have innovated in the past will have the best opportunity to innovate in the future, a tendency that increases with the strength of IPRs. Although economic theories suggest that this may be the most "efficient" outcome, it may fail to attain political considerations of fairness and equality.

Innovation and Growth

Technological innovation has achieved prominence within the past 20 years among economic studies concerning growth and human welfare. Neoclassical economic theory considers growth an outcome of capital accumulation, holding technology to be an exogenous factor. Endogenous growth theory holds technological development to be fundamental to the growth process and describes the market for technology using incentives of supply and demand.[3]

According to Romer (1990), the three underlying presumptions of endogenous growth theory claim that (1) technological development lays at the heart of economic growth; (2) technological development is excludable and can respond to market incentives; and (3) technology is a non-rival good that builds on existing levels of technology. Under this theory, profit-maximizing firms invest in R&D to gain innovations that yield temporary monopoly rents (or economic profit) that recover, on the aggregate, the original investment.

Open-economy endogenous growth models extend this framework to include North–South differences in innovative capabilities. The specification on technology assumes that the North carries much lower costs for innovation, while the South enjoys lower costs in production.[4] Grossman and Helpman (1991) develop a series of endogenous growth models that adhere to the principles of non-rival knowledge that can be excludable through intellectual property protection, usually via trade secrets or patents, and Helpman (1993) includes patent protection as a tool of government policies. He offers strong theoretical support to the popular notion that IPRs tend to favor Northern welfare at the expense of the South. In Helpman's model, strong IPRs grant considerable market power to innovating Northern firms, meaning that they can raise prices much higher than their static cost. This market power is sufficient to *decrease* the rate of innovation because Northern firms find a better return in shifting their resources to the production of previous innovations.[5]

Helpman's results are not robust to assumptions on the resource constraints of each region, and developments in the theory show that they do not necessarily hold if innovating Northern firms can shift production to the South.[6] That is, they produce goods using the new technology in the South, taking advantage of lower relative labor costs, yet maintaining control over the technology through IPRs. This scenario leads to the classic implication that each region specializes according to its comparative advantage: the North in R&D, the South in production. In this situation, a concentrated R&D core derives from optimal activity, and both the North and the South gain from the division of labor.

With technology transfer across regions, the optimal allocation of resources leads to the North engaging primarily in R&D and the South in manufacturing, a result that parallels the neoclassical doctrine of specialization according to comparative advantage. This efficient allocation of resources demonstrates welfare gains from the concentration of R&D among the industrialized countries. However, this outcome faces political controversies when the specialized goods are not Ricardo's famous wine and cloth, but non-rival technologies that have dramatic effects on future growth. Developing countries may be inclined to argue that such a result leaves them perpetually lagging behind the technology frontier. Such a public discussion highlights the important issues of the impact technology transfer actually has on the productive and innovative capabilities of the South, and the extent to which that transfer entails a shift in the control of proprietary knowledge.

TECHNOLOGY TRANSFER

Foreign Direct Investment and Licensing

In the models of open-economy endogenous growth outlined above, the methods of interregional technology transfer include imitation, foreign direct investment (FDI), licensing, and a "natural adaptation" where diffusion occurs according to an exogenous process. These modes carry different characteristics depending upon whether the transfer is intentional by the firm. Since technology is non-rival and only partially excludable, quite a lot of diffusion occurs outside voluntary market transactions. With regard to North–South technology differences in the context of intellectual property rights, attention must be paid to whether a shift in control of the proprietary asset occurs. Since stronger protection of intellectual property generally increases the value of existing innovations, significant wealth shifts may occur if the actual control of the knowledge is not transferred. The simple presence of new technology within its national borders may not benefit that country without some shift in control.

One aspect of this that carries particular interest is how and why technology gets transferred across national borders. An important distinction is whether an innovating firm *chooses* to transfer the technology. The impact of technology transfer on the recipient country can be different depending on whether the transfer occurs without intent of the firm, such as by "leakages" or spillovers.[7] The firm's incentive to make an intentional transfer relates to the non-rival nature of technology.

Theoretical approaches regarding intentional firm-specific technology transfers differ according to whether the activity crosses national borders. Baldwin and Scott (1987) distinguish between technology distributed by intent of that firm, "dissemination," and diffusion by competitive activities, or "imitation." The difference is not determined by legal or ethical claims, but rather by the appropriation rights of the innovator. Imitation often carries a negative connotation, but the activity itself, as generally defined, entails substantial cost and creativity by the competing firm, and the new products developed via imitation add value to economic growth (Hobday, 1995).

Profit-maximizing firms choose to distribute their technology in order to obtain the optimal returns on their proprietary assets. Firms may choose to do this in a variety of ways, but the simplest depiction regards a firm choosing to expand its in-house operations or to contract with an appropriate third party. This is the economic theory of "internalization" and follows Coase's (1937) classic explanation of integration as the optimal outcome when the transaction costs of using the price mechanism exceed the cost of organizing through direct control. This decision is at the heart of a firm's organization, from whether to hire janitors or to contract with a janitorial service, to whether AOL chooses to enter traditional media outlets by starting its own magazine or by acquiring Time Warner.

After firms succeed in innovation, under traditional assumptions they will seek to maximize the returns of that innovation and will naturally transfer the technology anywhere in the world where the return exceeds the cost. The decision to make this transfer, and the manner in which they do it, depends on many things, including the extent to which the innovation carries protection in a foreign country. The choice of integration or licensing reflects market imperfections, technological interdependence within the sector, the returns to economies of scale, and short-run differences in productive capacity or operating costs. These, in turn, relate to the strength of intellectual property protection.

Markusen (1995) details how these parameters extend to multinationals, and the popular eclectic theory of the multinational enterprise grants a central role to this decision.[8] Once firms have decided to service a foreign country with proprietary technology, they will do so by FDI if the returns to internalization exceed those of licensing. A large literature focuses on the actions by multinationals that have made this decision to engage in FDI. These theories usually assume that the firms undertake a fixed cost to develop a proprietary asset that can support production over multiple plants. Downstream production plants may shift overseas for multiple reasons, including lower factor costs or avoiding high transport costs.

However, relatively little attention has been paid to firms that choose *not* to undertake FDI—firms that choose to engage in cross-border licensing. Presumably, the impact of a technology transfer will differ depending on whether the technology arrives through direct licensing or by wholly owned subsidiaries of a foreign-based multinational. The value of technology transfer to the recipient economy depends on the extent control is shifted. This decision plays a crucial role in the transfer of technology across borders.

Hobday (1995) argues that diffusion of technology across borders primarily occurs by firm-level learning, both between and within firms. Mansfield and Romeo (1980) show that except for process innovations, FDI is the principle channel of technology transfer, and technology transferred through multinationals tends to be newer than in licensing or joint venture agreements. According to Dunning (1981), since the mid-1960s the dissemination of technology via multinationals has accelerated as various countries in Latin America and Asia, following Japan, have become adept at acquiring technologies. Technology is now being transferred by multinationals through their advice and assistance to local firms. Dunning suggests that the proprietary advantage of multinationals and the source of their market power may no longer be due to their possession of superior knowledge but to their ability to transfer it.

Technology transfer is a complicated process, however, and does not necessarily translate into immediate benefits for the recipient country. By definition, FDI entails the firm maintaining control of its proprietary innovation. In order to understand the nature and extent of technology being transferred by multinationals and the benefits conferred upon the host countries, Mansfield and Romeo (1980) surveyed executives from 31 United States based firms.[9] They found that the mean time of leakage of technology by multinationals was about 4 years after its initial introduction into the United States, but in most cases the technology may have leaked even without such transfer. Executives reported that in about one-quarter of the cases leakage was advanced 2½ years because of FDI, but in general leaks would have occurred as quickly. Seventy large British firms said those leaks are how they gained some technological capabilities, but only 20% said they were "important." Most competitors obtaining the technology were not usually headquartered in the host country; in the sample, the majority were headquartered in West Germany. Technology primarily leaked through reverse engineering, with important roles also played by information revealed in patents and defection in personnel. For 80% of leaked technologies leaks led to cost reductions for users, and for 40% of the technologies they led to supply savings.

In a detailed study, Hobday (1995) shows that FDI in East Asia includes stages of technology transfer that may lead to innovation by the recipient countries. He interviewed company directors and engineers to determine learning mechanisms for firms in the region, specifically concerning late-comer firms with technological disadvantages and in undeveloped local markets. These firms engaged in numerous channels in order to acquire technology and overcome technical barriers to entry. His theoretical stages of technology development include (1) basic production capabilities, (2) reverse engineering, (3) product design capability, (4) product innovation capability, and (5) advanced product/process innovation. Some examples of this development carried to fruition include Samsung and Goldstar of South Korea and ACER and Tatung of Taiwan.

A substantive economic literature on the impact of foreign direct investment on the host countries provides weak support that FDI generates positive spillovers. A review by Hanson (2001) suggests that domestic plants in industries with a large multinational presence realize lower rates of productivity growth. In particular, Haddad and Harrison (1993) found that weak plant-level productivity growth negatively correlated with foreign presence in Morocco, and Aitken and Harrison (1999) found similar results for Venezuela. Kokko (1994) found that spillovers are less likely in "enclave" sectors with large technology gaps and high foreign shares. The overall results imply that FDI may reflect a technology transfer that fails to generate technological development for the recipients.

No similar studies exist for the impact of cross-border licensing on host countries. The most relevant may be Scherer (1977), who discusses the impact compulsory licensing may have on innovation and patenting. Multinationals often patent their products in many jurisdictions but have production facilities in few, forcing developing countries to pay high prices for imported patented products while they are unable to use frontier technology. The response is that some governments use "working" requirements on patents, or subject these technologies to compulsory licensing.

Scherer shows that licensing can have the same financial consequences as exclusive exploitation if no scale economies are present in the production or distribution of goods, the cost of license enforcement is minimal, and a competitive royalty rate can be assigned. However, a royalty rate equivalent to potential monopoly rents can be difficult to obtain, and thus mandatory licensing may be analogous to weakening, or shortening, a patent. His study does not discuss the impact of licensing on the host country. The relative roles of FDI and licensing are being developed in a burgeoning literature on the impact of TRIPs on technology transfer. The economic evidence with regard to the impact of IPRs on technological dissemination suggests that stronger protection of intellectual property increases international transfers. This research has not yet been applied to unintentional transfers, but some ancillary results support theoretical projections that they are prevented by stronger protection. Intellectual property rights are an important determining element for both FDI and licensing, as well as with other forms of technology transfer.

Maskus (1998) observes that stronger IPRs have a positive and statistically significant impact on the sales, exports, and assets, for affiliates of United States based multinationals respond positively to increased IPRs in developing countries, although the number of patent applications declines. Smith (2001) finds support for theoretical suggestions that stronger IPRs enhance the advantages of licensing relative to FDI, especially in countries with strong imitative capabilities that pose a greater threat to leakages. Yang and Maskus (2001b) show that stronger IPRs have a positive and statistically significant effect on both the royalty fees from unaffiliated foreigners and for the licensing of industrial processes, but not on the royalty and licensing fees from affiliated subsidiaries.

While IPRs may tend to increase direct forms of transfer, they also leave proprietary control in the hands of the innovators. Consequently, the ultimate impact of stronger international IPRs may be a diminishing in the transfer of *usable* technology. That is, IPRs provide an incentive for the physical transfer of technology, but with proprietary control (and the associated rents) remaining with Northern-based multinationals. This may, of course, lead to the allocation of resources optimal for overall welfare—that is, the scarce resources in the world are arranged according to market-based incentives in a manner that best benefits the global community. This Pareto efficient outcome is, perhaps unsurprisingly, the prediction of the economic models discussed above.

These studies are particularly interesting in lieu of TRIPs, the multilateral agreement to strengthen IPRs in developing countries. The literature suggests that although IPRs may tend to increase the rate of FDI between the North and the South, the greater impact may be on licensing. This supports the theoretical idea that strong IPRs increase the incentives for licensing by weakening the internalization motivation for FDI, and highlights the need for future studies of the impact of cross-border licensing on domestic countries. Overall, the implementation of TRIPs likely increases the tendency toward a greater concentration of technological development, but with accompanying increases of technology transfer to the absolute benefit of both the North and the South. However, the relative benefits may accrue to the R&D core, leaving the rest of the world at a relative disadvantage, stuck behind the eight ball.

CONCLUSION

The concentration of industrial research and development in the North can be explained by traditional economic incentives regarding specialization according to comparative advantage. Open-economy endogenous growth models suggest that a dynamic shift toward further concentration of industrial R&D in the Northern countries may be an efficient allocation of resources, and multilateral agreements strengthening international intellectual property rights may achieve a dynamic optimum by encouraging this concentration. Multinational firms play a fundamental role in the diffusion of technology from the innovative core to the rest of the world, with an impact on the recipient country that differs according to whether technology transfer occurs through wholly owned subsidiaries or by cross-border licensing.

However, this efficient outcome may be detrimental to the technological advancement of developing countries. Such countries can expect an increase in incoming FDI as they accede to the requirements of TRIPs, but the very structure of intellectual property protection may prevent knowledge spillovers from this FDI to generate technological development in the South. In the face of this apparent paradox, developing countries have a number of policy options. They can fight the economic incentives toward dynamic specialization via subsidies to domestic industrial R&D, a politically popular option that could yield negative short-term results with an uncertain possibility of future benefits. Alternatively, they can work within existing market conditions with an eye toward licensing and acquisition of appropriable technology. This option is more likely to appeal to the countries with domestic industries just removed from the technology frontier. And, of course, they can work outside the legal framework in appropriating technology.

The terms *endogenous* and *exogenous* were introduced above; exogenous factors are taken as a given by the agent and endogenous factors are subject to a choice. In this framework, a policymaker could consider the inequalities in wealth as exogenous, an outcome of forces extending back centuries, and the manner in which the means of economic development shift between regions an endogenous choice. This would pose the question, "Given that some countries are rich and heavily investing in R&D, and given we care about the poorer countries, what policy variables could encourage development in the rest of the world?"

Multilateral negotiations through the WTO and similar institutions have yielded a multifaceted—and legally complex—agreement on IPRs. One's reaction to such agreements as a favored policy for development likely reflects one's prior assumptions about the benefits or harms of trade, multinational firms, and globalization itself. Appropriate policy must take into account the incentives for both static and dynamic efficiency, the means for technological innovation and development, and the actual benefits that may be accrued by technology transfer. This chapter has discussed the various options; the optimal policy for any particular country depends on its ultimate objectives relating to efficiency and equity.

ACKNOWLEDGMENTS

The views expressed in this chapter are those of the author, and do not necessarily represent the views of the Federal Trade Commission or any individual commissioner. I am grateful for the comments by participants at the April 2002 "Responding to Globalization" conference at the University of Colorado, and I thank Jason Green, Edward Greenberg, John O'Loughlin, Jim Russell, Lynn Staeheli, Julie Sherman, and Steven Tenn for helpful comments and suggestions.

NOTES

1. The Trade-Related Aspects of Intellectual Property Rights agreement (TRIPs) was adopted by the WTO in the mid-1990s. Compliance with the agreement was staggered according to relative levels of national wealth. Developed countries were to comply by 1996, developing countries by 2000, and least-developed WTO member nations by 2006.
2. The exceptions were the two city-states Hong Kong and Singapore.
3. Economists use the terms "endogenous" to describe factors that are determined by behavior within a model or a market and "exogenous" for factors with given values. The price of a gallon of gasoline is endogenous to the operators of a gas station (they get to set it) but exogenous to a driver (who cannot negotiate). The choice to purchase the gasoline is endogenous for the driver.
4. A specification introduced by Krugman (1979).
5. Glass and Saggi (2002) call this the "crowding out" of research and development.
6. Lai (1998) shows that if production can shift via FDI, then an increase in the strength of IPRs will lead to a higher innovation rate. Yang and Maskus (2001a) show this also holds if production shifts via licensing.
7. Grossman and Helpman (1991) define technological spillovers occurring when firms acquire information created by others without paying for it in a market transaction, and the creators (or current owners) have no effective recourse under prevailing laws to prevent other firms from utilizing the information so acquired. This quality highlights the partial excludability of knowledge goods, and serves as a primary theoretical benefit of FDI (Wang & Blomstrom, 1992). Unintentional spillovers, or "leakages," can occur in a variety of ways across national borders. Reverse engineering, similar to imitation, describes the process of developing a competing product using the original. This method is generally employed in fields in which innovative qualities can be easily mimicked, such as certain chemicals. Leakages also occur through defection, when a former employee begins working for a competitor, through the emigration of high-skilled workers, or when a firm breaks a licensing contract to produce its own good absent of royalties.
8. Dunning's (1981) eclectic theory, also known as "OLI," describes the ownership, location, and internalization conditions for a firm to engage in foreign

direct investment. An innovating firm seeking to shift production overseas already meets the first two criteria.

9. Their survey primarily includes transfers between developed countries, such as from the United States to Western Europe, likely due to the availability of such data. Eaton and Kortum (1996) discuss results for transfers carried out around the world.

REFERENCES

Aitken, B., & Harrison, A. (1999). Do domestic firms benefit from foreign direct investment? *American Economic Review, 89*, 605–618.

Baldwin, W. L., & Scott, J. T. (1987). *Market structure and technological change.* New York: Harwood.

Coase, R. H. (1937). The nature of the firm. *Economica, 4*(16), 17–31.

Correa, C. M. (2000). *Intellectual property rights, the WTO and developing countries: The TRIPs agreement and policy options.* New York: Zed Books/Third World Network.

Diamond, J. (1997). *Guns, germs, and steel: The fates of human societies.* New York: Norton.

Dunning, J. (1981). *International production and the multinational enterprise.* London: Allen & Unwin.

Eaton, J., & Kortum, S. (1996). Trade in ideas: Patenting and productivity in the OECD. *Journal of International Economics, 40*, 251–278.

Gallup, J. L., Sachs, J. D., & Mellinger, A. D. (1998). Geography and economic development (NBER Working Paper No. 6849). Boston: National Bureau of Economic Research.

Glass, A., & Saggi, K. (2002). Intellectual property rights and foreign direct investment. *Journal of International Economics, 56*, 387–410.

Grossman, G., & Helpman, E. (1991). *Innovation and growth in the global economy.* Cambridge, MA: MIT Press.

Haddad, M., & Harrison, A. (1993). Are there positive spillovers from foreign direct investment? *Journal of Development Economics, 42*, 51–74.

Hanson, G. H. (2001). *Should countries promote foreign direct investment?* (G-24 Discussion Paper Series No. 9). Geneva: United Nations Conference on Trade and Development.

Helpman, E. (1993). Innovation, imitation, and intellectual property rights. *Econometrica, 61*(6), 1247–1280.

Hobday, M. (1995). *Innovation and East Asia: The challenge to Japan.* Brookfield, VT: Edward Elgar.

Kokko, A. (1994). Technology, market characteristics, and spillovers. *Journal of Development Economics, 43*, 279–293.

Krugman, P. (1979). A model of innovation, technology transfer, and the world distribution of income. *Journal of Political Economy, 87*, 253–266.

Lai, E. (1998). International intellectual property rights protection and the rate of product innovation. *Journal of Development Economics, 55*, 133–153.

Mansfield, E., & Romeo, A. (1980). Technology transfer to overseas subsidiaries by U.S.-based firms. *Quarterly Journal of Economics, 95*(4), 737–750.

Markusen, J. (1995). The boundaries of multinational enterprises and the theory of international trade. *Journal of Economic Perspectives, 9*, 169–189.

Maskus, K. (1998). The international regulation of intellectual property. *Weltwirtschaftliches Archiv, 134*, 186–208.

Maskus, K. (2000). *Intellectual property rights in the global economy.* Washington, DC: Institute for International Economics.

Romer, P. (1990). Endogenous technological change. *Journal of Political Economy, 98*, S71–S102.

Ryan, M. P. (1998). *Knowledge diplomacy: Global competition and the politics of intellectual property.* Washington, DC: Brookings Institution Press.

Scherer, F. M. (1977). *The economic effects of compulsory patent licensing.* New York: New York University Graduate School of Business.

Smith, P. J. (2001). How do foreign patent rights affect U.S. exports, affiliate sales, and licenses? *Journal of International Economics, 55*(2), 411–439.

United Nations. (1999). *Human development report.* New York: Author.

United Nations. (2001). *Human development report.* New York: Author.

Wang, J., & Blomström, M. (1992). Foreign investment and technology transfer. *European Economic Review, 36*, 137–155.

Watal, J. (2000). *Intellectual property rights in the World Trade Organization: The way forward for developing countries.* London: Kluwer Law International.

Yang, G., & Maskus, K. (2001a). Intellectual property rights, licensing, and innovation in an endogenous product cycle model. *Journal of International Economics, 53*, 169–188.

Yang, G., & Maskus, K. (2001b). Intellectual property rights and licensing: An econometric investigation. *Weltwirtschaftliches Archiv, 137*, 58–79.

7

The Global Culture Factory

ANDREW KIRBY

This chapter provides an examination of the way that popular cultural products are produced and consumed. The first section examines the ways in which technologies have advanced and how the "digital revolution" has facilitated the creation of global corporate structures within the media industries in recent years. The second examines the complexities of this culture factory; particular attention is given to the diversity of corporations and products emerging from the different continents. The third section then evaluates the impacts that popular culture, with its many forms and origins, has on those who consume it around the world.

There is a received wisdom on globalization and culture—a highly pessimistic perspective—that is summarized in the following quotation:

> Globalization conjures up in many minds a spectacle of instantaneous electronic financial transfers, the depredations of free-market capitalism, the homogenization of culture, and the expansion of Western, by which is usually meant American, political hegemony. Hardly an attractive prospect—especially when accompanied by evidence of widening economic inequality, worsening ecological degradation, intensified ethnic rivalry, spreading militarism, escalating religious nationalism and other ills. . . . (Gunn, 2001, p. 19)

From such a standpoint, popular culture is regarded as trivial and disposable, something that reflects all that is wrong with contemporary

133

capitalism, something that stands alongside the hot dog, the hamburger and the handgun as America's least noble contributions to the world of commerce. Yet these are common misperceptions. First, popular culture is often disposable but it is hardly trivial. It is produced by hundreds of thousands of individuals and consumed by billions. It is, if nothing else, a large economic sector. Second, it is socially significant, beyond the narrow worth of its content. In societies that appear to be going through accelerating change, the relations between state, economy, and civil society are continually under negotiation. These conversations were once accomplished by relatively small numbers of elite citizens, but are now subject to almost universal participation that is facilitated by the media of mass communication. In short, it is popular culture that constitutes a subliminal public space within which social norms are presented and debated—these have typically been national, but are increasingly transnational conversations, about identity, sex, religion, and virtually any topic one can imagine (e.g., Kirby, 2002). Third, the production of popular culture is a genuinely global process, and it is important to make this statement absolutely transparent. While many use "Hollywood" as a synonym for popular culture, nothing could be further from the truth. Many more motion pictures are made outside the United States than within it—and many more sound recordings, many more television and radio programs. It must be emphasized, of course, that American media corporations are highly adept at global marketing—indeed, this sector has led virtually all others based in the United States in creating global reach. But they do not have anything like a monopoly, and this sector is fascinating precisely because there are European and Asian corporations that are also shipping their cultural products around the world. Consequently, we must start this chapter with a restatement of accepted wisdoms, characterized so dolorously by the quote above; accordingly, this chapter argues the following:

- First, popular culture is important precisely because it is disposable; it is a shifting and contemporary medium in which ideas are constantly offered, debated, accepted, and rejected.
- Second, the media of mass communication are important economic sectors, with global reach.
- Third, although U.S. corporations are powerful players in this sector, they do not have a monopoly, and thus any examination of the mass media involves a genuine study of globalization.

In reading this chapter, we can see that the distribution of media products fits into this book's organizing principles without any stress or strain. First, we are focusing on *cultural flows and mass communication* that encompass migration, the diffusion of popular culture, and the rise of a global mass communications industry (see O'Loughlin, Staeheli, & Greenberg, Chapter 1, this volume, p. 3). As noted, the transfer of media images is made possible by large conglomerates that have been practicing *economic globalization* for several decades. In turn, because of its combustible content, popular culture can be linked to broader social questions, such as *political change and democratization*, linked in turn to the transformations in institutions, attitudes, and behaviors associated with democratization and de-democratization (O'Loughlin et al., Chapter 1, p. 12).

THE TECHNOLOGICAL TERRAIN

The media of communication have seen intense innovation over the past decade. When the Berlin Wall collapsed, most messages were still transmitted in analog forms that would not have seemed unfamiliar to those who had built the wall back in the 1950s. Television was essentially a national medium, in contrast to radio, which was broadcast via short-wave bands around the world. Telephony depended on landlines for transmission of voice or fax messages; music was sold on large vinyl frisbees or fragile tapes; movies were distributed on bulky reels or videotapes with a finite life.[1] These were all highly inefficient and imperfect ways to transmit messages, although their ability to transgress national boundaries—such as from West to East Germany—was fully demonstrated by the extent to which popular culture was familiar to those behind the Iron Curtain.

In the intervening years, the move from analog to digital transmission has been dramatic. The consolidation of the Internet is perhaps the most perfect illustration of this shift, but it has occurred in all media. Movies can be distributed globally to theaters in digital form, with the result that the same film can open in multiple markets on the same day; the 2003 *X-Men* sequel opened in over 90 different countries (thus, in turn, limiting the possibility of prerelease pirate copies being sold in some of those markets). Satellite television also permits news and entertainment to be distributed across nations on an instantaneous basis, creating, almost as an afterthought, whole new regions of shared programming. Music can also be broken up into digital packets, sent via satellite or Internet connections, and reassembled into an array of perfect and permanent forms—compact disks, CD-ROMs, MP3 files—and this can be done by authorized and unauthorized users alike. Pocket-size wireless telephones can now send voice or text messages virtually anywhere, with the option to include digital video images as required, so that a lone reporter can now replace an entire news bureau.

A glance in a catalogue or airport electronics store would suggest that these devices are ubiquitous. This inference is both true and somewhat disingenuous. There is what is often termed a "digital divide" between the affluent and the impoverished. However, what is interesting about this differential access to technology is that it has little to do with geography and all to do with disposable income. There are parts of the United States and of Europe that have marginal communications connections; in contrast, there are parts of the developing world that have seen spectacular communications growth in recent years. For instance, the number of cellular telephones has risen to approximately 1 billion, which matches the number of landlines on the planet. The latter figure took a century to reach; there were, essentially, no cell phones prior to 1989. Senegal went from 100 wireless subscribers in 1995 to 390,800 in 2001, Egypt from 7,400 to 3 million. In many postcolonial nations where infrastructure has traditionally been in the grip of parastatal monopolies, cell phones have allowed more subscribers to circumvent the rigid control of communication enjoyed by the state (Rheingold, 2002; Shirky, 2002).

Without descending into a rigidly technological argument (for, after all, this hardware did not invent itself—it was developed within an evolving business sector, or, in the case of some important components, for the military), it is now clear that the creation of digital technologies has utterly altered the way in which information is moved about the globe. A piece of visual information once descriptively termed a "newsreel"—say, of Amelia Earhart landing in Europe after her first transatlantic flight—might have taken up to 2 weeks to appear in American theaters, as the image first had to be returned across that same expanse of ocean. Film had to be developed, a vocal soundtrack had to be added, musical interludes had to be recorded, copies had to be printed, and the reels had to be distributed to movie theaters (Hall, 1998). Now, in contrast, the information transfer is immediate; nor do we have to go to a theater in order to receive news. It enters our homes, offices, and even the mall via multiple media. We can attend a sports event and receive commentary on the same game via pocket radios; we can watch television and participate via cellular phone to vote someone off a talent show (and with greater efficiency than many political elections).

Quite what this means for us in social terms is a complicated issue. Numerous commentators have weighed in on the topics of time–space compression, or what it means to live now—not so much in a world without boundaries, or in a world without geography—but more literally in a world, as opposed to a neighborhood or a region. McLuhan, of course, famously coined the term "global village" (McLuhan & Powers, 1989), but while that phrase has entered the popular consciousness, it does so in an ambiguous and often unclear manner—is a global village an expansion of our horizons, as the geographical term implies, or a shrinkage to a prior set of social values, as in the phrase "It takes a village"? It is also the case that many critics do not even accept the implications of such insights, for, as Gitlin notes, the village is fragmented by having both "mansions and huts" (2001, p. 176), while the very process of global information transfer is a complex question of economic and intellectual domination (Mitchell & Schoeffel, 2002; Giddens, 2000). Much more will be said on this topic below.

What this all means for the business of communication has, however, proved to be much simpler. Even something as rudimentary as a movie clip once lay atop a very complicated pyramid of production, which involved many disparate activities, different trades and professions, and was thus hard to integrate. As Peter Hall (1998) has shown in detail, new movie technology was initially developed in New York: that was where the Broadway entrepreneurs and the finance were.[2] But very quickly there was a need to develop a new productive location, where all the newer and disparate trades could be assembled, and that is why the U.S. movie industry decamped for Los Angeles. The movie business in the early 20th century was, in that sense, not so very different from the jewelry or furniture trades of the 19th century, insofar as they prospered via proximity. Skilled labor was available, the business services understood exactly what was expected of them, as did the police and the courts, and this all added up eventually to an integrated regional economy (Scott, 2000).

Technological change has, subsequently, allowed the inevitable centrifugal forces to exert themselves. There has always been an urge to create popular cultural products in cheaper locations in order to save on labor costs, but this still involved the assembly of the finished product and its marketing at home base. In movie terms, this might be exemplified by filming *Anthony and Cleopatra* in England and Italy during 1960–1962, before assembling the film in Hollywood and then premiering it in Los Angeles and New York. What, though, if the entire project could have been created in cheaper locations, assembled in some virtual technological space, and then marketed concurrently throughout the world? This would be much more efficient, and is exactly how contemporary products such as *The Matrix* and the *Lord of the Rings* trilogies were created. Most of the six movies in the two series were filmed in the South Pacific, and many of the special effects were developed there too. The films have been released globally, although a significant proportion of receipts is based on the marketing of nontheater merchandise: a soundtrack, toys and action figures, then a rapid release of home video (with special features on DVD). (Paramount Pictures, of whom more below, proclaims proudly, if ingenuously, on its website, that it is now the sole major motion picture studio still based in Hollywood).

Implicit, then, in this new commercial age, is integration in terms of the ways that a product is assembled, marketed, and consumed. The same digital "thing" can be shown in theaters, in aircraft, on satellite television, on computers, and in homes or dentists' offices, which makes the distribution process much simpler. The same can be said of advertising, which too can reach across all these media.[3] And that is only the beginning. Integration can be extended almost infinitely. Why not link up film studios and television networks, so that movies that have entered their rental phase can be shown on television, but with the advertising revenues returning to the studio? Or link studios and theme parks—or cruise ships, or hotels in New York—so that visual icons can provide fun and comfort to those on vacation in varied locations (Hannigan, 1998)? Implicit in these deliberations is a seamless commodification of images. The modern entertainment corporation looks to create an integrated web of consumption that transcends geography. That is to say, just as it is no longer focused on the theater, it is no longer limited to any region, country, or continent. For example—and this is a complicated sequence that should be read slowly—DVD rental dispensers are now being installed in the same fast-food restaurants, that sell the toys, that are associated with the movies, that are shown on the DVDs. These will be the same fast-food restaurants that will be found in the theme parks, that feature the full-size action figures, which are also displayed in the movies. The potential for this kind of "synergy," as it has been termed in the business, is enormous—Disney's 1994 animated movie *The Lion King* is a canonical example. Grossing three-quarters of a billion dollars in global box office, it has subsequently spawned a soundtrack, a stage show, cartoons shown on the Disney Channel, an attraction at Disney parks, and numerous toys. The net profit has been more than a billion dollars (Weber, 2002).

From these examples, we can see that a fundamental convergence has occurred in recent years with regard to the transmission of information. It was once the case that there were recognizable differences between the electronic media (film, television, radio), and thus differences in terms of what each medium had to offer. Crucially, they saw themselves involved in a zero-sum game; for instance, the rise of radio was once thought to be the death of recorded music. This is less and less the case. The media are now more often seen as complementary, in the sense that they handle the same basic products. Put another way, a single commodity can be cross-marketed across different media—hence the proud claim of Howard Stern to be the "king of all media." Beginning as an FM disk jockey, Stern developed his "shock-jock" persona over a number of years and in several radio markets. Slowly, he achieved national recognition, as his program was syndicated, a process that was facilitated by the growth of national radio chains in the 1980s and 1990s.[4] Based on this recognition—and the notoriety of his run-ins with the Federal Communications Commission (FCC)—Stern then wrote a best-selling memoir titled *Private Parts*. In turn, this was made into a Hollywood movie; in addition, the found-music soundtrack from that was released, and became a successful CD (e.g., Kassabian, 2001). Concurrently, Stern was then able to move his radio show into a television format (initially by simply filming what went on in the radio studio), and subsequently to develop a production company. His television show *Son of the Beach* aired on FX for several seasons; he is currently developing remakes of successful movies from the 1980s.

Howard Stern might be taken as an example of the postmodern condition, in the sense that his fame cumulates but is based upon nothing but his prior fame. This example has, though, a very solid material base. Seen from that direction, Stern is not so much "king of all media" as much as he is "master of his own domain," which he has expanded tirelessly. He has one product: himself. And he has been able to commodify himself so extensively precisely because of the interconnections within his industry. As Figure 7.1 indicates, all of Stern's campaigns have been made possible by the way in which media corporations have morphed into one another in recent years, at what is sometimes wryly termed "the bleeding edge." Record companies, book publishers, and movie distributors routinely acquire each other via takeover or merger, with the result that products are easily moved from market to market. Infinity Broadcasting Corporation (owner of WXRK radio, where Stern is physically located when he broadcasts his show), Simon & Schuster (owners of the Pocket Books imprint), CBS (which initially tried to place him on network television), and Paramount Pictures (which made his movie) are all part of Viacom, one of the "Big Seven" media giants (see Figure 7.1 and Table 7.1).

The example of Howard Stern indicates quite well the manner in which the corporate structure facilitates cross-marketing, or synergy (Klein, 1999). Moreover, it also indicates the limits to his work. As his radio and television shows are dependent upon a certain type of language-based humor, they have not been a global marketing success; while the radio show is broadcast in Canada, and the television show

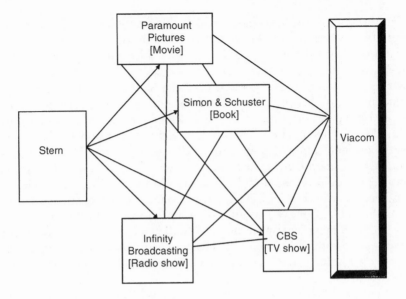

FIGURE 7.1. Howard Stern and his relation to various media corporations.

has been shown on Bravo in the United Kingdom, both lack the universal appeal that is central to global marketing. One other aspect of this is worth repetition—namely, that media products like Stern are very dependent upon functional integration within the parent corporation. That may seem obvious, but, as we will see, there are examples of conglomerates that have not prospered precisely because they have not learned that the media must be integrated and not made to compete. When we return to the AOL Time Warner case, we will see the implications of this.

TABLE 7.1. Corporate Activity of Global Media Conglomerates

Corporation	Sales (millions)	Income (millions)	Employees
Sony	$57,117	$115	168,000
Vivendi–Universal	$51,125	($12,119)	381,504
AOL Time Warner	$41,065	($98,696)	89,300
Disney	$25,329	$1,236	112,000
Viacom	$23,223	($224)	122,770
News Corporation	$16,344	($6,738)	33,800
Bertelsmann	$ 8,592	$825	83,000

Note. Data are the most recent available filings, compiled by the author, for 2001 and 2002.

THE GLOBAL ECONOMIC TERRAIN

In the period following the end of the Cold War, one of the most active economic sectors has been in the communications and entertainment field. Drawing on the technological opportunities discussed above, and reflecting the drive to expand activities across a now-undivided planet, corporations began to integrate their holdings. Interestingly, this was by no means a uniquely U.S. activity, as Table 7.2 indicates. The table indicates how the larger corporations in this field are distributed across the Americas, Europe, and Asia. In addition to the seven major players in the field—Sony, Viacom, News Corporation, Disney, AOL Time Warner, Bertelsmann, and Vivendi Universal, which are themselves representative

TABLE 7.2. Regional Distribution of Media Corporations

Corporate name	Base	Activities/holdings	Sales (millions)
NAFTA			
AOL Time Warner Inc.	US	Diverse media	$41,065
Bell Globemedia Inc.	Canada	Diverse media	$ 756
Cisneros Group of Companies	Venezuela	Electronic media	$ 3,800
Cox Enterprises, Inc.	US	Print, broadcast	$ 8,693
Grupo Televisa, S.A.	Mexico	Diverse media	$ 2,147
The Hearst Corporation	US	Print, broadcast	$ 3,300
The Thomson Corporation	Canada	Electronic media	$ 7,237
Viacom Inc.	US	Diverse media	$23,223
The Walt Disney Company	US	Diverse media	$25,329
EU			
Bertelsmann AG	Germany	Diverse media	$ 8,592
Bonnier AB	Sweden	Diverse media	$ 1,586
Pearson	UK	Print	$ 6,152
Reed-Elsevier	Netherlands	Print	$ 6,617
Roularta Media	Belgium	Diverse media	$ 388
Lagardère SCA	France	Print, web media	$11,777
Vivendi Universal S.A.	France	Diverse media	$51,125
Asia			
The News Corporation Limited	Australia	Diverse media	$16,344
Nippon Hoso Kyokai	Japan	Broadcast media	$ 5,042
Shochiku Co., Ltd.	Japan	Broadcast media	$ 529
Sony	Japan	Diverse media	$57,117

of those three major geographical groupings—a number of other corporations have transnational reach. Taken together, their annual sales amount to an astonishing one-quarter of a trillion dollars.[5]

We can begin to explore this evolution by a closer examination of the News Corporation, whose history cannot be understood except in terms of the manner in which CEO Rupert Murdoch manufactured a business plan for creating transnational media (Chenoweth, 2002). He was the first in the field to extend his business across more than one continent (into Europe, Asia, and ultimately the United States), while at the same time expanding across traditional divides between radio, television, and print media. From a base in the Australian newspaper market, Murdoch pushed in two apparently unrelated directions. He moved into Asia and developed a key presence in the expanding television sector, first with Sky and then with the Star channel, now seen by 300 million people in 53 countries.[6] Concurrently, he moved into the British print market, buying across the social-class spectrum with newspapers as disparate as the *Times* and the *Sun* (to which he has subsequently added the *New York Post*). His biggest purchase has been in terms of the Fox group, also in the United States, which extends from the studios of 20th Century Fox to Fox Television to FX and Fox Sports cable. The full extent of News Corporation's holdings is shown in Figure 7.2, which follows the company's own emphasis upon "communication platforms." As can be seen, the structure builds upon the complementarity of the sectors, ranging from sports teams and retail outlets at one end to satellite television at the other. Note that relatively few of the dozens of holdings have a geographic suffix (cf. Fox Sports Middle East), insofar as it assumed that these all have transnational, or even global, potential.

One of the things that is particularly interesting about this sector is the way in which globalization has not been driven solely by corporate logic (as we might find in deals such as the merger of Daimler and Chrysler) or corporate raiding (exhibited by Tyco) but by old-fashioned "intrepid entrepreneurialism" (Pavlik, 2000, p. 230). Key figures in the acquisitions and mergers of the past decade have been larger-than-life CEOs, who had aspirations to dominate and decimate their neighbors. Besides Murdoch, these have included Ted Turner (Time Warner), Sumner Redstone (Viacom), Jean-Marie Messier (Vivendi), and Michael Eisner (Disney). One of the apparent results of this "big man" process has been a subsequent series of spectacular failures in the industry, based on hubris and poor decision making. The near collapse of Vivendi in 2002 is one example, and the corporate problems of AOL Time Warner during 2001–2003 are another. In fact, the near-demise of the latter is a particularly important insight into the manner in which technological change has had an impact within the industry. When the component

TELEVISION

Broadcast
FOX Broadcasting
FOX Sports
Fox Television Stations

Cable
Canal Fox
Fox News Channel
Fox Sports Middle East
Fox Sports Latin America
Fox Sports Net
Fox Sports World
Fox Sports World Español
FOXTEL
FX
Fox Movie Channel
National Geographic Channel
Speedvision
STAR
TV Guide Channel

Satellite
Canal Fox
Fox Sports Middle East
Fox Sports Latin America
FOXTEL
Sky PerfecTV!
STAR

Digital Television
Sky
Sky Latin America

Syndication and Distribution
Twentieth Century Fox Television Distribution
Twentieth Television

PRINT

Magazines
INSIDEout (News Magazines)
Phoenix Weekly (STAR)
SmartSource Magazine (News America Marketing)
TV Guide

Newspapers
News Limited Newspapers
New York Post
The News of the World (News Group Newspapers)
The Sun (News Group Newspapers)
The Sunday Times (News International Newspapers)
The Times (News International Newspapers)

Publishing
HarperCollins Publishers
HarperCollins Australia/New Zealand
HarperCollins Canada
HarperCollins UK

INTERNET

newscorp.com
foxppv.com
harpercollins.com
foxinteractive.com
foxmusic.com
fox.com
foxsports.com
dodgers.com
pac-10.org
fxnetwork.com
thefoxmoviechannel.com
foxsportsworld.com
fswe.com
nationalgeographic.com
foxnews.com
smartsource.com
newsamerica.com
tvguide.com
nypost.com
mundofox.com
mundofox.com.br
lossimpson.com
ossimpsons.com.br
foxsportsla.com
sky.com
thetimes.co.uk
sunday-times.co.uk
the- sun.co.uk
newsoftheworld.co.uk
foxsportsme.com
startv.com
espnstar.com
channelv.com
phoenixtv.com
news.com.au
careerone.com.au
realestate.com.au
insideout.com.au
foxtel.com.au

Interactive Television
Fox News Virtual Channel
Fox News Extra
Fox News Active
Fox Sports Virtual Channel
Fox Sports Extra
Fox Sports Active
GUIDE Plus + Gold (TV Guide)
Sky Active
TV Guide Interactive

RADIO

Fox News Radio
FOX Sports Radio
Sky News Radio

PRODUCTION

Theatrical
20th Century Fox
20th Century Fox Animation
Blue Sky Studios
Fox 2000 Pictures
Fox Searchlight Pictures

Television
20th Century Fox Television
Foxstar
Fox Foundry
Fox Television Studios
Productions
Fox TV Pictures
Fox World
New Zealand Regency Television
The Greenblatt–Janollari Studio

RETAIL

In-Store
News America Marketing-In-Store

Home Video
20th Century Fox Home Entertainment
20th Century Fox Home Entertainment International

Consumer Products
Fox Music

Licensing and Merchandising
20th Century Fox Licensing and Merchandising

Interactive
Fox Interactive
SmartSource iGroup (News America Marketing)

OUTDOOR
News Outdoor

PLACE BASED

Sports Team
Los Angeles Dodgers

Themed Attractions
Dodger Stadium
Fox Sports Sky Boxes
Fox Studios Baja

FIGURE 7.2. News Corporation: Global "communication platforms."

parts of the current company merged in 2000, it was expected that the new power of the Internet would transform the Time Warner group, which was still based on relatively traditional products such as magazines and movies. The merger was hailed as creating "an operating system for everyday life" (Rose, 2000, p. 158), but proved to be an expansion that only produced more fiefdoms. Frequently reported struggles ensued between Time Warner and Turner Broadcasting, for example, over access and content to prime cable markets, despite their own corporate links. After reporting what may be the largest ever U.S. corporate loss by a nonbankrupt firm in 2002, the conglomerate has shed executives and is developing a new business model. Its experiences suggest a number of lessons. First, the Internet has proved to be less profitable than first thought; AOL actually has fewer clients than broadband services within the conglomerate as a whole. Second, a lack of desirable content has driven down advertising revenues, despite the supposed sexiness of the medium—an outcome that seems at odds with McLuhanist arguments. And third, the state has proved to be more powerful than is sometimes thought in discussions of regulation in a global era—FCC regulations have hamstrung the AOL merger in the United States and will be one more factor promoting the ultimate breakup of the group (Rose, 2002).

While it is often argued that larger corporations have greater power and thus global reach, it is interesting in this context to compare the AOL corporate map with that of its rival, the News Corporation (compare Figures 7.2 and 7.3). While the latter shows a tightly integrated structure, it will be seen that AOL shows much more segmentation. Analysts frequently point adoringly to the independence of HBO, which is headquartered in New York and has little to do with the other parts of AOL, even those related to television, which are based in southern California (e.g., *The Economist*, 2003, pp. 73–75). This, it is argued, results in high-quality program material, but for the purposes of this chapter it is important to make two rather different inferences. First, these two cases indicate that corporate strategies can be, and often are, very different; and second, the way that different conglomerates have responded to technological challenges, such as the digital revolution and the arrival of broadband, is also varied. For our purposes, we can collapse this into the following insight: globalization is not proceeding according to any particular playbook. It is not a smoothly evolving stage of capitalist development; if anything, it is a case of "two rapid steps forward, one step reeling backward." It is certainly not the case that companies are rolling out cultural products with any specific social or political intention—their only goal is an economic return, and, as is frequently observed, most records, most films, and most television shows fail, because audience tastes are hard to predict.[7]

GLOBAL—OR AMERICAN?

The emergence of global corporations raises interesting questions about how the cultural impacts of the phenomenon are evolving and may de-

AOL

Interactive Services Group
AOL, CompuServe, Netscape Navigator, AOL TV

Interactive Properties
ICQ, Digital City, AOL MovieFone, Spinner, Winamp

AOL International (joint ventures)
AOL Latin America, AOL Australia, AOL Hong Kong

Home Box Office
HBO, HBO Plus, HBO Signature, HBO Family, HBO Comedy, HBO Zone, HBO en Español, Cinemax, MoreMAX, ActionMAX, ThrillerMAX

Joint ventures
Comedy Central, HBO Ole, HBO Brasil, HBO Asia, HBO Hungary, HBO Czech, HBO Poland, HBO Romania

Trade Publishing
Little, Brown; Warner Books; Book-of-the-Month Club

Time, Inc.
Asiaweek, eCompany Now, Coastal Living, Entertainment Weekly, Fortune, Life, In Style, Money, Mutual Funds, People, Teen People, People en Español, Progressive Farmer, Real Simple, Small Business, Southern Living, Southern Accents, Sports Illustrated, Sports Illustrated for Kids, Sports Illustrated for Women, Sunset, This Old House, Time, Time Digital, Time for Kids, Wallpaper, Parenting, Family Life, Baby Talk, Cooking Light, Health

Turner Broadcasting System
CNN, CNNfn, CNN Headline News, CNN Interactive, TBS Superstation, Turner Network Television, Turner Classic Movies, Cartoon Network

Turner Sports
Atlanta Braves, Atlanta Hawks, Atlanta Thrashers, World Championship Wrestling, Goodwill Games

Joint ventures
Cartoon Network Japan, Court TV

Time Warner Cable
12.6 million customers connected
Five local news channels, including *NY1 News*

Joint ventures
Road Runner, Time Warner Telecom

Warner Bros.
Warner Bros. Pictures, Warner Bros. Television, Warner Bros. Animation, Looney Tunes, Hanna-Barbera, Castle Rock Entertainment, Telepictures Productions, Warner Home Video, Warner Bros. Consumer Products, Warner Bros. Studio Stores, Warner Bros. International Theatres, Warner Bros. Online, *DC Comics, Mad Magazine*

New Line Cinema
New Line Cinema, New Line Television, New Line New Media, New Line Fine Line Features, New Line Home Video, New Line International, New Line Cinema Studio Store

Warner EMI Music (joint ventures)
Warner Bros./Reprise, Atlantic, Elektra, EMI, Virgin, Capitol, London/Sire, Maverick, Rhino, Qwest, Tommy Boy Music, Giant (Revolution), Priority, East/West, Angel, Blue Note, Columbia House

Music Publishing
Warner/Chappell Music, EMI Music Publishing, Jobete Music

The WB
Television network, including *Buffy the Vampire Slayer, Dawson's Creek, Felicity, Sabrina, Roswell, Pokémon*

FIGURE 7.3. Corporate divisions within AOL Time Warner, 2000–2002.

velop in the future. Before tackling this theme, though, some clarifications are in order. First, the relationship between corporate branding, capitalism, and culture is not addressed in any depth here, although it is certainly the case that the creation of brands, as distinct from products, has many repercussions (Klein, 1999). A consumer taught via repetition to revere a Big Mac is, evidently, in no position to objectively assess the demerits of a lump of fatty animal parts, produced in conditions so disgraceful that they necessitate that the animal be pumped full of antibiotics, while the slaughterhouse worker faces constant on-the-job injuries (Ritzer, 1996; Schlosser, 2001). The same is true—in principle at least—of the music aficionado who is led to believe that the Spice Girls could sing, that Madonna can act, and so forth. This is the outcome of a global phenomenon, in which utterly disparate products legitimate each other simply because they share a brand and a logo. Disney and Virgin, to take two instances, are branding chains that extend from records, to films, to theme parks, through to railways, airlines, and even a municipality; it no longer matters that these corporations began, respectively, with a cartoon mouse and an obscure long-playing record titled "Tubular Bells." Similarly, in Japan, the Hello Kitty brand is "an icon that doesn't stand for anything at all" which has been affixed to products as disparate as vibrators and automobiles (Anderson, 2001, p. 121). In short, successful capitalism is not merely about commodification, but is increasingly about creating a web of products that become linked in the consumer's mind via a logo—a symbol that can transcend both time and space, insofar as it does not depend upon local languages and other referents in order to be valued. It should not be news that indigenous cultures are not nurtured under such circumstances. There are commentators who are apparently still surprised by this, but that is beyond the scope of this discussion (for more on what can be called the "Rigoberta Menchú syndrome," see, e.g., Arias, 2001).

Here, instead, the focus is upon a more focused question. Put simply, do branded goods (media products included) fit into a web of national domination, one country over another? Cultural commentators have been keen to lament the trend to globalism on two counts: first, with the argument that anything global cannot be local, and second with the assertion that this involves an Americanization of the planet. With regard to the first, Poster argues:

> Consumption is also planetary in scope bringing across borders alien cultural assumptions as embodied in commodities. The popular need no longer be local. Although foreign goods are inflected with community values, and easily adapted to local conditions, they remain indexes of otherness. What is more dramatic still . . . nation states are losing their

cultural coherence by dint of planetary communications systems.
(2002, p. 98)

A summary of his argument is that foreign is not local, and in all likelihood means American. The extent to which globalization is a process
driven by explicitly American corporations and American values has
been examined by disparate sources. There are certainly complaints
about American cultural imperialism embedded in France's efforts to
maintain the integrity of the French language, for instance (and Disneyland Paris was once dismissed as a "cultural Chernobyl"; see Weber,
2002). More systematically, Sassen (1996) has pointed to the ways in
which international trade with U.S. corporations tends to result in U.S.
legal codes being imposed upon other nations, at the expense of their
own legal values. Intellectual property rights clauses have been incorporated into the General Agreement on Tariffs and Trade (GATT) agreement, also to the primary benefit of U.S. software and drug corporations. In numerous contexts, in ways both obscure and obvious, dozens
of other examples can be invoked.

Yet this manages to miss two key dimensions to this story. The first
is that the global corporate model works well for more than one party—
as Gitlin puts it in his study of the media, "the cultural gates are poorly
guarded and swing both ways" (2001, p. 188). A global corporation
cannot—by definition—be based exclusively in one country, nor can it
peddle only one type (or source) of content if it wants to be successful.
As was noted earlier, many more movies (to take one medium) are made
outside the United States than within it. This ensures that there is more
than a one-way flow from Hollywood (Table 7.3), or, to put it in corporate jargon, ownership of the pipe does not automatically define the content.

It is important to recognize how the strategic arrangement of media
corporations around the world facilitates complex flows of cultural
information. Let us take a specific example, the 1999 movie *Crouching
Tiger, Hidden Dragon*. This production won four Oscars© in 2001, including Best Foreign Language Film and Best Original Score. Based in
the *wuxia pian* tradition of premodern chivalry, and shot on the mainland of China (including distant Mongolia), it was partially financed by
Columbia TriStar Asia (a Hong Kong-based division of Sony), and Sony
Classics Pictures in the United States. "This $15 million movie is a mi-
crocosm of Chinese ambitions in the world market. Filmmakers from
the three Chinas are trying to stay true to their heritage while appealing
to the West" (Corliss, 1999, p. 21). By "three Chinas," the commentator
refers to the People's Republic of China (PRC), to Hong Kong, and to
Taiwan, all represented in the production. The director, Ang Lee, grew

TABLE 7.3. Numbers of Films Produced Globally in 2001

Producer country	Number of films released, 2001 (%)	Investment, $billions, 2001 (%)
India	809(43)	0.07(00.6)
US	720(38)	9.50(84.2)
France	170(09)	0.85(07.5)
UK	85(4.5)	0.81(07.1)
China	85(4.5)	0.04(00.3)
Total	1,869(100%)	$11.27(100.0%)

Note. Data compiled from The Economist (2002).

up in Taiwan, while the two main stars are from Malaysia (Michelle Yeoh) and Hong Kong (Chow Yun-fat), although neither speaks Mandarin as a first language. The film was released in both English and Mandarin versions, in order to compete in two of the largest global markets. In addition, there is a soundtrack (also marketed by Sony) that features PRC composer Tan Dun, Chinese American Yo-Yo Ma, and Hong Kong singer CoCo Lee, whose contribution is offered twice, once in English and once with Mandarin vocals.

Crouching Tiger offers no hint of U.S. values; its representation of a preindustrial society involves only collective responsibility, reflects rigid social and gender roles, and fails to display a Hollywood feel-good ending (by the close, two of the three protagonists are dead). Moreover, the original theatrical release was in Mandarin with English subtitles, as the dubbing was ready only for the tape and DVD release. Yet it was a commercial and critical success (grossing approximately $300 million), indicating that the distribution mechanisms for contemporary entertainment do work for films that do not recycle Hollywood's norms. European films, too, are increasingly visible in the global marketplace, and Hindi-language films are making an impact for the first time (e.g., Lagaan and Monsoon Wedding). It is not wishful thinking to argue that this is globalization and not Americanization (see also Ching, 2000).

Nor are these the only examples. Consider the impacts that Japanese corporations have had in various fields. It is not hyperbole to state that their consumer electronic innovations have revolutionized entertainment, first with the Walkman, then with the Betamax video recorder, the compact disk, game technology, and subsequently the most advanced aspects of wireless telegraphy. All have had important social impacts on how we define personal space, what constitutes privacy, and the amount of time we spend in the home consuming electronic entertainment. In terms of popular culture, the Japanese have been global leaders in both comic books and video games, and their aesthetic has been inordinately influential. The *manga* style of comic books has influenced Western productions such as *The Little Mermaid* and *The Lion King*. The game market is a vast one (pushing $20 billion annually), and Japan has led in both the technology (Nintendo) and the aesthetic (arguably, much of Hollywood's action output now borrows from Japanese game dramas). The *anime* genre (now finding full expression as digital animation) has also been an influential aesthetic that is found across the world's cartoon networks. The Pokémon cartoon has been broadcast in 65 countries, in more than 30 languages; between 1996–2002 it grossed $15 billion.

None of this amounts to an open global marketplace in which the consumer can pick and choose freely. It has long been recognized that the media of communication have been used to pass (mis)information across borders, while internally, corporate, bureaucratic, and citizen interests serve to police the public sphere. As mentioned with the French example, censorship is often recast as cultural resistance to foreign threats, although it rarely seems to occur in ways that one might predict. British newspapers with naked women on display simply don't appear on the newsstands in theocratic countries—so there is nothing to debate. But material that we might expect to pass safely "under the radar," insofar as it seems to contain no explicit political or sexual message, may still be troubling to those in authority. For example, Hollywood action heroes represent values of individuality that are both attractive, for what they offer in terms of social or business innovation, and yet troublesome to leaders in countries such as Singapore, where communal values and obedience are culturally rooted (for an extended editorial quotation on this topic, see Kirby, 2001).

As efforts to police the Internet have typically been only partially successful, any cultural defense must usually concentrate upon traditional media (e.g., Castells, 2001). As noted earlier, popular culture is often a forum for nonelite debate, and social anxieties are constantly confronted there (Kirby, 2002). I have already introduced the Chinese film *Crouching Tiger*, which has at its heart a story of unconsummated love—an unusual theme for American audiences. Rather than receiving accolades from religious commentators, the movie was, however, criticized for "offenses to God," including "spiritual meditation without God," and "walking on water," a reference to a scene in which characters spring across a lake. The fact that these are characters from the Qing dynasty, representatives of another religion, does not diminish a perceived threat to American sensibilities.[8] Another example also reveals something of the self-referential world of popular culture. In the second *Harry Potter* film, it was suggested that Dobby (an unlikable computer-generated character with an inordinately large nose) was deliberately created to look like, and thus be disrespectful of, Russian leader Vladimir Putin. It was widely reported on the Internet that Russians were upset by the portrayal and were calling for the movie to be banned. This episode says much about the ease of communication via the Internet, and the willingness of individuals to be offended by cultural symbols, be they British, American, or Chinese in origin.

The transfer of images across borders reminds us that popular cultural product appears almost everywhere, although that point needs to be examined at more than face value. While it is the case that capitalism is adept at branded selling, it is also the case that the cash register opens for anyone, to extend Gitlin's phrase. A significant proportion of the product that flows around the world may be brands, labels, or artists recognizable as American, Japanese or European, but which have not been licensed and are not supported by advertising. In this instance, then, we should ask whether such pirate production and trafficking enhances or undermines globalization? Certainly, the purchase of a fake Brittany Spears disk reflects some aspect of Americanization—but only to the extent perhaps that the purchase of cocaine on an American street corner reflects some Colombian influence in the United States (Boulware, 2002).

In trying to find balance on this topic, we should recognize that capitalism is typically successful because it is adept at the nuances of giving consumers what they want, and in the most palatable manner. As an instance, a 2002 U.S. CD release by Shania Twain contains 19 songs; on one disk the songs are offered in country mode, on a second disk the same 19 are presented in a light rock format. This guarantees that one or the other part of the package can be successful in the two main U.S. contemporary music radio formats. And, in turn, globalization offers an opportunity to find another, larger audience; the non-U.S. release also has two disks, but in this instance the 19 songs are presented in a light rock and in a world music format. Perhaps the most adept example of this flexible cultural production is the Disney Corporation, which is a definitive case of the complexities of global marketing. Precisely because its characters come with implicit social messages, the company has rolled out its products in ways that are both tentative and culturally sensitive.[9] Initial moves into the Chinese market, for instance, were led by local pirate entrepreneurs (as noted above), which deflected political concerns. In 1997, the release of *Kundun*, with its pro-Tibet message, threatened Disney's position in China, but the company was also producing a traditional animated folktale, *Mulan*, which rehabilitated its image. One result was a successful bid to create Hong Kong Disneyland, which will open in 2005. As with its Japanese counterpart, Disney is working with local government (and will receive over $2 billion worth of subsidies), and uses local corporations, such as Sea Rainbow, to handle key tasks, such as management of the Mandarin website (Weber, 2002).

DISCUSSION: GLOBALIZATION AND THE FUTURE

Globalizing corporations are not only doing new things in terms of business, their activities are also having interesting social repercussions. In the past decade, for instance, satellite television in Asia has presented its viewers a glimpse of an everyday reality that includes a new creation: the transnational popular cultural region. Cable television (such as the BBC or Sky) now offers the weather report, or sports results, for large mixes of countries. Have the residents of Abu-Dubai, of India, and of Japan thought about themselves as having much of anything in common on a daily basis? Unlikely. Does the repetition of such a regional construct change social realities? Perhaps. The collapse of time–space borders in the presentation of information creates the opportunity for people to think about themselves in different ways. The European experience seems to indicate that it is possible to manufacture cultural cohesion across national divides, and popular culture contributes to that process.

In a post-9/11 world, we now ask if this is a good or a bad thing, or indeed whether this contributes to cultural anxieties. As we have seen, popular culture can be a touchy subject. Todd Gitlin, who styles himself as "one of America's foremost public intellectuals," makes the following fatuous observation: "What do American icons and styles mean to those who are not American? We can only imagine—but we can try" (2001, p. 179). We should be able to empathize with others' dismay over the erosion of the local, but if we cannot, then it is not hard to read about it, for, after all, more prominent public intellectuals—Baudrillard and Calvino, Lefebvre and Mao—have all written extensively on these perceived threats. And it is hardly difficult to think of reverse examples that are closer to home. The uncoordinated arrival of rock music from Great Britain in the United States in 1964 was nonetheless perceived as the "British Invasion," and records were smashed or burnt by community leaders. Was the music immoral and socially incendiary? To a degree, but the threat was in reality the arrival of "race music" in American homes, and this was exacerbated by the fact that it arrived so brazenly via JFK Airport (Inglis, 2000). More contemporary examples occur daily, and implicit in such examples is the reality that even the most disposable product is a "sugar-coated bullet," in Mao's phrase. That is why, in this new global era, the purveyors of culture succeed by being subtle, working with governments, and generally keeping the focus on the entertainment that the consumer is willing to consume—knowing all along that if governments restrict the supply, then much of this product is easily pirated anyway and a demand will arise of its own accord.

We should, though, restrain any urge to make this account more dramatic than it actually is. The vast majority of the world's consumers still spend most of their time listening to programs on local radio, in their own language, and watching local television, and sometimes seeing foreign films—and in that sense, the virtual world predicted by Meyrowitz (1985) 20 years ago is still under construction. The most basic impediments to homogeneity still persist—for example, a DVD calibrated for European or Japanese televisions will not work with American televisions, which use different displays. And many countries remain committed to controlling information flows, a practice that includes the United States.[10]

And the future—will that be so different? That is more complex. The evolution of wireless technologies and the availability of cheaper electronics are set to make the dissemination of information even simpler. Where people want to mount cultural resistance, it will be cheap and easy to manufacture and distribute local products, be it music or poetry recorded inexpensively and copied on computers, or films recorded on cheap digital video. Where people want to consume from the mainstream, that too will proceed, though digital file sharing is decimating the "big seven" players and their smaller counterparts. Despite last-ditch efforts to sue individual "pirates" for billions of dollars, no one really believes that the monopolistic global music industry will survive the decade. Thus we see the same technologies facilitating very different outcomes—namely, more standardization and also more local product—in all media. So there will probably be two parallel evolutions—one global, the other highly fragmented, which is, after all, exactly why popular culture is popular.

CONCLUSIONS

This chapter has provided a snapshot of a complex industrial sector that has equally complicated social and cultural implications. We can say with some precision that this is a field in which corporations from several countries have developed global reach, and in doing so have led their counterparts in other industries. They have emerged from the past decade with complex structures of interlocking holdings that extend across the continents. By exploiting new digital technologies, they have produced some very interesting social and economic possibilities. It is this aspect of the future that is harder to predict. As technologies have been made cheaply and widely available, it offers the possibility of both uniformity and diversity within popular culture.

Does this constitute globalization? Yes, to an important degree, that is exactly what this is. The corporations span the globe, although it is easily forgotten that even the big corporate players are still only making limited returns on their global investment. Moreover, the flow is chaotic in terms of origins and destinations, and there is a complicated picture of production and consumption that varies by medium (and would get even more complicated if we were able to address all the aspects of popular culture, including sports). Moreover, because popular culture tends to be very much a generational product, there also tends to be fragmentation between adolescents or young adults, who have the most disposable income and the most technological competence, and older consumers, who tend to enjoy more traditional fare. So it must be recognized that claims of a tidal wave of American hegemony—a cultural McDonaldization of the planet—are wildly overstated. Indeed, globalization is creating a truly dynamic popular culture, a mixing of many streams of information, and there are many more exciting changes ahead—both technological and social—in the next decade.

NOTES

1. Movies were available on digital laser disks, but these were to become another example of an adequate technology that did not succeed, alongside the Betamax and the eight-track tape player. The size of the disks and the weight of the players militated against their widespread diffusion.
2. As he also points out, the technology evolved in numerous places, and there was a burgeoning film industry in France in 1910 that supplied over half of the movies seen in the United States that year, reminding us that globalization is a cyclical process that reached one peak in 1913 (see Kirby, 2001).
3. Digital transmission works both for entertainment and for advertising, and rapidly blurs the lines between them. Commercial product placement was visible as early as the 1950s (see the 1959 movie *Pillow Talk*), but is reaching its apotheosis in a digital age, when products can be stripped in and out of digital presentations at low cost but with high return for advertisers.
4. This had been impossible in earlier decades due to monopoly restrictions, but was then facilitated by the intentional weakening of FCC regulations during the Reagan administration, a process still continuing in 2003 (e.g., Rose, 2002).
5. It should be pointed out that this figure includes, in the case of a corporation like Vivendi, some nonmedia activities, but that lack of integration also goes a long way toward explaining the imminent collapse of the company at the time of writing.

6. Data from News Corporation, 2001.
7. It is impossible to fully understand how idiosyncratic different markets can be. In 1997 German television decided not to syndicate *Seinfeld* and returned instead to the popular standby *Hogan's Heroes*—which, it will be remembered, is set in a wartime German POW camp.
8. The analysis can be seen at *www.capalert.com*. The web page is titled "Media Analysis of American Culture," but seems to make no allowances for products that are imported. We might mention that there have also been strong fundamentalist criticisms of the *Harry Potter* series for promoting witchcraft, which reflects the different religious sensibilities in Europe and the United States. It is another indication of how cultural resistance flairs in unlikely ways.
9. As Americans are used to thinking of cartoon characters as innocuous and childish, it is initially hard to conceive that others react in different ways: to the portrayal of childhood itself, for instance, or of attractiveness, or of the assertiveness of girls (Ossman, 2002).
10. The United States has led the world in attempting to place legal controls on the Internet, although all this legislation has been deemed unconstitutional to date.

REFERENCES

Anderson, C. (2001, October). Ichiban. *Wired*, pp. 120–125.
Arias, A. (2001). Authoring ethnicized subjects. *Publications of the Modern Language Association of America, 116*(1), 75–88.
Boulware, J. (2002, November). Pirates of Kiev. *Wired*, pp. 110–15.
Castells, M. (2001). *The Internet galaxy.* New York: Oxford University Press.
Chenoweth, N. (2002). *Rupert Murdoch.* New York: Crown Books.
Ching, L. (2000). Globalizing the regional, regionalizing the global: Mass culture and Asianism in the age of late capital. *Public Culture, 12*(1), 233–257.
Corliss, R. (1999, November 29). Back to China. *Time Asia*, p. 21.
The Economist. (2003, January 18). How to manage a dream factory, pp. 73–75.
Giddens, A. (2000). *Runaway world : How globalization is reshaping our lives.* London: Routledge.
Gitlin, T. (2001). *Media unlimited.* New York: Metropolitan Books.
Gunn, G. (2001). Introduction: Globalizing literary studies. *Publications of the Modern Language Association of America, 116*(1), 16–31.
Hall, P. G. (1998). *Cities in civilization: Culture, technology and urban order.* London: Weidenfeld & Nicholson.
Hannigan, J. (1998). *Fantasy city.* New York: Routledge.
Inglis, I. (2000). The Beatles are coming: Conjecture and conviction in the myth of Kennedy, America and the Beatles. *Popular Music and Society, 24*(1), 31–85.
Kassabian, A. (2001).) *Hearing film: Tracking identifications in contemporary Hollywood film music.* New York: Routledge.
Kirby, A. (2001). What in the world? *Political Geography, 20*, 727–744.

Kirby, A. (2002). Popular culture, academic discourse and the incongruities of scale. In A. Herod & M. Wright (Eds.), *Geographies of power* (pp. 171–192). New York: Blackwell.

Klein, N. (1999). *No logo.* New York: Picador.

McLuhan, M., & Powers, B. R. (1989). *Global village: Transformations in world life and media in the 21st century* . New York: Oxford University Press.

Meyrowitz, J. (1985). *No sense of place.* New York: Oxford University Press.

Mitchell, P., & Schoeffel, J. (Eds.). (2002). *Understanding power: The indispensable Chomsky.* New York: New Press.

Ossman, S. (2002). *Three faces of beauty.* Chapel Hill, NC: Duke University Press.

Pavlik, J. V. (2000). The structure of the new media industry. In A. Greco (Ed.), *The media and entertainment industries* (pp. 214–247). Boston: Allyn & Bacon.

Poster, M. (2002). Digital networks and citizenship. *Publications of the Modern Language Association of America, 117*(1), 98–103.

Rheingold, H. (2002). *Smart mobs: The next social revolution.* New York: Perseus.

Ritzer, G. (1996). *The McDonaldization of society.* Thousand Oaks, CA: Pine Forge Press.

Rose, F. (2000, September). Reminder to Steve Case: Confiscate the long knives. *Wired,* pp. 156–172.

Rose, F. (2002, March). Big media or bust. *Wired,* pp. 88–97.

Scott, A. (2000). *The cultural economy of cities: Essays on the geography of image-producing industries.* Newbury Park, CA: Sage.

Schlosser, E. (2001). *Fast food nation.* Boston: Houghton Mifflin.

Shirky, C. (2002, October). Sorry, wrong number. *Wired,* pp. 97–98.

Weber, J. (2002, November). The ever-expanding, profit maximizing, cultural imperialist, wonderful world of Disney. *Wired,* pp. 71–79.

Part III

GLOBALIZATION'S OUTCOMES

Human Well-Being

8

Globalization's Impact on Poverty, Inequality, Conflict, and Democracy

MICHAEL D. WARD
KRISTIAN SKREDE GLEDITSCH

This chapter explores the impact of globalization on inequality, economic growth, conflict, and access to democracy. Globalization became the new buzzword for academics, politicians, activists, and pundits in the 1990s and into the 21st century. Stanley Fischer (2003) points out that while globalization was never mentioned in the *New York Times* during the 1970s, there were more than 1,200 references to it between 2000 and 2002. A search for "globalization" on the Google Internet search engine in September 2003 yielded over 2.6 million links, a million more than the figure reported in the lecture by Fischer, only 1 year earlier.[1] One editor of this volume identified 20 articles about winners and losers from globalization appearing in the *Economist* over the past few years, each with a different twist and many with diametrically opposed conclusions about the good, bad, and ugly consequences of globalization.

Although it is difficult to characterize fairly all the work that has been done over the last five decades on globalization and its effects, it is undoubtedly the case that most of it has been impressionistic and normative. From the first critiques of the Schumann Plan (the precursor to the European Union [EU]) to the latest analyses of the antiglobalization riots and protests known as the "Battle of Seattle," a wide-ranging set of deleterious effects have been identified, ranging from destruction of labor unions, diminished productivity, and decimated local industries, to declining terms of trade, pauperization of developing societies, and cultural hegemony.[2] Recent works in this vein include Martin and Schumann's (1998) *The Global Trap*, Stiglitz's (2002) *Globalization and Its Discontents*, and Rodrik's (1997) now-classic query: *Has Globalization Gone Too Far?* At the same time, proponents of globalization generally have lined up a range of benefits, which focus on increased firm productivity, diffused throughout the globe, with supply chains that efficiently extract resources at the point of greatest leverage. Evocative titles in this genre include Bhagwati's (2002) *Free Trade Today*, Bhalla's (2002) *Imagine There's No Country*, Irwin's (2002) *Free Trade under Fire*, and Norberg's (2001) *In Defense of Global Capitalism*.

The major enduring questions concerning globalization can be easily summarized: Does globalization help to promote economic productivity and growth qua development? Does globalization promote inequality? How does globalization affect the role and efficiency of state-based institutions? How does globalization affect the functioning of the market and more broadly of the mechanisms of political economic governance? Does globalization foster greater levels of conflict? In this chapter, we take stock of the empirical record for some of the questions regarding the effects of globalization that have been prominent in the popular debate.[3] We examine these questions by looking at outcomes and the impacts of globalization in these arenas, not by unraveling the actual processes involved.

Democratization has gained central ground in discussions of governance at the state as well as the global level. Under the Clinton administration (1993–2001), the United States actively promoted democracy as a state-building mechanism. At the Millennium Summit in September 2000, member states of the United Nations resolved to "spare no effort to promote democracy and strengthen the rule of law, as well as respect for all internationally recognized human rights and fundamental freedoms, including the right to development".[4] Apart from its inherent benefits, democracy is often argued to promote both peace (viz. the democratic peace literature, which asserts that democracies do not wage war against each other) and economic development, both directly through more favorable policies and indirectly by avoiding the deleterious effects on health and productivity caused by conflicts and extreme autocracy. About 50% of the independent countries in the world presently have political systems that can be characterized as "relatively democratic" in the sense of having widespread electoral processes as well as some functional separation of powers in different branches of government. Countries with electoral democratic institutions represent the preponderance of world economic activity, with some estimates as high as almost 90%. At the same time, few states with an Islamic or Arab majority have open competitive elections with leaders regularly chosen by popular mandate, and other regions also have "democratic deficits."

One of the problems that face most analyses of globalization and democratization is that it is easy to assume that both forces occur in a vacuum. This permits many analysts to assume that any identifiable change or stasis can be attributed, for better or for worse, to one of these two "facts." It seems clear that many changes are afoot around the world, not all of them a direct result of globalization or of democratization. Nor can the present state of affairs always be regarded as a consequences of globalization, without a serious effort to examine whether problems existed prior to globalization and whether changes, no matter how slow they might be, appear to be improving or undermining the prospects for democracy and development. We return to both of these points in our review.

EXPANSION OF GLOBAL PRODUCTIVITY

We define *globalization* to mean the increasing internationalization of processes previously confined to nation-states. Globalization may in principle encompass economic, political, and social processes. The current globalization debate is in many ways a reincarnation of previous controversies over the virtues and vices of the capitalist market. Polyani (1944) was perhaps the first to highlight the putative evils of a globalizing force in the world, particularly as it is thought to erode the networks of social interactions that give meaning to local interactions.[5] However, contemporaries such as Hayek (1944) highlighted the virtues of self-organizing markets and emphasized the negative consequences and assaults on freedom inherent in efforts by the government to intervene in market mechanisms.

Polanyi and Hayek were writing at a time when communism as well as the idea of a "third way," balancing the tyranny of planned economies and the unchecked market, were seen as viable alternatives. The wave of privatization by avowed socialist governments in countries such as France and Greece in the 1980s and the transitions of Eastern Europe in 1989, however, signaled a general acceptance that only markets—in some form—provide effective provision of goods and services. There is no longer an epic clash of ideologies, but rather an attack on what is viewed as a dominant single perspective, that is, a so-called neoliberal globalism.

Expansion of Trade and Gross Domestic Product

There is no argument that both the economic output of the world and the volume of international commerce have grown dramatically over the past five decades. In Figure 8.1 data from the World Trade Organization[6] are used to portray these two important facts. The length of each line segment in these two spirals corresponds to a year's worth of trade or gross domestic product (GDP). If trade were to stay constant over time, the spiral would simply be a circle. If there were a downward trend in trade, the spiral would fall inward, with the earliest date being on the outer edge, the most recent being in the center. However, both graphs show that there has been an upward, increasing spiral in global trade as well as productivity.[7] Not only has economic activity spiraled upward, but so too has international commerce. The general trend is clear: not only has there been a fairly constant growth in the global economy over the past five decades, but there has been pervasive expansion of the interdependency of that global economy, with the last decade or so show-

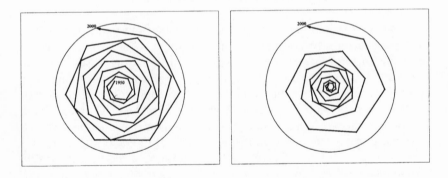

      (a) GDP 1950–2000   (b) Export Volume 1950–2000

FIGURE 8.1. There can be little doubt that global economic productivity has seen constant expansion during the period from 1950 to the present. As seen in panel (a) global productivity, in terms of an index of global gross domestic product (1990 = 100), continues its expansion since 1950. Even more striking is the growth of the volume of global exports (1990 = 100), which shows in panel (b) both growth and (especially during recent periods) acceleration. Although the expansion has not been linear, the postwar period has seen unprecedented progress in economic productivity without setbacks at the scale of the great depression in the 1930s.

ing overwhelming evidence of a rapid acceleration in that economic activity that is categorized as international commerce.

Obviously, economic globalization includes more than an accumulation of the net amounts of commerce (i.e., international trade); it also includes flows of other economic resources, ranging from human capital, through so-called foreign investment, on to the flow of financial instruments—private and governmental—throughout the globe. For better or worse, however, most visible and notorious are the statistics collected by governments and international agencies to assess international trade. Figure 8.2 shows the plots of trade and GDP for the entire world. Alesina, Spolaore, and Wacziarg (2000) point out that this understates the expansion of trade because it ignores the changing population of states. Many former colonies have gained independence since 1945, and these tend to have less trade integration. Alesina et al. (2000) have compiled time series data based on the average trade-to-GDP ratio for eight countries in continuous existence from 1870 (France, the United Kingdom, Denmark, Italy, Norway, Portugal, Australia, Brazil, and Sweden). Their data suggest that the trade-to-GDP data was stable up to the global depression of the 1930s, and did not recover until after World War II. Beginning in the 1970s, however, there was a notable expansion of trade that was not just a continuation of previous trends. We take this as evidence that "globalization" is a meaningful term to describe increasing internationalization, exemplified by the unprecedent expansion of economic exchange across boundaries. Taken in simplest terms, globalization and economic expansion will be two easy summaries of economic history with respect to the last five decades.

Economic expansion under globalization does not play out identically over time or space. Although GDP per capita differs tremendously between different areas of the world, the *relative* growth rates in income are essentially the same for developing and developed societies. However, since the initial GDP per capita for developing and industrialized societies were so different, the gap in income that existed at the outset has increased even further. Moreover, there is quite a bit of variation between different geographical regions of the world. As can be seen in Figure 8.3, whereas countries in Latin American and Asia overall have experienced considerable growth, countries in Africa have fared much worse. The Middle East, after a period of initial growth, has experienced a severe economic recession since the oil crisis in the 1970s. The lack of progress in many developing countries is all the more surprising because economic growth theory suggests that countries with lower GDP should have a potential for higher growth rates (this is sometimes known as the "convergence thesis").

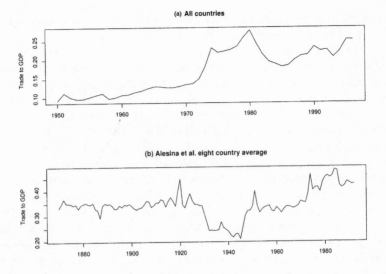

FIGURE 8.2. Growth of trade as a share of GDP over the post-1950 period is shown in panel (a), while panel (b) illustrates the stability of this index for eight countries throughout much of the modern industrial period. Only since World War II has this gone up for the set of eight longest existing countries: France, United Kingdom, Denmark, Italy, Norway, Portugal, Australia, Brazil, and Sweden.

Globalization, Economic Openness, and Growth

To what extent have trade and integration with global markets helped countries achieve higher growth rates? Recent research has consistently shown that a higher trade-to-GDP ratio appears to be associated with higher *recent* growth rates (Levine & Renelt, 1992). Sachs and Warner (1995) estimate that countries with open economies on average experience 2% more annual growth than countries with closed economies. Ireland provides an excellent example of a country that has benefitted from internationalization, even if the boom of the 1990s is fading. Indeed, Dollar and Kraay (2001b) note:

> The post-1980 globalizers have seen large increases in trade, and significant declines in tariffs over the past 20 years. Their growth rates have accelerated from the 1970s to the 1980s to the 1990s, even as growth in the rich countries and the rest of the developing world has declined. The post-1980 globalizers are catching up to the rich countries while the rest of the developing world is falling farther behind.

Dollar and Collier (2001) present an overview of recent empirical evidence for the benefits of increasing trade and growth generally, as well as against claims of the putative antiequalitarian effects of globalization. Bhalla (2002) finds that poor individuals have had the fastest relative growth in income. Dollar and Kraay (2001a) further establish that average income in the poorest sectors of societies tend to increase proportionately as average income increases, thus suggesting that growth not only does not hurt, it actually helps those who are poorest in a growing economy. This finding seems to hold for different regions, different eras, and within different countries at different pregrowth levels of economic productivity. The regions that display the poorest economic performance, as measured by the growth of GDP per capita, in Figure 8.3, are also the regions with the least globalized countries. We conclude that the evidence overwhelmingly suggests that countries with more globalized economies do better and experience sustained growth in income both for the richest and for poorer sectors of society. Indicators of development other than GDP per capita, such as life expectancy, lead to similar conclusions. Differences between regions preceded the recent economic expansion, and there is little basis for connecting globalization either with the lack of development in Africa or with the economic contraction in many Middle Eastern countries.

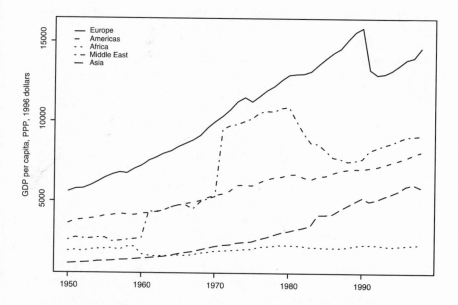

FIGURE 8.3. There is differential, but globally positive, growth in GDP across most regions of the world. Africa remains the outlying region in terms of development, much like it has since the beginning of the industrial age.

Challenge: Inequality as an Outcome of Globalization

Many critiques of economic globalization have argued that although it may generate wealth and prosperity in the aggregate, globalization benefits only those countries that already are wealthy, and tends to harm more fragile groups, such as workers, even in the wealthy countries.

It has been known for a long time that economic growth tends to promote inequality in the short run, though there are debates about its long-term effects (see, e.g., Gottschalk & Smeeding, 1997). Critiques of globalization have brushed aside increases in economic productivity to argue that the real consequences have been to reinforce and accentuate existing inequalities.

How has inequality changed in the past five decades? Inequality is a very difficult concept to grasp, in part because it is fractal. Economists began studying inequality by looking at distributional patterns of income within sectors of modern economies. These first studies assumed that wages would serve as a good surrogate for income on a sectorwide basis. So initially inequality was calculated on sectoral bases within individual countries. At the same time, economists and others became interested in how income was distributed around the globe, to get at international inequality. For the most part these studies used national income statistics to determine the extent to which individual countries shared equally in the world's income. There are some significant problems with this approach, but what does the evidence show in broad strokes? If we use GDP adjusted in terms of purchasing power parity (PPP) (Summers & Heston, 1991), it is not clear that there has been any remarkable increase in global inequality over the past several decades.[8] Several studies have been prominent in arguing for increasing inequality (United Nations Development Programme, 1999; Korzeniewicz & Moran, 1997; Milanovic & Yitzhaki, 2001), but those studies that suggest a large increase in global inequality tend to be based on currency-adjusted, national-level income data. Indeed, using foreign exchange rates instead of PPP indices does suggest a growth in global inequality over time. But these studies are offset by a sheaf of critiques that point to declining inequality if PPP data are employed, or at least to stable patterns of inequality (Chotikapanich, Valenzuela, Prosada, & Rao, 1997; Castles, 1998; Deininger & Squire, 1996; Firebaugh, 1999; Melchior, Telle, & Wiig, 2000; Schultz, 1998). Both the PPP data and the currency-adjusted data are actually different versions of the same underlying data on national economic outputs.

Perhaps the most convincing study is Dowrick and Akmal (2001), which suggests that between 1980 and 1993 (the years most typically studied) there is a very modest increase in inequality of slightly less than

2%, about one-fourth the estimates typically found in studies of inequality.

Studies that use national-level data are criticized by Milanovic and Yitzhaki (2001)—among others—since they homogenize each country, even when they are weighted by population. Milanovic and Yitzhaki (2001) suggests that there is really no middle class, globally. Rather, they find a world of very rich and very poor. Unfortunately, this study does not address the issues of change and whether inequality is increasing.

There are three major problems yet to be overcome with the extant studies of global inequality. First among them is a reliance on national-level data that does not include China. Chinese productivity has been growing rapidly, often by as much as 10% per annum in recent years. Since China has a population of more than a billion, this almost certainly will *reduce* estimates of global inequality, even if it will take China until the second half of the current century to catch up with per capita levels of economic production in Northern Europe and North America. Second, relying on PPP data is probably a good idea, but it has the consequence of typically overstating the true income figures in especially poor countries since the purchasing power goods are global, not local. The third real problem is that most studies of change are mired in the last century, typically stopping with the 1993 data. Clearly, missing out on the last decade of global economic change *and* the case of China, which alone holds over 20% of the world's population, are pretty serious omissions. Both of these influences may well work in opposite directions, but it is unlikely that they will cancel each other out. Overall, we conclude that good evidence for a massive and important shift in global inequality has yet to be discovered.

It is certainly possible that inequality can be retreating at global as well as local levels, but that poverty is still climbing because the really poor are falling even further behind, despite the widespread advance of the rest of the population. However, if we look at absolute level of poverty, defined by the World Bank as living on an income of less than U.S. $1 per day, rather than at measures of inequality, then it is clear that there are many encouraging trends suggesting that poverty is retreating. Although it is hard to find good data over time, Bourguinon and Morrisson (2000) estimate that poverty rates—the share of the population that is poor—have declined impressively from about 55% in 1950 to 23.7% in 1992. For many years the World Bank's *World Development Report* gave data on household income surveys for only a handful of developing countries, but recently the World Bank has begun publishing comparable data drawn from almost 300 income surveys for over 80 developing countries (Chen & Ravallion, 2000). Table 8.1 displays estimates for different regions for 1990 and

TABLE 8.1. Population Living below U.S. $2.15 per Diem, Based on the
World Bank Surveys of Household Income Dynamics in over 80 Countries

	Poverty rate (% below $2.15 per diem)	
	1990	2000
East Asia with China	68.5	48.3
Exclude China	64.9	50.8
China	69.9	47.3
East Europe and Central Asia	6.8	21.3
Latin America	27.6	26.3
Middle East and North Africa	21.0	24.4
South Asia	86.3	77.7
Sub-Saharan Africa	76.0	76.5
Total	60.8	53.6
Total (minus China)	57.5	55.7

Note. These data provide the percentage of the population that fall below the U.S. $2 a day
poverty line set as a benchmark for assessing global poverty on a local level, and are explained
and amplified in Chen and Ravallion (2000).

(2000). The data show progress in many parts of the world, while also illustrating the absence of change in certain regions. Eastern Europe and Central Asia are the only regions with a substantial augmentation in poverty in the 1990s. Here, the data largely reflect the newly independent former Soviet Union republics, such as Azerbaijan and Ukraine, that have joined the community of nations with high (~ 30%) poverty levels in the 2000 data. Bhalla (2002), combining the distribution of consumption over household surveys with the increases in consumption over national accounts, estimates that global poverty was 30% of world population in 1987 and only 13% in (2000). Moreover, large countries that have liberalized have had particularly large decreases in poverty. In India, poverty rates declined from about 40% in 1987–1988 to 26% in 1999–2000 (Deaton, 2000). Recent World Bank data suggest that the number of poor in China fell by 150 million from 1990 to 1999. Although progress in reducing poverty has been painfully slow, the picture is fairly encouraging in that the proportion of the world's population living below the U.S. $1 to $2 a day level is declining. Again, the positive trends appear to be the strongest in the economically open countries, while problems of poverty have persisted to a greater extent in countries that are the least globalized.

Challenge: Are the Details Devilish?

Globalization is often held to create winners and losers. In an article entitled "Winners and Losers," Wade (2001), writes that

technological change and financial liberalisation result in a disproportionately fast increase in the number of households at the extreme rich
end, without shrinking the distribution at the poor end. Population
growth, meanwhile, adds disproportionately to numbers at the poor
end. These deep causes yield an important intermediate cause that
makes things worse: the prices of industrial goods and services exported
from high-income countries are increasing faster than the prices of
goods and services exported by low-income countries, and much faster
than the prices of goods and services produced in low-income countries
that do little international trade. (p. 72)

To see whether globalization in and of itself can carry the blame in cases
that have fared poorly in the recent decades, it is instructive to examine
how globalization and the expansion of economic openness have played
out in specific countries.

If we had to look for a typical country that we expect to would be
harmed by globalization, Peru would seem a good candidate for a "loser."
It is dependent on agricultural products, including illegal drugs; has significant corruption; and has a fairly unskilled workforce. Its population
is about 25 million inhabitants, growing by about 2% per annum, with
a growth in the labor force of almost 3%. About one-half of the population is below the national poverty line. Life expectancy is about 70
years, but infant mortality remains high, at about 40 per 1000.[9]

Since the 1980s, Peru has become increasingly market-oriented, especially following significant privatization in mining, electrical, and telecommunications. Helped not only by significant foreign private investment but also by major investments and loans from the International
Monetary Fund (IMF) and the World Bank, its economy has begun to
emerge from decades-long stagnation. Adult mortality has been reduced
by about one-third in the last two decades. Immunizations (e.g., diphthe-
ria, pertussis, and tetanus toxoid [DPT] and measles) have gone from
10–20% in 1980 to essentially 95 out of 100 infants. Literacy is high
(90%), illiteracy having been decreased by 50% in the last 20 years, and
school enrollment is also high. GDP has grown by about 5% in the last
decade, and by a factor of 2 from 1980 to the present ($20.7 to $53.5
billion). As a result of robust growth in GDP over the past two decades,
gross domestic investment as well as exports of both goods and services
have all fallen as a percentage of GDP. However, exports of goods and
services grew by about 9% during this period, a trend that is continuing.
The economy has seen its service sector grow to about 60% from less
than 50% in the early 1980s. Copper exports have increased only modestly, but manufactures have grown threefold in the last decade (and
terms of trade are only about 80% of their 1990 levels).

It is hard to argue for Peru as a poster child of globalization, and many significant problems remain. The growth in GDP per capita was overtaken by population growth, and the total debt has grown by a factor of 4 and now approaches U.S. $30 billion. At the same time, it is hard to ignore that Peru has benefitted significantly in recent years from an increase in economic productivity as well as in international commerce. Many of the problems in Peru seem to stem more from the unfortunate policies of the government toward the poor, in particular its disregard for the informal sector, rather than form the consequences of globalization (for an interesting discussion of Peru, see De Soto, 1989). While even critics of globalization have admitted that it might be a boon for high- or even middle-income countries, Peru would not be widely recognized as falling into either category.

Even if some benefits of globalization may fall disproportionately in Northern lands, it is hard to ignore that there is significant progress to be found, even in the poorest countries. Indeed, the main areas that have not experienced growth or fallen behind—notably Africa and the Middle East—tend to include the least globalized countries.

THE POLITICAL REALM

Although important, economic wealth is not the only criterion for human progress. Broader conceptions of human rights also include elements such as security from violence, political freedom, and the provision of human rights (United Nations Development Programme). In this section, we review the status of war and peace and democracy in the contemporary world, and explore how globalization may influence these factors.

Globalization and Conflict

Despite the pessimism expressed by many scholars and pundits alike (for an amusing review, see Mueller, 1994), most empirical data on war and peace suggest that the end of the cold war has been accompanied by a decrease in the number of violent armed conflicts. Conflict can come in two main types, namely, interstate conflict between states and intrastate conflict within nations. Figure 8.4 displays the number of countries involved in wars, defined as conflicts involving more than 1,000 casualties, based on the data described in Gleditsch (2003). The solid line indicates proportion of independent states in the international system involved in an interstate war, whereas the dashed line indicates the proportion of states involved in an intrastate, or civil war, for the period

1816 to (2001). The frequency of interstate war has declined consistently since the end of the Vietnam War. After a brief flare-up after the end of the cold war and the breakup of the Soviet Union, the incidence of civil war has declined to the lowest level ever seen after World War II.[10] The Uppsala armed conflict data, with a much lower casualty threshold of only 25 deaths, finds a similar, strong decline in the number of conflicts after the cold war (see Gleditsch, Wallensteen, Eriksson, Sollenberg, & Strand, 2002). Preliminary data also suggest that the absolute number of annual casualties from conflict has declined markedly since the end of the cold war.[11] Although this decrease in violence probably is due to many factors other than globalization, there is substantial evidence suggesting that globalization may have positive effects in decreasing the risk of conflict. In this section, we review the evidence for a positive relationship between globalization and interstate and civil peace.

In a series of articles, Oneal and Russett (Oneal & Russett, 1997; Oneal & Russett, 1999; Russett & Oneal, 2001; Russett, Oneal, & Davis, 1998) have presented empirical evidence that increasing trade decreases the likelihood of conflict among states. There is also evidence

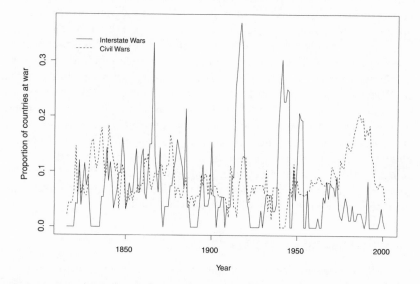

FIGURE 8.4. The solid line indicates the proportion of independent states in the international system involved in an interstate war, whereas the dashed line indicates the proportion of states involved in an intrastate or civil war, for the period 1816–2001. As can be seen, the end of the Cold War is associated with a dramatic decrease in the proportion of states involved in wars.

that increasing trade and interdependence appears to go together with a lower incidence of civil war. The State Failure Task Force Report finds that states with more trade relative to their GDP were substantially less likely to experience state failures.[12] Gleditsch (2002) finds that a greater degree of intraregional trade decreases the likelihood of state failure. Many have expressed concerns over the effects of globalization on income inequality, which in turn may increase the risk of conflict. Even supposing that globalization tended to increase inequality (which is, as noted above, disputed), there is little evidence of a link between income inequality and conflict. Collier and Hoeffler (2001b) find no evidence that greater income inequality is associated with a higher risk of civil war. If we accept a link between globalization, wealth, and the prospects for democracy (we address this issue in the next section), the prospects for sustaining peace should improve with further globalization. Many have noted that two democratic regimes have never waged war on one another, and a wealth of recent research illuminates how democratic institutions may reduce the risk of violent conflict between states (see, e.g., Russett & Oneal, 2001; and Gleditsch, 2002). Moreover, Hegre, Ellingsen, Gates, and Gleditsch (2001) find that democratic states are particularly unlikely to experience interstate conflicts, and that the risk of conflict is the highest among "mixed regimes" that are not sufficiently autocratic to fully deter conflict, but at the same time not sufficiently open to allow groups to pursue their political interest by nonviolent political means in a democratic political system. Thus mixed regimes encourage conflict by their absence of both genuine openness and repression.

Nonconventional forms of warfare such as terrorism has received considerable attention in the wake of September 11, 2001. Many have seen this as evidence of a clash of civilizations, which may be accelerated by the perceived threats to other cultures and traditional values under globalization. Recent studies, however, suggest that more globalized countries tend to have a lower incidence of terrorist activities (Li & Schaub, 2003). Again, the evidence suggests to us that the problem may emanate more from areas that have fallen behind in globalization than from the pernicious effects of globalization itself.

Globalization and Democracy

Figure 8.5 illustrates the important trends in democratization over the 19th and 20th centuries. Panel 8.5a shows that from a low of 5% at the outset of the 19th century, currently about half of the countries in the world are considered democracies by the Polity data.[13] As can be seen, the period of globalization, defined as the expansion of global economic exchange, has clearly coincided with a spread of democracy to a large number of countries. Doorenspleet (2000) has argued that since measures based on all states reflect not only states that change but also new states entering the system, what Huntington (1991) infers to be "waves" are merely artifacts of changes in the population of states over time rather than changes in prevailing types of governance. As in the case of trade, this may lead observers to exaggerate changes in regime types from global measures. Panel 8.5b, displays alternative measures based on a sample of only 21 states that have been in continuous existence from 1816. As can be seen, the trends over time are similar, and here too there is a dramatic increase in the extent of democracy following the emergence of increasing global economic exchange in the postwar period.

The traditional literature on democratization has related the prospects for democracy to a society's wealth (e.g., Lipset, 1959, 1994). Przeworski, Alvarez, Cheibub, and Limongi (2000) argue that democracies never break down when they reach a sufficient level of per capita income. If globalization enhances productivity and wealth, the prospects for democracy should improve with globalization. As of yet, few studies have examined linkages between globalization and democratization in much detail, although it is generally the case that the factors that make countries more likely to be open to trade (e.g., high per capita income, lower levels of conflict) also tend to be associated with democracy. However, it is well known that many of the transitions to democracy in the third wave of democratization has taken place in cases that would seem to be unlikely candidates. Gurr, Marshall, and Khosla (2000) note that some states have had surprising success with democratic transition, such as Bangladesh and Benin.

The traditional focus on social requisites would lead us to expect that democratic institutions may be rather fragile in many newly democratic states, given that the presumed prerequisites are not in place. Here, we would like to draw attention to some of the regional effects on democracy, and their implications for democracy in a globalizing world. Figure 8.6 plots the probabilities of transitions to and from democracy as a function of the proportion of democracies among neighboring countries. The likelihood that countries will become democracies and remain democracies is greater for countries that have a large proportion of neighboring states that are democracies. This relationship between regional context and the prospects for transitions and democratic survival does not disappear when controlling for the effects of plausible domestic and international influences. This suggests that the prospects for survival may be better than could be expected from a purely economic view, since many of the recent transitions in developing countries have tended to take place among regional clusters of countries. Many Latin American

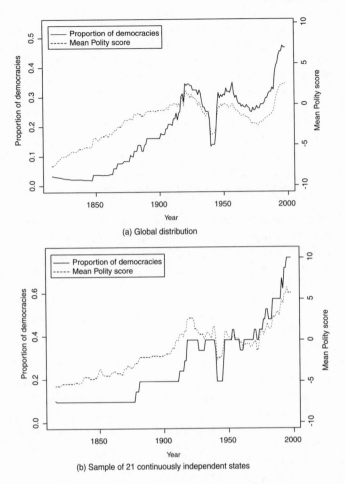

FIGURE 8.5. The share of the world's population of countries that are democratic has shown strong, but not constant, growth over the past several decades, measured by the proportion of states that are democracies as well as by the global mean of the Polity democracy score.

states experienced transitions to autocratic regimes in the 1950s, 1960s, and 1970s, when these states were experiencing high growth. Despite the current economic problems the region faces, however, the fact that all states in the region currently are democracies and that democracy is regarded as an essential prerequisite in regional organizations such as the Organization of American States and the MERCOSUR (Mercado Común del Sur) can help strengthen democratic institutions. As Mueller (1999) has pointed out, even if systematic evidence may fail to show

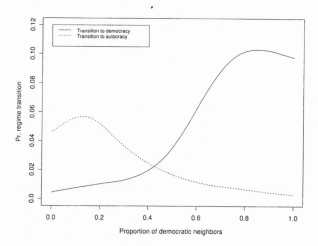

FIGURE 8.6. The diffusion of democracy is seen in terms of the increased likelihood of transitions to democracy for countries located in countries with a high proportion of neighboring democracies and the decrease in risk of transitions to autocracy for states with more democratic neighboring countries.

conclusively that democracy helps foster good economic policies and performance, widespread popular beliefs in such a link can in themselves help sustain democracy. Overall, we conclude that globalization appears to be associated with economic trends that should be beneficial to democracy, and that democratization under globalization has proceeded in geographical patterns that should make the prospects for democracy relatively good in most areas, except for the least globalized areas.

CONCLUSION

Evidence suggests that economic globalization has brought with it considerable prosperity, spread fairly broadly around the globe. It is certainly not everything its staunchest proponents have envisioned. There are many places around the globe with horrendous poverty, conflict, bad governance, and endemic corruption. In this sense globalization has not necessarily eliminated the barriers and imperfections that plague many countries. Markets are still "messy" in the sense that they exist within social and institutional contexts. As a result, they have many unintended consequences that are often ascribed either to the market mechanism itself, or to the contexts, or—worse—to both. Not everyone will benefit equally; indeed, some may not benefit at all. But at the same time, with the major general exceptions of Africa and the Middle East, many, many people around the world have benefitted from the expansion of economic activity and commerce in the last five, and especially the last two, decades. If a proper accounting of China were to be undertaken, it seems likely that global inequality as well as poverty will have been dramatically reduced.

At the same time, while progress has been made in terms of democratization, the world is still plagued by corruption and authoritarian regimes. Progress in this regard has been slower than might have been envisioned. Democratic institutions may not by themselves be a panacea for eradicating many forms of bad governance or policy failures. However, we are reminded of Winston Churchill's famous claim that "democracy is the worst form of government, except for all the others that have been tried." We believe that globalization, with deeper and more widespread global transparency, stronger prospects for civil society, and improved communications, eventually may help make a wider range of democratic institutions more accountable and effective. Keck and Sikkink (1998) document the rise of global activists in world politics. A search on Google on "Shell+Nigeria," for example, yields over 180,000 hits, the vast majority to activists' pages, and very few located in Nigeria itself. This is more than the 164,000 hits for "globalization+inequality" or the 84,700 hits for "antiglobalization". In this sense, global activism is just as much a part of globalization as increasing economic productivity and international commerce.

If globalization is generally associated with good things, why is there so much dissatisfaction with globalization? We believe that some of the allegations of the vices of globalization stem from unfounded myths (e.g., Micklethwait & Woolridge, 2003). Much of the criticism against globalization highlights the consequences for countries that are unable to participate and reap the benefits of globalization. This is a real problem, but trying to "un-/deglobalize" the world is unlikely to provide a feasible solution.

The globalization movement is not irreversible, and there is a possibility that the backlash against globalization may slow down the processes. Micklethwait and Woolridge (2003) argue that the anti-WTO demonstrations and riots in Seattle made U.S. president Bill Clinton back away from another round of trade liberalization. Another contraction of the world economy may well lead to a resurgence of protectionism and government intervention. Nonetheless, we believe that the virtues of globalization augur well for the future.

Norberg (2001) points out in his anti-antiglobalization tome *In Defense of Global Capitalism* that contemporary Congo is at about the same level of development as his native Sweden in 1870, with widespread poverty, low levels of productivity, and staggeringly high infant mortality rates. Globalization—in the form of international trade as well as the increased, specialized productivity it encouraged—changed all of that for Sweden, now thought of as one of the wealthiest countries in the world. The real question is whether in a few generations a Congolese book will be able to tell the same story. Both theory and evidence suggest that elimination of internal and external barriers—political and economic—should permit enormous progress to occur even where we find it most absent today. Developing countries with open economies grow faster than those with closed economies. Poverty is on the retreat among the majority of the world's population, especially in China and South Asia, and according to most figures the developing world is growing more rapidly than the developed world, in parallel to what happened in Sweden during the Industrial Revolution.

The world has never been in better shape. There is less poverty, greater wealth, more democratic governance, and many fewer civil and international conflicts than at any point in history. Globalization surely has its downside, but it is hard to blame it for impoverishment, inequality, and strife, since these appear all to be on the downturn exactly at the moment that globalization is reaching new heights.

ACKNOWLEDGEMENTS

We thank Ed Greenberg, John O'Loughlin, and Lynn Staeheli for their robust stewardship of the April 2002 Conference on Responding to Globalization and this ensuing volume. Michael E. Shin and Keith Maskus also helped along the way.

NOTES

1. Searching by the British spelling "globalisation" yields another 1.4 million hits.
2. The Schumann Plan was the basis for the European Coal and Steel Community (ECSC) established in 1952 among Belgium, France, Italy, Luxembourg, the Netherlands, and West Germany. The ECSC eventually grew into the European Economic Community and then transformed into today's European Union.
3. One question that is also very controversial is whether globalization is good or bad for the environment. Space and time do not permit an examination of this issue here, but see Lofdahl (2002) for a pessimistic perspective and Lomborg (2001) for a more optimistic account. The relationship between the natural environment and conflict within and between societies is examined in Diehl and Gleditsch (2000).

4. The full text of the declaration is available online at *http://www.un.org/millennium/summit.htm*.

5. Earlier studies of social change, including those by Marx, also emphasized the market as a modernizing force, but tended to see this as a virtue.

6. Data are taken as indices, with 1990 being set to 100 in each series.

7. Wassily Leontief used similar graphs to study the contraction under the global depression of the 1930s.

8. The theory behind purchasing power parity (PPP) is based on the idea that identical products and services should cost the same in all countries, at least in the long run. Exchange rates will adjust currency values to ensure this characteristic. The PWT data are available online at *http://pwt.econ.upenn.edu/aboutpwt.html*.

9. See *http://www.worldbank.org/data/countrydata/aag/per_aag.pdf* for the World Bank Group's summary data on Peru.

10. Assessing the incidence of civil war prior to World War II is somewhat complicated due to the issue of colonialism. Many colonial wars have elements of both interstate and civil war, and the true extent of violence outside independent states in the 19th century is probably underreported. See Gleditsch (2003) for further discussion.

11. Personal communication, Kristine Eck, Department of Peace and Conflict Resolution, University of Uppsala.

12. The State Failure Task Force was established in 1994 in response to U.S. vice president Albert Gore's interest in establishing a capability within the U.S. government to accurately forecast the collapse of regimes around the globe. This task force was led by a team of distinguished academics and policy analysts who worked as consultants to the U.S. intelligence community. This project—still running as of 2003—has received much attention. For an interesting introduction, see "CIA Study: Why Do Countries Fall Apart?: Al Gore Wanted to Know," in *US News and World Report*, March 12, 1996. More scholarly discussion can be found in King and Zeng (2001).

13. The Polity democracy score is an additive index of scorings on the general openness of political institutions, notably executive recruitment, constraints on the chief executive, regulation of political participation, and the competitiveness of political competition. This scale is widely used in comparative studies of democracy. The construction of the measure is explained in Gleditsch and Ward (1997) and Jaggers and Gurr (1995).

REFERENCES

Alesina, A., Spolaore, E., & Wacziarg, R. (2000). Economic integration and political disintegration. *American Economic Review, 90*(5), 1276–1296.

Bhagwati, J. (2002). *Free trade today*. Princeton, NJ: Princeton University Press.

Bhalla, S. (2002). *Imagine there's no country: Poverty, inequality, and growth in the era of globalization*. Washington, DC: Institute for International Economics.

Bourguignon, F., & Morrisson, C. (2002). Inequality among world citizens, 1820–1992. *American Economic Review, 92*(4), 727–744.

Castles, I. (1998). The mismeasure of nations: A review essay on the human development report 1998. *Population and Development Review, 24*(4), 831–845.

Chen, S., & Ravallion, M. (2000). *How did the world's poorest fare in the 1990s?* (Working Paper No. 2409). Washington, DC: World Bank Development Research Group. Available online at *http://econ.worldbank.org/resource.php.*

Chotikapanich, D., Valenzuela, R., Prasada Rao, D. S. (1997). Global and regional inequality in the distribution of income: Estimation with limited and incomplete data. *Empirical Economics, 22*(4), 533–546.

Collier, P., & Hoeffler, A. (2001). *Greed and grievance in civil war.* Washington, DC: World Bank, Development Research Group.

Deaton, A. (2000). Is world poverty falling? *Finance and Development, 39*(2), 4–7.

Deininger, K., & Squire, L. (1996). A new data set measuring income inequality. *World Bank Economic Review, 10*(3), 565–591.

de Soto, H. (1989). *The other path: The invisible revolution in the third world.* New York: Harper & Row.

Diehl, P. F., & Gleditsch, N. P. (Eds.). (2000). *Environmental conflict.* Boulder, CO: Westview.

Dollar, D., & Kraay, A. (2001a). *Growth is good for the poor.* Washington, DC: The World Bank, Development Research Group.

Dollar, D., & Kraay, A. (2001b). *Trade, growth, and poverty.* Washington, DC: The World Bank, Development Research Group.

Doorenspleet, R. (2000). Reassessing the three waves of democratization. *World Politics, 52*(3), 384–406.

Dowrick, S., & Akmal, M. (2001). *Contradictory trends in global income inequality: A tale of two biases.* Department of Economics, Australian National University, Canberra, Australia. Available online at *http://ecocomm.anu.edu.au/economics/staff/dowrick/world-inequ.pdf.*

Firebaugh, G. (1999). Empirics of world income inequality. *American Journal of Sociology, 104*(6), 1597–1630.

Fischer, S. (2003). Globalization and its challenges. *American Economic Review, 93*(2), 1–30.

Gleditsch, K. S. (2002). *All international politics is local: The diffusion of conflict, integration, and democratization.* Ann Arbor: University of Michigan Press.

Gleditsch, K. S. (2003). A revised list of wars between and within states, 1816–2001. Department of Political Science, University of California, San Diego. Available online at *http://weber.ucsd.edu/ kgledits/papers/exp_wars.pdf.*

Gleditsch, K. S., & Ward, M. D. (1997). Double take: Reexamining democracy and autocracy in modern polities. *Journal of Conflict Resolution, 41*(3), 361–383.

Gleditsch, N. P., Wallensteen, P., Eriksson, M., Sollenberg, M., & Strand, H. (2002). Armed conflict 1946–2001: A new dataset. *Journal of Peace Research, 39*(5), 615–637.

Gottschalk, P., & Smeeding, T. M. (1997). Cross-national comparisons of earnings and income inequality. *Journal of Economic Literature, 35*(2), 633–687.

Gurr, T. R., Marshall, M. G., & Khosla, D. (2000). *Peace and conflict 2001: A global survey of armed conflicts, self-determination movements, and democ-*

racy. College Park: University of Maryland, Center for International Development and Conflict Management.

Hayek, F. A. (1944). *The road to serfdom*. Chicago: University of Chicago Press.

Hegre, H., Ellingsen, T., Gates, S., Gleditsch, N. P. (2001). Toward a democratic civil peace? Democracy, political change, and civil war, 1816–1992. *American Political Science Review, 95*(1), 33–48.

Huntington, S. (1991). *The third wave of democracy*. Norman: Oklahoma University Press.

Irwin, D. (2002). *Free trade under fire*. Princeton, NJ: Princeton University Press.

Jaggers, K., & Gurr, T. R. (1995). Transitions to democracy: Tracking democracy's "third wave" with the Polity III data. *Journal of Peace Research, 32*(4), 469–82.

Keck, M. E., & Sikkink, K. (1998). *Activists beyond borders: Advocacy networks in international politics*. Ithaca, NY: Cornell University Press.

King, G., & Zeng, L. (2001). Improving forecasts of state failure. *World Politics, 53*(4), 623–658.

Korzeniewicz, R. P., & Moran, T. P. (1997). World economic trends in the distribution of income, 1965–92. *American Journal of Sociology, 102*(4), 1000–1039.

Levine, R., & Renelt, D. (1992). A sensitivity analysis of cross-country growth regressions. *American Economic Review, 82*(4), 942–963.

Li, Q., & Schaub, D. (2003). *Economic globalization and transnational terrorist incidents: A pooled time series cross sectional analysis*. Paper Presented at the annual meeting of the International Studies Association, Portland, OR.

Lipset, S. M. (1959). Some social requisites of democracy: Economic development and political legitimacy. *American Political Science Review, 53*(1), 69–105.

Lipset, S. M. (1994). The social requisites of democracy revisited: 1993 Presidential Address. *American Sociological Review, 59*(1), 1–22.

Lofdahl, C. L. (2002). *Environmental impacts of globalization and trade: A systems study*. Cambridge, MA: MIT Press.

Lomborg, B. (2001). *The skeptical environmentalist*. Cambridge, UK, and New York: Cambridge University Press.

Martin, H.-P., & Schumann, H. (1998). *The global trap: Globalisation and the assault on democracy and prosperity*. London: Pluto.

Melchior, A., Telle, K., & Wiig, H. (2000). *Globalisation and inequality: World income distribution and living standards, 1960–98* (Studies on Foreign Policy Issues Report 6[B], 1–42). Oslo, Norway: Royal Norwegian Ministry of Foreign Affairs.

Micklethwait, J., & Wooldridge, A. (2003). *A future perfect: The challenge and promise of globalization* (rev. ed.). New York: Random House.

Milanovic, B., & Yitzhaki, S. (2001). *Decomposing world income distribution: Does the world have a middle class?* (Working Paper No. 2562). Washington, DC: World Bank.

Mueller, J. (1994). The catastrophe quota: Trouble after the cold war. *Journal of Conflict Resolution, 38*(3), 355–375.

Mueller, J. (1999). *Capitalism, democracy, and Ralph's pretty good grocery.* Princeton, NJ: Princeton University Press.

Norberg, J. (2003). *In defense of global capitalism.* Washington, DC: Cato Institute.

Oneal, J. R., & Russett, B. M. (1997). The classical liberals were right: Democracy, interdependence, and conflict, 1950–1985. *International Studies Quarterly, 41*(2), 267–295.

Oneal, J. R., & Russett, B. M. (1999). Assessing the liberal peace with alternative specifications: Trade still reduces conflict. *Journal of Peace Research, 36*(4), 423–442.

Polyani, K. (1944). *The great transformation: The political and economic origins of our time.* New York: Rinehart.

Przeworski, A., Alvarez, M., Cheibub, J., & Limongi, F. (2000). *Democracy and development: Political institutions and well-being in the world, 1950–1990.* Cambridge, UK: Cambridge University Press.

Rodrik, D. (1997). *Has globalization gone too far?* Washington, DC: Institute for International Economics.

Russett, B. M., & Oneal, J. (2001). *Triangulating peace: Democracy, interdependence, and international organizations.* New York: Norton.

Russett, B. M., Oneal, J. R., & Davis, D. R. (1998). The third leg of the Kantian tripod for peace: International organizations and militarized disputes, 1950–1985. *International Organization, 52*(3), 441–468.

Sachs, J., & Warner, A. (1995). Economic reform and the process of global integration. *Brookings Papers on Economic Activity, 1,* 1–95.

Schultz, T. P. (1998). Inequality in the distribution of personal income in the world: How it is changing and why? *Journal of Population Economics, 11*(3), 307–344.

Stiglitz, J. F. (2002). *Globalization and its discontents.* New York: Norton.

Summers, R., & Heston, A. (1991). The Penn World Table (Mark 5), An expanded set of international comparisons, 1950–1988. *Quarterly Journal of Economics, 106*(2), 327–368.

United Nations Development Programme. (1999). *Human development report 1999.* New York: Oxford University Press.

Wade, R. (2001, April 26). Winners and losers. *The Economist,* pp. 72–74. Available online at *http://www.economist.com,StoryID=587251.* Accessed on September 12, 2003.

9

Some Measurable Costs and Benefits of Economic Globalization for Americans

J. DAVID RICHARDSON

The mushrooming debate over the promise and perils of globalization outcomes is almost universally impressionistic and anecdotal. Artful evocation is the norm, careful amassing of evidence the exception. The Institute for International Economics's Globalization Balance Sheet (GBS) family of projects aims first to redress this imbalance, then to quantify some of the most important economic benefits and contentious costs for U.S. citizens at the grassroots level. Follow-on projects may apply the GBS template methods to other regions and may also try to expand the types of benefits and costs that can be evaluated.

GBS projects seek primarily to bring depth, accuracy, and balance to economic aspects of the globalization debate in the United States. They attempt to measure some of American globalization's underappreciated "fitness" benefits and underquantified distributional costs. It is rare for economists to construct comprehensive distributional accounts of a country's winners and losers from global integration. Yet the identification of such winners and losers helps in many ways. It helps evaluate whether the American incidence of globalization is equitable from various social justice perspectives. It helps explain apparent gaps between social perceptions and the realities of American global exposure. It helps inform the political debate over the North American Free Trade Agreement (NAFTA) and the World Trade Organization (WTO) and over the tax and regulatory treatment of cross-border corporate operations. Distributional accounts are a necessary input to the ethical and political debate and are the most important objective of the GBS projects.[1]

The GBS template methods exploit the growing availability of genuinely microeconomic data: censuses and surveys of real-life households, workers, voters, migrants, firms, plants, and communities.[2] Six projects have been completed, or are nearly completed, with several more underway. Each defines globalization uniquely: the range of conceptions includes trends in export intensity, import penetration, immigration density, and foreign investment, both inward and outward.

The American focus of the GBS projects was not arbitrary. Rich American data availability was one reason (though similarly rich data are increasingly available around the world). An even more important reason is the strong, persistent "globalization backlash," even in the world's leading economy. Cynical American public opinion has often identified globalization with a narrow "American *corporate* agenda," even during the 1990s "good times." As the global economy has turned down, skepticism has intensified about the benefits of deeper global economic relations to mainstream U.S. citizens. Such skepticism is ironic from the perspective of the rest of the world, which identifies "globalization" with "Americanization," and sees all Americans as winners and themselves as losers.

In the completed projects, there is a surprising resonance with anxious American perceptions. The projects' clear consensus is that deeper global integration is a mixed blessing for the United States. Globalization of most types enhances the economic "fitness" and opportunities of a large number of Americans who in turn rejuvenate their workplaces, unions, firms, and industries. Yet it actually worsens prospects for many Americans who are unwilling or unable to engage globally.[3] And if this bifurcation of benefits and costs is true in the United States, it is very likely true elsewhere as well. And if it is ubiquitous, then the policy challenge for *every* country, broadly speaking, is how to continue to capture the economic gains from global integration—GBS and other research shows them to be positive in the aggregate—yet simultaneously to diffuse them internally through ancillary domestic policies that insure and empower national subpopulations who bear most of the economic costs.

GBS research so far has revealed the following specifics. Americans with average skills, women, blue-collar union members, and insular communities (those that are neither cosmopolitan nor mobile) are among those who are disproportionately represented in the subpopulation who are on the periphery of globalization's gains. Some even lose from it. The research also reveals, however, that these same socioeconomic groups face even worse outcomes from changing lifestyles and technology—globalization definitely contributes, but only modestly, to their problems.

The policy implications of this research consensus involve empowerment and security in mobilizing to meet *all* these challenges, not just globalization. Empowerment entails mobility and educational objectives that are rewarded in an increasingly globalized world (e.g., through the portability of health and pension benefits, or via language and cultural training). Security entails creative types of insurance against new and sometimes accentuated risks arising from global linkage. The GBS researchers propose policies that aim at more broadly diffusing globalization's gains and opportunities across American society, and at better insuring Americans against those trends that are unavoidably adverse.

An ultimate vision for a global family of GBS projects is the capability to weigh credible measures of globalization's traditional and new costs and benefits against each other in many of the most important regions of the world. Since many of the newer costs and benefits are distributional (i.e., some individuals and groups gain at the expense of others), this ultimate aim will involve judgment that goes beyond mere measurement.[4] Even more ambitious extensions of projects like these involve subtle, judgmental assessments of the many effects of all-encompassing globalization—economic, cultural, and sociopolitical—that cannot be quantified at all.

In the rest of this chapter I distill the insights and policy implications of the completed GBS family of research projects. I focus especially on their surprising consensus about Americans who bear burdens from U.S. global integration of various types, including immigration, or who perceive that they do. I close with a discussion of our growing consensus about American policy evolution.

GLOBALLY DISLOCATED AMERICAN WORKERS

One of Kletzer's (2001) projects tries to measure the effects of exceptional import penetration on American manufacturing workers—in isolation from trends in technology, outsourcing, unionization, and other shocks during the last quarter of the 20th century—up to 1999. The import effects on worker dislocation and job/earnings recovery turn out to be modest, and hard to disentangle from the special burdens that women workers face in the entire American labor market. But the effects on manufacturing workers of generalized dislocation from *both* global and local shocks are by no means modest, and manufacturing workers bear them especially heavily compared to other American workers. This conclusion leads Kletzer, together with GBS and other coauthors, to propose a radical reshaping of U.S. labor market policies for *all* dislocated workers (Kletzer & Litan, 2001). Among other benefits from her proposal for earnings-and-benefits insurance would be less severe distributional costs, and wider dissemination of the gains, from greater U.S. global integration. For example:

• Import-displaced U.S. manufacturing workers have similar profiles (in terms of age, education, job tenure, etc.) to other displaced manufacturing workers except that they are disproportionately female. Forty-four percent of import-displaced manufacturing workers are women, compared to only 36% of normally displaced manufacturing workers. Import-displaced workers are 2% less likely to become reemployed 3 years after displacement than normally displaced workers, but that difference could be ascribed entirely to their being disproportionately women.

• There are almost no differences in the earnings recovery patterns of import-displaced and other displaced manufacturing workers: on average, both have lost about 17% of their former earnings 3 years after displacement, with roughly one-quarter losing 30% or more, yet a third losing nothing (or even gaining). For both groups of workers, half are reemployed in manufacturing, and only one-tenth are reemployed in retail trade. Earnings losses of the second group are indeed large, but losses of the first group are small.

Other Kletzer projects (2002, 2003) update her studies to 2002 and hone in on female (and male) workers in the most import-sensitive sectors of all, textiles and apparel, featured also in the Kletzer, Levinsohn, and Richardson (in press) project discussed below. Early results of these projects suggest that dislocation increased across the board in the American economy in 2001–2002, but that men's reemployment probabilities in textiles and apparel, always lower than elsewhere, declined less precipitously than men's reemployment probabilities declined elsewhere. For women, on the other hand, the 2001–2002 decline in reemployment probabilities for all import-sensitive manufacturing, including textiles and apparel, was sharper and steeper than elsewhere.

GLOBALLY ENERGIZED AMERICAN WORKERS— AND FIRMS AND COMMUNITIES

Lewis and Richardson (2001) synthesize a growing body of microeconomic research that suggests that there are significant opportunities for American workers, firms, and communities to prosper, as long as they are able to engage the global economy in some fashion. Globally engaged Americans earn more, grow those earnings faster, and fail less frequently than comparable yet insular Americans. Global engagement includes not only export engagement, but imports of inputs (e.g., machines, components, managers, technologies) and the welcoming of inward and outward investment. As globally engaged population shares grow, the microlevel gains have the potential to rejuvenate workplaces, industries, regions, and the entire American economy.

For example, globally engaged American firms, workers, and communities enjoy significant advantages compared to otherwise-identical counterparts (twins): comparable firms, workers, and communities that are not globally engaged. Productivity, growth rates, wages, and job stability differ between the twins by significant amounts, usually from 5 to 15%. Another generation's growth in global commitment at the same rate that the past generation globalized would raise American standards of living by a quarter.

But there is another face to these potentially beneficial outcomes, a distributional downside. Workers, firms, and communities that cannot or will not or fail to engage the global economy lose heavily over time. They are displaced by the globally engaged. The challenge is to empower them, or their children, to be able to engage—to engage change of all types, especially when their skills, education, history, and willingness to move make such engagement unlikely or very costly.

GLOBALLY SKEPTICAL AMERICAN VOTERS—AND THE NEGLECTED IMPORTANCE OF RESIDENTIAL WEALTH

This failure of certain segments of American society to engage with the global economy helps explain why Scheve and Slaughter (2001) find so much skepticism in their American voter surveys about globalization in the early and mid-1990s, especially among a large number of voters with median (or less) skills and education. Such typical (median) voters do not favor freer trade, nor freer immigration. They fear—perhaps correctly, as Scheve and Slaughter show—that they have little or nothing to gain from further U.S. global integration, nor do their communities. For example, in import-sensitive communities, homeowners strongly oppose further trade liberalization, presumably because they fear that rising imports will cause a steep decline in community property values. Because immigration (imports of people) has more favorable effects on home prices, however, homeowners in "gateway" communities for migrants show no greater opposition to immigration than other voters do.

By contrast, American voters with extensive skills and education do favor additional trade and immigration, and seem to gain from it. In fact, the impact of voter skills and education is so strong that expected additional correlates of voter globalization preferences, including the industry in which they work (and its global exposure), contribute *nothing* to explaining the remaining variation among the voters! That industry does not matter after taking voter skills and education into account is quite surprising; most research traditionally identifies individual interest with the interests of an employer from a specific industry. The overwhelming central dominance of skills and education resonates also with the findings of Lewis and Richardson (2001) and Baldwin (2003). The "haves" (who gain from American globalization) come predominantly from upper socioeconomic strata; the "have nots" from lower. The bifurcated benefits and costs have more of the feel of "class" distinctions than economics, ethnicity, ideology, or region of residence. The issue that this research dramatically highlights is whether skill building and stronger education could turn typical American voters from global skeptics to global enthusiasts.

GLOBALLY VICTIMIZED AMERICAN LABOR UNIONS AND UNION-FRIENDLY REGIONS—OR NOT?

Baldwin (2003), in his study of American deunionization in both the goods and the services sectors, finds resonance with Scheve and Slaughter's (2001) results. Unionists with only basic education fared far worse than other unionists. For example:

• Between 1977 and 1997 there was a precipitous decline in the proportion of workers with median education (12 years or less) who were represented by a labor union, from 29% to 14%. By contrast, unionization proportions declined much less among workers with above-median education, from 19% to only 13%. The union wage premium also declined for basically educated workers, from 58% to 51%, whereas it rose slightly for better educated unionists, from 18% to 19%.

• For basically educated workers the surge in imports during the decade from 1977 to 1987 contributed a little more than 20% to unionization's decline across all sectors, controlling for technological change and shifts in demand. But during the second decade (1987–1997) a surge in *exports* was correlated with *declining* demand for basically educated workers, controlling for technological change and shifts in demand, entirely concentrated on unionists.

The correlation of export *growth* with *declines* in unionized employment is quite surprising. Yet increased export competitiveness can indeed cause reduced employment of the less educated as firms upgrade technology and skill requirements and sell their output in more intensely competitive global markets.

During the first decade, by contrast, Baldwin (2003) finds that rising exports helped basically educated unionists very modestly, and during the second decade, basically educated unionists were relatively sheltered from imports—rising imports were associated with disproportionately large employment losses among basically educated *non*unionists and, curiously, *more*-educated unionists.

These conclusions reveal the value of microdata research. For American union workers as a whole, without regard to education, globalization helps explain employment declines only very modestly (technological change and shifts in demand likewise contributed only modestly). In both decades, but especially the first, ubiquitous trend declines in union affiliation—within each goods and services sector, within each region, within each worker educational group—explain half to almost all the fall in the overall unionization rate.

In sum, any safety net that American unions provided was disproportionately lost by the less-educated workers who arguably need it most. The problem is no less severe when globalization's role in this story is slight.

AMERICAN PLANT-LEVEL DISLOCATION AND SURVIVAL IN TEXTILES AND APPAREL

Kletzer et al. (in press) has profiled the mixed blessings of globalization for American textile and apparel plants from the early 1970s to the mid-1990s. The most surprising conclusion of this research is that there are any blessings at all! But they find that some plants turn out to be surprisingly resourceful in this slumping sector. For example, rates of productivity growth among incumbent plants over this period, and rates of entry and job creation by high-productivity new plants, all turn out to be quite strong (the typical plant's overall productivity increased by 68% in textiles, 26% in apparel).

On the other hand, plant shutdown is also quite prominent, along with job destruction and dislocation of workers. Kletzer (2002, 2003) documents the significantly lower wages and reemployment probabilities of dislocated textile and apparel workers—especially women—through the year 2002.

Resonating with Baldwin (2003) and with Scheve and Slaughter (2002), Kletzer et al. (in press) find that shutdown and worker dislocation are concentrated on plants with average productivity and below— whose workers are also often less skilled.

On the other hand, this bad news can be better news for the industry on average, via the rejuvenation effects described in Lewis and Richardson's (2001) work. On balance, the infamous "churning" in the textiles and apparel sector still leaves open many sectoral opportunities for workers and investors because of active entry, strong productivity trends, and job creation. For example, both industries feature successful outsourcers who pay their workers more in annual salaries than they would otherwise earn—$1,000 more (per year) in textiles and $200 more in apparel—for every 10% of their inputs that they outsource. In apparel, successful outsourcers also have more stable jobs (higher plant survival rates), consistent with the general results in Lewis and Richardson (2001).

Counterbalancing these (mixed) blessings on productivity and *new* job opportunities, however, the authors find unremittingly gloomy plant-level trends in wages and *net* job growth. Both are strongly negative— wage declines are especially negative for white-collar (nonproduction) workers, job declines for blue-collar (production) workers.

EMERGING POLICY IMPLICATIONS FROM THE GBS PROJECTS

Taken together, these research projects suggest a radical reorientation of familiar American training, adjustment, and "safety-net" policies toward policies that are consciously adopted to complement and to enable global liberalizing initiatives, by spreading the gains from global integration more widely within American society and stabilizing the uneasy popular support for them.

To be more specific, GBS authors increasingly agree that American policies should:

- Move away from specific industry and job-based relief and toward worker empowerment.
- Move toward education and skill-building experience, including on-the-job training.

- Move toward insurance programs that preserve an individual's lifetime earnings potential.

One such set of policies would allow older and less mobile dislocated workers to avoid having to change sectors by facilitating mobility among younger workers—for example, through tax-favored job search and corresponding programs, or benefit portability enhancements. Such programs, in turn, help stabilize individuals' lifetime wealth and make them less fearful of global initiatives that expose them to capital losses (e.g., on their homes).

Still another program is the institutional creation of new incentives for labor unions and other "worker agents" to provide their members with better global opportunities and insurance, as envisioned in Elliott and Richardson (in press).

The most fully developed GBS reform program, however, is the combination of wage insurance and subsidized health insurance for dislocated workers, outlined by Kletzer and Litan (2001), and actually implemented on a limited basis in recent American legislation. Kletzer and Litan (2001) use the U.S. Displaced Worker Surveys to estimate the costs of a two-part proposal to ease the burdens and shorten the duration of U.S. worker dislocation. They recognize that dislocation is the distributional dark side of the energy and change that drives much of the U.S. economy, but consider globalization as merely one element of that dynamism, not a special element, and hard to isolate from technology, education, and other institutional and social change. The spirit of their proposal is to make *all* change less burdensome to American workers, and therefore more acceptable to them, as well as more defensible to broad public opinion. Both they and Scheve and Slaughter (2001) cite polls that show how voters shift their preferences in favor of further trade liberalization when it is tied to programs that assist workers to adjust.

In this spirit, the most arresting conclusion of their study is how cost-effective their proposed policy seems to be. Their program would cost the public budget only $3–$3.5 billion dollars and would entail:

- Wage insurance to replace 50% of a full-time dislocated worker's earnings loss in a new job, with up to $10,000 of total annual reimbursement for a 2-year period of eligibility.
- Health insurance premium subsidies (50% of the displaced worker's contribution) to maintain affordable medical insurance for 6 months after a worker's dislocation.

Even more generous provisions bring the public cost no higher than $6.5 billion (e.g., no dollar cap on wage insurance reimbursement, 70% reimbursement of earnings losses and worker contribution to medical insurance). Kletzer and Litan make no calculation of the savings to the conventional unemployment insurance system from sharpening incentives to become reemployed. Such unemployment insurance savings only makes their proposal *more* cost-effective.

NOTES

1. Smeeding and Rainwater (2000), along with others, have pioneered cross-country comparisons of the welfare of similar groups at similar positions in their own country's size distribution of income, but there is not yet enough time depth in the underlying data to do cross-country studies of response to trends such as globalization. See Heckman, Smith, and Clements (1997) for reasons and methods to do distributional accounting, especially accounting for the impacts of policy evolution.
2. There is also growing *global* availability of such "microdata," and not only in the richer countries. Nobel Prizes in Economics in 2000 were awarded to James J. Heckman and Daniel McFadden for pioneering techniques in handling such data and their widespread applications. There is a great deal to learn from worker-level, plant-level, voter-level, and census-tract-level data that cannot be seen in aggregates and averages. Some of it validates what we think we already know about the benefits and costs of globalization, but much does not, as this family of projects has revealed.
3. In cross-country historical research using more aggregated data, there is a growing consensus that the same conclusion holds for countries. Prospects are worsened for those who cannot or will not engage—even the status quo is unsustainable. See Bordo, Taylor, and Williamson (2003), Lindert and Williamson (2003), and World Bank (2002).
4. The forthcoming Richardson (in press) monograph aims to summarize the GBS family of projects and their policy implications, and possibly to lay the groundwork for a more ambitious project that would likely involve globalization scholars beyond the GBS family. A derivative aim of the GBS projects is a template for similar exercises abroad and possibly on a global scale. The Institute is cooperating with the Deutsche Gesellschaft für Auswäartige Politik in Berlin and the Institut Francais des Relations Internationales in Paris on GBS and parallel projects being undertaken there. These European projects are focused more on the backlash to globalization than on measuring its effects.

REFERENCES

Baldwin, R. E. (2003, June). *The decline of U.S. labor unions and the role of trade.* Washington, DC: Institute for International Economics.

Bordo, M. D., Taylor, A. M., & Williamson, J. G. (Eds.). (2003). *Globalization in historical perspective.* Chicago: University of Chicago Press.

Elliott, K. A., & Richardson, J. D. (in press). *Free trade in worker agency services.* Washington, DC: Institute for International Economics.

Heckman, J. J., Smith, J., & Clements, N. (1997). Making the most out of programme evaluations and social experiments: Accounting for heterogeneity in programme impacts. *Review of Economic Studies, 64,* 487–535.

Kletzer, L. G. (2001, September). *Job loss from imports: Measuring the costs.* Washington, DC: Institute for International Economics.

Kletzer, L. G. (2002). *Gender and globalization: Job loss from apparel and textiles.* Unpublished manuscript, Institute for International Economies, Washington, DC.

Kletzer, L. G. (2003). *Workers at risk.* Unpublished manuscript, Institute for International Economics, Washington, DC.

Kletzer, L., Levinsohn, J., & Richardson, J. D. (in press). *Responses to globalization: The U. S. textiles and apparel industries.* Washington, DC: Institute for International Economics.

Kletzer, L. G., & Litan, R. E. (2001, March). *A prescription to relieve worker anxiety* (Policy Brief No. 01-2). Washington, DC: Institute for International Economics.

Lewis, H., & Richardson, J. D. (2001, October). *Why global commitment really matters!* Washington, DC: Institute for International Economics.

Lindert, P. H., & Williamson, J. G. (2003). Does globalization make the world more unequal? In M. D. Bordo, A. M. Taylor, & J. G. Williamson (Eds.), *Globalization in historical perspective.* Chicago: University of Chicago Press.

Richardson, J. D. (in press). *Global forces, American faces: U. S. economic globalization at the grass roots.* Washington, DC: Institute for International Economics.

Scheve, K. F. ., & Slaughter, M. J. (2001, March). *Globalization and the perceptions of American workers.* Washington, DC: Institute for International Economics.

Smeeding, T. M., & Rainwater, L. (2002, February). *Comparing living standards across nations: Real incomes at the top, the bottom, and the middle.* Unpublished manuscript, Maxwell School, Syracuse University, Syracuse, NY.

World Bank. (2002). *Globalization, growth, and poverty: Building an inclusive world economy.* Washington DC: Author.

10

Globalization and Health

MICHAEL E. SHIN

Interactions and exchanges between the people and places of the world have occurred for centuries, if not millennia. As described in *Guns, Germs, and Steel* by Jared Diamond (1999), it is precisely such interactions, mixed in with a bit of geographical possibilism, that led to the diffusion of beliefs (e.g., religion), innovations (e.g., irrigation) and institutions (e.g., government) that have shaped and continue to shape the world. What sets contemporary globalization apart from the interactions and exchanges of the past are its intensity and extent, and the degree to which people and places are now interdependent. Contemporary globalization comes with several conveniences (e.g., instantaneous communication, access to worldwide goods, services, information, etc.), but globalization has also generated several consequences (e.g., increasing social and economic inequalities). Of all the costs and benefits related to globalization, those concerning health are among the most important and compelling. They are the focus of this chapter.

Like globalization, health is understood and defined in many different ways. For instance, according to the constitution of the World Health Organization (WHO), "Health is a state of complete physical, mental and social well-being and not merely the absence of disease or infirmity." Though succinct, this definition has fueled many debates and discussions about whether health should be considered an unattainable end or a means to an end (e.g., see Noack, 1987). Furthermore, such a broad definition of health that combines both absolute and ambiguous terminology (e.g., "complete" and "well-being") makes measuring health a difficult task. Health is also understood in biomedical terms (see Curtis & Taket, 1996). Biomedical definitions of health are based largely upon Western medicine and revolve around "fixing" what is "wrong" or "abnormal" with the patient. This point of view is characterized by a high degree of specialization and presumes that a "normal" level of health exists. Medical specialists diagnose diseases, illnesses, and injuries, among other things, and also determine what cure, remedy, or treatment is necessary to restore "normal" health. There is, however, an increasing interest in and awareness of alternative approaches to Western medicine. Holistic, Eastern, and Asian approaches to health and medicine not only provide diverse frameworks to understand, improve, and treat health, but they are also a reflection and function of globalization itself.

The relationship between globalization and health is clearly multifaceted. This chapter focuses upon two specific and complementary dimensions of the globalization and health nexus. The first part of the chapter looks at the pathway between economic integration and health at the global scale. Gauging whether globalization plays a role in increasing income is of particular interest. This relationship is important because economic growth is believed to improve health, which in turn contributes to more growth (United Nations Development Programme, 2003). Though health is itself an important precondition for productivity and growth, poverty reduction may serve as an important stimulus for this joint and synergistic relationship, especially in developing countries. The second part of the chapter focuses upon the sociogeographical linkages and structures that enable and facilitate economic integration. Rather than examine the direct economic benefits of such linkages and structures, several health externalities and consequences are discussed within the context of diffusion at the beginning of the 21st century. The chapter closes with some thoughts about the potential influence that health has upon interdependence and integration, and the overall significance of the globalization and health nexus.

INCOME, INTEGRATION, AND HEALTH

Over the course of the 20th century, the overall health of the world's population improved dramatically. It is estimated that in 1900 the global average for life expectancy at birth was between 30 and 40 years (Livi-Bacci, 1997). This statistic is currently 67 years and continues to increase (Population Reference Bureau, 2002). Though health cannot be entirely captured by a single number, life expectancy presents itself as an accessible measure of health that is widely recognized and used. Increases in life expectancy during the last century are generally attributed to improved nutrition and higher standards of living. Since 1990 alone, the World Bank reports a 20% decrease in the global poverty rate, measured as the proportion of a country's population that lives on less than U.S. $1 a day.[1] Despite such dramatic increases in life expectancy and poverty reduction, such figures conceal the uneven geography of health and poverty across the world, as well as the acute disparities that exist within many countries.

Country-to-country variations in life expectancy and infant mortality rates are frequently understood within the context of the epidemiological transition model (see Omran, 1971, 1983; Wilkinson, 1996). This model documents a society's pathway to the reduction of death rates, particularly those associated with communicable disease. Demographic characteristics of countries that have yet to complete the epidemiological transition include high levels of mortality due to infectious disease, high levels of fertility, and a predominantly young population that is particularly vulnerable to infectious disease. Once the detrimental effects of communicable disease can be overcome by a society, mortality and fertility rates gradually decrease and stabilize. As deaths due to communicable disease decrease, those attributed to noncommunicable diseases (NCDs) such as heart disease, stroke, diabetes, and cancer become more and more common as the population ages (Wilkinson, 1996). It is estimated that NCDs contributed to nearly 60% of all deaths across the world at the end of the 20th century, and in 20 years this figure is expected to reach nearly 75% (World Health Organization, 2002a).

Of the several factors that lead to the completion of the epidemiological transition, increases in income are perhaps the most significant and tangible. Reducing poverty and increasing income can have several positive impacts upon health (United Nations Development Programme, 2003, pp. 67–70). At the household level, and especially in poor countries, increases in income can raise standards of living by increasing savings and investment. Such surplus income can be used to offset shortages in times of crisis (e.g., drought, flood, war, etc.), and it can be invested in resources that create additional opportunities to sustain and generate income (e.g., livestock, seed, tools, machinery, etc.). Economic growth can also improve population health by stimulating government expenditures on education, critical infrastructures, and health care. Prichitt and Summers (1996) show that increases in income and growth are indeed related to lower infant and child mortality rates, broader access to education, and higher life expectancies.

How, then, can the incomes of the poor be sufficiently increased in order to propel a society through the epidemiological transition? Evidence suggests that policies that promote economic integration, such as trade liberalization and tariff reductions, may encourage economic growth and increase income, especially for poorer nations (World Bank, 2002; Dollar, 2001; Edwards, 1998; Frankel, 1997). As technological developments facilitate more and more exchanges between firms and economies, policies that inhibit such transactions (e.g., tariffs, quotas, barriers, subsidies, etc.) reduce competitiveness and may stifle growth. Liberal policies arguably do just the opposite. Freedom to participate in the global market permits economies to access and exchange ideas, choices, capital, and labor. Access to multiple sources of capital, to a broad array of resources, and to skilled and unskilled labor stimulates competition, innovation, and economic growth that can in turn lead to global, national, regional, and local improvements in health. In fact, no study has ever identified a country that was less open to trade in the 1990s than in the 1960s and that concurrently improved its standard of living relative to other countries (World Bank, 2002, p. 37).

To visualize the global relationship between globalization and health, examine Figures 10.1a and 10.1b, which provide aggregate portraits of the relationship between changes in trade and life expectancy, and changes in trade and changes in life expectancy between 1980 and 1999, respectively.[2] Though the relationship between change in trade and life expectancy is positive and statistically significant, it needs to be interpreted carefully for several reasons. First, the relationship between globalization and health presented and discussed thus far is indirect (i.e., trade leads to growth, which leads to better health), and Figure 10.1a presents it as a direct causal relationship. Second, it is not entirely clear whether changes in trade lead to high life expectancy or whether higher life expectancies lead to increased changes in trade. Within the context of this simple aggregate analysis, life expectancy may be serving as a proxy for economic development. In other words, over the last two decades healthier and wealthier countries were probably better positioned to participate in and benefit from globalization than developing countries. Figure 10.1b supports these reservations and shows a slightly negative but statistically insignificant relationship between changes in trade and changes in life expectancy since 1980.

Several other factors and policies, not included above, are certainly related to economic growth, integration, and health (e.g., geography, history, etc.). Aggregate analyses of globalization patterns and processes are often criticized for such omissions and their overall utility questioned due to this abstraction of reality (see Dollar, 2001). Among the concerns of using cross-national measures of globalization and health are the incorrect specification of variables and relationships, the complexity of multiple causal relationships, and the problems with inferring subna-

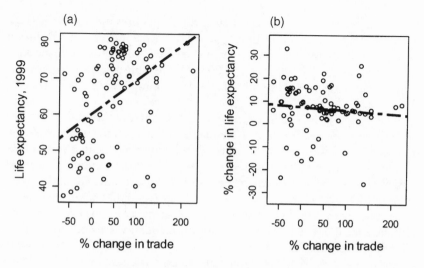

FIGURE 10.1. Trade and health, 1980–1999.

tional health and socioeconomic circumstances from aggregate data. Finding consistent and agreed-upon measures of globalization and health is difficult, and the externalities that emanate from the processes and structures of globalization are often remarkably complex and are manifest at multiple scales from the global to the local. Clearly, the linkage between economic integration and health is one such relationship, and it is also probably reciprocal in nature. Complicating matters even more are incomplete and inconsistent data sets that render disputable insights into global patterns and processes.

Consider, for example, the omission of China from many aggregate data sets and subsequent analyses of globalization processes, effects, and outcomes. In many respects China represents the ultimate globalization success story, yet it also embodies and encapsulates the many challenges that globalization presents to societies across the world at the beginning of the 21st century. For decades China had a closed economy. It was not until the 1980s that the communist leadership in China began to experiment with economic reforms such as privatization, rural and agrarian reforms, and the creation of special economic zones (SEZs) where foreign capital, technology, and know-how were welcome (Knox & Agnew, 1998, pp. 356–359). These changes stimulated the growth of domestic capital and facilitated the influx of capital from abroad, which in turn spurred localized industrialization. Firms of varying size, with ownership from all around the world, have since replaced the huge state-owned companies and now dominate manufacturing output from China.

According to World Bank estimates, this transformation from a state-controlled to a hybrid economy was reflected in China's gross domestic product (GDP) growth rate, which averaged 10% annually between 1990 and 2001 (World Bank, 2003). Since 1980, life expectancy in China also registered a notable increase from 66.8 years to slightly over 70 years.

Underlying the recent expansion of the Chinese economy, however, are extreme sociogeographical inequalities. For instance, significant differences in wealth exist between the rural provinces located in the interior of China and those provinces that are situated along the coast, which serve as gateways for trade and investment, and are active locations within the global economy (e.g., see Coughlin & Segev, 2000). Moreover, as the rural poor are drawn to coastal cities in search of work and a better life, urbanization and urban inequality are additional globalization outcomes that China and the world must face. Such sociogeographical disparities are important to note because in addition to the relationship between wealth and health, there is a growing body of evidence that suggests relative income is also related to health, or more precisely, that poorer health is linked to greater disparities in income within societies that have completed the epidemiological transition (e.g., Kawachi, Kennedy, Lochner, & Prothrow-Stith, 1997; Waldmann, 1992; Wilkinson, 1996). The pathway between income inequality and health remains debated, but one plausible explanation is based upon the argument that more egalitarian societies provide better social, welfare, and health services than less egalitarian societies, which in turn leads to better health.

A strong case can be made that increases in income and economic growth can improve health, especially in developing countries. Economic integration as a strategy for development, however, is not without its critics. For instance, the antiglobalization movement contends that such policies tend to overlook and exploit the poor and exacerbate socioeconomic inequalities. The uneven and diverse impacts of economic integration represent one dimension of the discussions and debates that are concerned with determining whether globalization is good for your health (e.g., Feachem, 2001). Yet such discussions fail to capture some of the most important and intriguing aspects of the globalization and health nexus, namely, the nature of the linkages that permit globalization to occur. Though official statistical databases suggest that globalization (e.g., trade as a share of GDP, foreign direct investment, etc.) and health (e.g., life expectancy, infant mortality rates, etc.) are discrete phenomena that can be measured on a country-by-country basis, the persistence, emergence, and diffusion of diseases such as tuberculosis, malaria, HIV/AIDS, the West Nile virus, and recently severe acute respiratory syndrome (SARS) highlight the importance of other scales of analysis, and the inescapable interdependencies and interactions that exist between the people and places of the world.

OF MICROBES AND "MACROBES" . . .

In the spring of 2003, anxiety spread across Asia and on to Europe, the Americas, and other parts of the world as a mysterious respiratory illness claimed the lives of more than 800 people and infected more than 8,000 people (World Health Organization, 2003a). Given the name "severe acute respiratory syndrome," or "SARS" for short, this new and potentially deadly flu-like disease led to the reintroduction of quarantine laws, the closing of schools, the cancellation of public events, and WHO's unprecedented use of travel warnings across Asia and Canada. Traced back to Guangdong Province around Hong Kong, the rapid spread of SARS is, in part, attributed to the highly connected air traffic routes that can shuttle people, and subsequently diseases like SARS, anywhere in the world in less than a day. Adding to the difficulties in identifying the source and stopping the spread of SARS was misinformation provided by the Chinese government, which in February 2003 reported that the mystery illness was under control.

The rapid and widespread diffusion of SARS is a telling example of a globalization consequence, and it embodies several of the health challenges that globalization presents to the world in the 21st century. This is not to say that the diffusion of disease is a new phenomenon, but that the diffusion of disease in the context of contemporary globalization reveals the complexity of the ever-changing networks that connect the people, the places, and subsequently the microbes of the world. Dealing with health issues that ignore social, economic, and political boundaries underscores the need for multilateral coordination and cooperation. Such health issues are not limited to the spread of germs, but include what I call "macrobes," or health matters that are somehow linked to globalization, such as the tobacco trade and the provision of affordable drugs. Therefore, in addition to understanding how a disease such as SARS can spread so quickly in an effort to contain and minimize its harmful effects, it is also necessary to appreciate the role that international institutions can play in facilitating and impeding the spread of disease, vaccines, better health practices, and even harmful habits.

The spread and survival of disease relies upon a variety of physical, social, economic, and political linkages at different scales of analysis, which in turn constitute a truly global network of connections. The diffusion of a disease typically begins at a point of origin or with someone who is infected, and through direct or immediate contact the disease spreads outward across space and over time. More connections to the infected provide more opportunities for the disease to spread. This is what is believed to have happened with SARS when persons from Singapore and Vietnam staying in a Hong Kong hotel came into close contact with another guest who was infected and staying on the same floor (World Health Organization, 2003b). Epidemics usually begin in densely populated environments such as cities, but as distance from the point of origin increases opportunities to become infected tend to decrease. This type of diffusion is referred to as "spatial contagion," and accounts for how the common cold, influenza, and SARS typically spread at the local level.

The infected, however, are not necessarily bound to a single place but can travel along the myriad connections that constitute the networks upon which globalization depends. What is more, advances in transportation technology have decreased travel times to the extent that the longest international flights are now shorter than the incubation period of any human infectious disease (Frenk & Gomez-Dantes, 2002). As the SARS outbreak illustrates, the emergence of an illness or disease in one location can result in a pandemic. Though the global spread of a disease such as SARS appears to disregard completely the geographical notion of distance, it still depends on the uneven geographies of the global networks upon which it travels. The spread of SARS was neither homogeneous nor isotropic across Southeast Asia; rather, it jumped from China and Hong Kong to Singapore and Vietnam, then made its way to Europe, Canada, and the United States, without affecting places in between. Figure 10.2 maps the diffusion of SARS using the cumulative number of deaths due to SARS reported to WHO in the spring of 2003. The size of each circle in the map roughly corresponds to the number of victims, and the shading of the circle corresponds to the month in which the deaths occurred. Note that China did not begin to report SARS cases to WHO until April 2003. The diffusion of SARS portrayed in Figure 10.2 is an example of "relocation diffusion," which occurs when a phenomenon spreads from one location to another without affecting places in between.

Though most diseases spread via spatial contagion at the local scale between individuals, relocation diffusion can account for the regional and global spread of diseases. Oftentimes an underlying hierarchy of places tends to determine the way in which, for example, a disease spreads, or "relocates," across the world. In the case of SARS, it is widely believed that international air traffic routes, and, in particular, the way in which airline traffic is organized in the Asia-Pacific region,

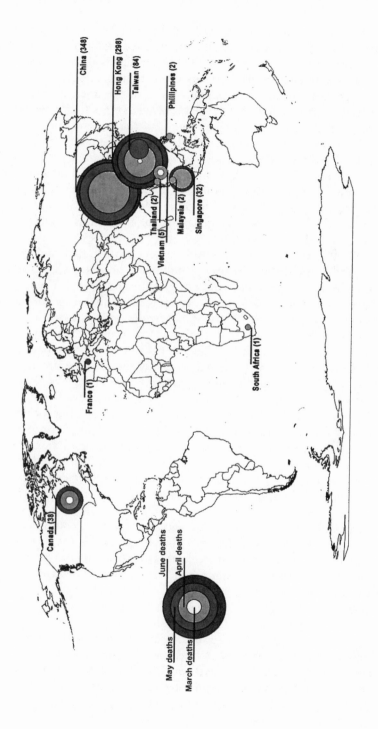

FIGURE 10.2. Deaths due to SARS, 2003. Data from World Health Organization (2003a).

facilitated the spread of this potentially fatal disease to some places but not others. In 2002, it is estimated that over 122,000,000 passengers departed, transited, and arrived at the Hong Kong, Bangkok, Singapore, and Beijing international airports (Airports Council International, 2003). If each passenger was considered a possible infectious contact, the number of potential infections that could have occurred at these four airports alone is staggering. Airline route maps found in the back of most in-flight magazines reveal the organizational hierarchy of air traffic, with most destinations served by a relatively small number of *hubs*, or locations that serve more destinations than others. Recognizing the relative importance of places within such hierarchies, and the linkages that exist between people and places, not only informs how a disease like SARS spreads but also sheds light upon the geographical unevenness of globalization processes and outcomes.

The recent emergence and spread of diseases such as SARS and the West Nile virus captured the attention of millions around the world, in part because there is an inherent fascination with and fear of the unknown, especially when it can kill. Accordingly, the global mass media taps into and responds to such fears and provides a disproportionate amount of coverage on such topics, which in some media markets is quite exaggerated and hyped. The real health risks posed by such pathogens, if measured in terms of those who actually die from contracting such viruses, remain negligible when compared to past outbreaks of influenza or the leading causes of death around the world. This is not to say that diseases such as SARS and West Nile virus are trivial, but for the sake of comparison, it is estimated that the 1918–1919 influenza epidemic killed more than 40 million people, which is greater than the total number killed in all the major wars of the 20th century combined (Brainerd & Siegler, 2002). Moreover, according to WHO estimates, HIV/AIDS, tuberculosis, and malaria continue to account for 25% of deaths around the world.

As the networks that support globalization grow more complex, it is a certainty that outbreaks of infectious disease will continue to occur. Yet microbes are not the only threats to health that benefit from globalization processes and networks. Macrobes, a term used to describe the health consequences related to the structures of globalization, as well as to social behaviors and lifestyles, such as the practice of smoking that kills 4.9 million people annually, also capitalize upon the innovations and utilize the networks that drive globalization (World Health Organization, 2003c). The roots of tobacco use, for instance, can be traced back several centuries, but it was not until the discovery of the New World by Europeans that the practice of tobacco use became widespread and eventually global in scale (Mackay & Eriksen, 2002; von Gernet, 2000). A complex mix of social, economic, physiological, and cultural factors play a role in the diffusion of smoking and nicotine addiction, but the advent of the rolling machine in the late 19th century ushered in the era of the mass-produced cigarette, which when linked to today's extensive trading networks, serve over 1.1 billion smokers around the world (von Gernet, 2000, p. 10). Table 10.1 provides an indication of the extent and scale of the global tobacco trade today.

It is clear from Table 10.1 that the tobacco trade and tobacco use are truly global phenomena of significant proportions. Self-reported revenues from the five largest tobacco multinational firms amounted to nearly U.S. $110 billion in 1999, which in effect amounts to an enormous amount of power and influence. When lawsuits emerge or when smoking restrictions and bans are imposed in one country, tobacco multinationals use their power, resources, and the international trade framework to look for other countries, usually in the developing world, with fewer or no regulations where they can distribute their profitable and addictive product. The increase in death due to noncommunicable disease (e.g., heart disease, cancer, obesity, etc.) in developing countries is probably related to such practices and the diffusion of unhealthy lifestyles (World Health Organization, 2002b).

In addition to creating opportunities for tobacco multinationals, such discrepancies in national regulatory structures and varied interpretations of international trading agreements pose several significant challenges to global health issues. Recognizing the possible health consequences associated with globalization, or more specifically how in

TABLE 10.1. The Global Tobacco Trade

Top 10 leaf importers, 1999 (thousands of metric tons)		Top 10 leaf exporters, 1999 (thousands of metric tons)		Top 10 cigarette importers, 2000 (billions of cigarettes)	
Russia	263	Brazil	343	Japan	83.5
United States	241	United States	191	France	67.2
Germany	190	Zimbabwe	164	Italy	56.5
United Kingdom	129	China	132	Malaysia	49.3
Netherlands	113	Turkey	129	United Kingdom	45.0
Japan	99	India	120	Germany	33.6
France	71	Greece	101	Spain	25.2
Ukraine	70	Italy	94	China	24.2
Poland	60	Malawi	93	Saudi Arabia	20.0
Egypt	55	Argentina	73	Slovakia	20.0

Note. Data adapted from Mackay and Eriksen (2002).

certain circumstances the relaxation of trade barriers could potentially threaten public health, the World Trade Organization (WTO) recently declared that member states "can put aside WTO commitments in order to protect human life" (WTO, 2002). Though this declaration speaks to issues surrounding the diffusion of infectious disease, it is primarily a response to concerns voiced by developing countries afflicted with diseases like HIV/AIDS, tuberculosis, and malaria (see Nicholson, Chapter 3, this volume). In such countries, it is feared that adherence to the terms of the WTO's trade-related intellectual property rights (TRIPS) and patent protection agreements would make potentially lifesaving drugs too expensive and put them out of the reach of those in need.

The health implications of trade liberalization are not limited to goods like cigarettes and drugs to combat infectious disease. Recent WTO discussions have focused upon the General Agreement on Trade in Services (GATS) that may also have several lasting health consequences (e.g., see Lipson, 2001). One of the central issues relates to the opening of health service markets to foreign competition. In theory, liberalizing health services markets will increase competition, improve efficiency, and broaden access to health services. In reality, where health markets have been liberalized and privatized, such changes tend to benefit the wealthy, shift resources (including physicians) to high-end care and elective treatments with high profit margins, and ultimately exacerbate health inequalities. Similar calls have been made to open markets and remove restrictions on health-related financial services such as health insurance. In some places this is problematic because it is common for insurance companies, many of which are multinational, to issue policies that are contingent upon good health, which in effect burdens further the public sector with high-cost patients (World Health Organization, 2000).

The outbreak of an infectious disease such as SARS, the diffusion of tobacco use and the modern tobacco trade, and the inclusion of health-related issues in WTO discussions reveal the complex and continually changing dimensions to the globalization and health nexus. Responses to such events, patterns, and trends already require a significant amount of coordination and cooperation not only at the local and national levels, but globally as well. Such coordination and cooperation must be universal because the risks and consequences of misinformation, underreporting, and nonparticipation far outweigh the benefits, if any. Just as globalization processes and networks facilitate the rapid spread of disease, greater efforts need to be made to use these same networks to promote better health practices, to prevent the conditions that can lead to the outbreak of infectious disease, and to create opportunities that can improve health.

AN APPLE A DAY KEEPS GLOBALIZATION AWAY . . .

Hong Kong is a ghost town. Offices are shut, conferences
cancelled, schools closed and businesses across the region
are reeling.
—*The Economist* (April 5, 2003, p. 13)

Many of the discussions surrounding globalization and health look at
how globalization can improve health (e.g., see Feachem, 2001). From
the externalities associated with trade liberalization to the diffusion of
disease, the arrow of causality usually points from globalization to
health. Following this causal path, the first half of this chapter reviewed
the argument that economic integration may stimulate growth, which in
turn may improve health. The pathways between globalization and
health, however, are not definitive or direct. Additional research is
needed to explore competing notions of globalization and health, and to
examine how globalization and its externalities, such as income inequalities, not only affect population health, but also the health of subnational
groups and individuals. Though data and measurement issues continue
to lead to differences in opinion about globalization's impact upon
health, efforts to improve such indicators are ongoing and promising
(e.g., see Asada & Hedemann, 2002; Gakidou & King, 2002).

The second half of this chapter looked at a few of the health-related
consequences of globalization, and also considered the influence that
health has upon economic integration itself. Arguably, the most important prerequisite for globalization through the ages is the health of its
participants. History provides examples of how civilizations and cultures—
the first globalizers—were stricken and affected by disease, and as we
enter the 21st century we should ask ourselves whether and how globalization itself is at risk. Put another way, as HIV/AIDS, tuberculosis, and
malaria continue to plague sub-Saharan Africa, as drug-resistant strains
of bacteria continue to be discovered in the developed world, and as
unfamiliar diseases like the West Nile virus and SARS spread at unprecedented rates over vast distances, is worldwide interdependence and integration at risk?

It is clear that in the short term the emergence of a disease such as
SARS can have notable economic effects. It is estimated that SARS cost
the Asia-Pacific region $30 billion (World Health Organization, 2003b,
p. 7). Travel to Asia all but ceased during the SARS crisis, exacerbating
problems for the global airline industry and creating several problems
for businesses throughout Asia and the world. For example, fear of
SARS is believed to have slowed production in many sectors as employees
stayed home. The repercussions of this slowdown were global because

many of the affected areas (e.g., Hong Kong, Singapore, Guangdong Province) are the "workshops" of the global high-tech industry, where foreign investment is often directed and whence information technology components are imported (Krugman, 2003). The long-term effects of SARS remain to be seen. One positive outcome is that the SARS crisis will likely lead to improved and even more efficient responses to future health threats. Though the use of travel warnings by WHO have been criticized, especially by the city of Toronto in Canada, failure to respond may have resulted in even greater losses, in terms of both life and money. This is one of the paradoxes of public health in this era of globalization: What is the cost of vigilance and is it possible to be too vigilant? These questions are difficult to answer. Looking to history, the 1918 influenza epidemic in the United States killed more than 675,000 people, but it is estimated that the effects of this epidemic later contributed to domestic economic growth until the Great Depression (Brainerd & Siegler, 2002). Since many victims in the United States were of working age, the influenza epidemic led to a rapid decrease in the labor pool, which in theory increased the wages of those who remained healthy and active.

Can the same effect be expected from the HIV/AIDS pandemic, especially in places like sub-Saharan Africa? Probably not. HIV/AIDS is a "slow plague" that erodes productivity, afflicts the poor, and has orphaned thousands upon thousands of children in Africa (Gould, 1993). This last point is perhaps the most compelling and most troubling aspect of the AIDS pandemic in Africa. When linked to poor and worsening economic conditions, food insecurity, civil unrest, armed conflict, and government corruption, state failure is a possible, if not, probable result in the future (see Collier et al., 2003). It is in failed states that disease spreads, terrorism takes root, human rights are violated, and history and culture are destroyed. In many circumstances, interdependence and integration can prevent state failure and may offer the best prospects to overcome the challenges of disease, despair, and conflict, but health and wealth, as well as disease and despair, exist beyond the scale of the state. Multilateral cooperation and efforts to coordinate health policy, relief efforts, and preventative strategies therefore must consider the diverse and multiple scales that are affected, from the local to the global. In the end, integration, interdependence, and globalization are probably not at risk of disease and poor health; rather, the greatest health risks may arise from the failure to include, participate, and integrate.

NOTES

1. The $1 figure is in 1985 U.S. dollars, adjusted for purchasing power parity.
2. Change in trade is measured as a share of total GDP for the 1980–1999 pe-

riod. All data were obtained from the World Bank. The correlation between change in trade and life expectancy in 1999 is .46, and is significant at the p < .01 level; the correlation between change in trade and change in life expectancy for the same period is –0.10 and is not significant at the p < .05 level.

REFERENCES

Airports Council International. (2003). *ACI passenger traffic reports.* Available online at *http://www.airports.org.* Accessed March 7, 2003.

Asada, Y., & Hedemann, T. (2002). A problem with the individual approach in the WHO health inequality measure. *International Journal for Equity in Health, 1,* 1–5.

Brainerd, E., & Siegler, M. (2002). *The economic effects of the 1918 influenza epidemic* (CEPR Working Paper). Washington, DC: Center for Economic Policy Research.

Collier, P., Elliot, L., Hegre, H., Hoeffler, A., Reynal-Querol, M., & Sambanis, N. (2003). *Breaking the conflict trap.* Washington, DC: World Bank.

Coughlin, C., & Segev, E. (2000). Foreign direct investment in China: A spatial econometric study. *World Economy, 23,* 1–23.

Curtis, S., & Taket, A. (1996). *Health and societies: Changing perspectives.* London: Arnold.

Diamond, J. (1999). *Guns, germs, and steel.* New York: Norton.

Dollar, D. (2001). Is globalization good for your health? *Bulletin of the World Health Organization, 79,* 827–833.

Edwards, S. (1998). Openness, productivity and growth: What do we really know? *Economic Journal, 108,* 383–398.

Feachem, R. (2001). Globalisation is good for your health, mostly. *British Medical Journal, 323,* 504–506.

Frankel, J. (1997). *Regional trading blocs.* Washington, DC: Institute for International Economics.

Frenk, J., & Gomez-Dantes, O. (2002). Globalisation and the challenges to health systems. *British Medical Journal, 325,* 95–97.

Gakidou, E., & King, G. (2002). Measuring total health inequality: Adding individual variation to group-level differences. *International Journal for Equity in Health, 1,* 1–12.

Gould, P. (1993). *The slow plague.* Cambridge, UK: Blackwell.

Kawachi, I., Kennedy, B., Lochner, K., & Prothrow-Stith, D. (1997). Social capital, income inequality and mortality. *American Journal of Public Health, 87,* 1491–1498.

Knox, P., & Agnew, J. (1998). The geographical dynamics of the world economy. In P. Knox & J. Agnew (Eds.), *The geography of the world economy* (3rd ed., pp. 64–120). London: Arnold.

Krugman, P. (2003, April). Guns, germs and stall. *The New York Times.* Available online at *http://www.nytimes.com.* Accessed April 4, 2003.

Lipson, D. (2001). The World Trade Organization's health agenda. *British Medical Journal, 323,* 1139–1140.

Livi-Bacci, M. (1997). *A concise history of world population* (2nd ed.). Oxford: Blackwell.

Mackay, J., & Eriksen, M. (2002). *The tobacco atlas.* Geneva: Author.

Noack, H. (1987). Concepts of health and health promotion. In T. Abelin, Z. Brzezinski, & V. Carstairs (Eds.), *Measurement in health promotion and protection* (pp. 5–28). Copenhagen: World Health Organization.

Omran, A. (1971). The epidemiological transition: A theory of the epidemiology of population change. *Milbank Quarterly, 64,* 355–391.

Omran, A. (1983). The epidemiological transition theory: A preliminary update. *Journal of Tropical Pediatrics, 29,* 305–316.

Population Reference Bureau. (2002). *2002 world population data sheet.* Washington, DC: Author.

Prichitt, L., & Summers, L. (1996). Wealthier is healthier. *Journal of Human Resources, 31,* 841–868.

United Nations Development Programme. (2003). *Human development report.* New York: Oxford University Press.

von Gernet, A. (2000). Origins of nicotine use and the global diffusion of tobacco. In M. Pope (Ed.), *Nicotine and public health* (pp. 3–15). Washington, DC: American Public Health Association.

Waldmann, R. (1992). Income distribution and infant mortality. *Quarterly Journal of Economics, 107,* 1283–1302.

World Health Organization. (2000). *World health report 2000. Health systems: Improving performance.* Geneva: Author.

World Health Organization. (2002a). *Noncommunicable diseases in south-east Asia region.* New Delhi: Author.

World Health Organization. (2002b). *World health report.* Geneva: Author.

World Health Organization. (2003a). *Cumulative number of reported cases of severe acute respiratory syndrome.* Available online at *http://www.who.int/.* Accessed April 10, 2003.

World Health Organization. (2003b). *Severe acute respiratory syndrome (SARS): Status of the outbreak and lessons for the immediate future.* Geneva: Author.

World Health Organization. (2003c). *Tobacco free initiative.* Available online at *http://tobacco.who.int/.* Accessed April 10, 2003.

Wilkinson, R. (1996). *Unhealthy societies: The afflictions of inequality.* London: Routledge.

World Bank. (2002). *Globalization, growth and poverty: Building an inclusive world economy.* New York: Oxford University Press.

World Bank. (2003). *World development indicators 2003.* Washington, DC: Author.

World Trade Organization. (2003). *WTO agreements and public health 2002.* Available online at *http://www.wto.org.* Accessed March 15, 2003.

11

Democracy and Social Spending

The Utility of Interregional Comparisons from a Latin American Perspective

GEORGE AVELINO
DAVID S. BROWN
WENDY HUNTER

The globalization of markets for capital, goods, and services that has taken place in the past two to three decades has few historical precedents. Against a wider context of international integration, the developing world has experienced the most dramatic change in its economic policy orientation since World War II. In the developing world, governments have instituted radical reforms aimed at integrating their economies into global markets. In the face of reform, large segments of the population are now exposed to market fluctuations that have important effects on employment and on government subsidies. Also at risk is government spending on social programs, including spending on health, education, and social security. Have governments become less generous toward citizens in response to the pressures generated by greater economic openness? Or have they created stronger safety nets and new forms of social assistance in order to meet the new social challenges of globalization? Answers to those questions depend, in part, on the constraints political institutions place on politicians; electoral laws, party systems, or institutional rules that govern executive and legislative relations (to name just a few) might all effect how much room politicians have to maneuver vis-à-vis the international economy and their constituents. In tandem with new economic reforms, many developing countries have undergone significant political liberalization (democratization) over the last two decades. Are democracies more likely than their authoritarian counterparts to provide the protection demanded by segments of the population put at risk by increasing trade and capital flows?

Heretofore a vast majority of the literature on globalization and social spending has been limited to the developed world (Cameron, 1978; Garrett, 1998; Hicks & Swank, 1992; Katzenstein, 1985; Pierson, 1996; Rodrik, 1998). Although some studies on the developing world exist, they have only appeared in the last few years (Garrett & Nickerson, 2001; Kaufman & Segura, 2001; Rodrik, 1999; Rudra, 2002; Rudra & Haggard, 2001). In this chapter, we argue that the questions posed above cannot be answered fully without "globalizing" the study of globalization: questions regarding the relationship between globalization, democracy, and social spending cannot be adequately addressed without making interregional comparisons. By "interregional comparisons," we mean comparing countries in one specific geographical region to countries in another: for example, comparing European countries to Latin American countries, or African countries to Asian countries. Specifically, we demonstrate how analyzing globalization as a truly global phenomenon leads to progress on three different fronts. First, interregional comparisons can help answer a wider range of questions than studies analyzing similar and more geographically proximate countries: interregional comparisons provide more variance in the dependent and independent variables. Using Latin America as a comparison, we identify a number of key variables that distinguish the Western European experience from that of the developing world. By making interregional comparisons, we not only learn more about globalization and its influence on the developing world, we can gain insights into the empirical patterns long established in the European context. Second, comparing developed with developing regions necessitates disaggregating the dependent variable. Social programs differ dramatically in both size and kind over such a wide range of countries. Some of the most important differences between the developed world and poorer nations may not simply exist in terms of quantity, they may reside in the kind of programs offered. Relying solely on aggregate data may obscure important empirical relationships. Third, making interregional comparisons requires more exacting measurement: interregional comparisons, for example, demand we pay more attention to differences that exist between the rich and the poor countries of the world. We show how traditional indices fail to account for important differences that exist between national economies, ultimately producing biased results.

The chapter proceeds as follows. In the first section, we describe the theoretical debate on globalization, and social spending in terms of efficiency, compensation, and democracy's impact on both. Next, we describe how interregional comparisons can help answer important questions that remain unexplored in studies that limit themselves to a single region. We then explore how disaggregating social spending into its component parts—a natural outcome of interregional comparisons—can help uncover previously obscured empirical relationships that exist between globalization and social spending. Finally, we show how interregional comparisons force us to revise previously used measures of globalization.

EFFICIENCY, COMPENSATION, AND DEMOCRACY

Before we proceed, it is important to set the stage. Let us briefly sketch a brief outline of the current debate. The literature addressing the interaction between globalization, domestic politics, and government spending is substantial, spanning several decades (Adserá & Boix, 2002; Cameron, 1978; Garrett & Mitchell, 1999; Hicks & Swank, 1992; Huber & Stephens, 2001; Katzenstein, 1985; Kaufman & Segura, 2001; Pierson, 1996; Rodrik, 1998, 2001). As both a function of ideology[1] and the empirical patterns observed by scholars, the question at the core of this literature asks whether governments respond to the challenges of globalization by cutting social spending (i.e., "efficiency") or by protecting people's welfare (i.e., "compensation").

Generally speaking, the majority of empirical work on globalization, democracy, and social spending focuses on government expenditures in health, education, and social security—the main components of social spending. Usually, social expenditures are expressed in per capita terms or as a percentage of a country's gross domestic product (GDP). The central notion of the efficiency approach is that governments will reduce taxes and social welfare expenditures that diminish corporate profits, discourage investment, and therefore threaten economic growth and international competitiveness. Social services burden business through the distortion of labor markets and higher taxes. If governments borrow to pay for these services, the higher real interest rates that result from this borrowing further depress investment. In short, the efficiency approach posits that economic openness places important constraints on welfare spending, leaving governments little choice but to restrict their social outlays.

The compensation argument recognizes the constraints imposed by economic integration on the social policy options of governments, yet accords greater weight to the countervailing demands imposed by citizens seeking protection from the state. It stresses the perception among top elected officials and bureaucrats that the social instability and political discontent engendered by internationalization could ultimately endanger the model of economic openness as well as their own careers. The core contention of the compensation thesis is that government officials should use the latitude they have to strengthen social insurance mechanisms and cushion citizens from the vagaries of the international economy.

Social expenditures are a clear general measure of the extent to which governments contract or expand their commitments to citizens.[2] Consequently, democracy's role in determining government expenditures on social programs is of particular relevance. Although many of the studies use different definitions, they are all based on a procedural conception of democracy. Most of the work in this literature defines *democracy* as a set of political institutions that allows for contestable elections in which politicians freely compete for a free vote. The focus on democratic procedures is important because it allows scholars to distinguish between processes (democratic institutions and their interaction with globalization) and outcomes (government spending on social programs).[3] Fluctuations in social spending, for example, can provide clues about the constraints facing public officials, their latitude for responding to those constraints, and the relative weight they place on competing priorities. As we noted earlier, a great majority of the literature's scope is limited to Western Europe, prohibiting scholars from examining the relationship between globalization, democracy, and social spending. Theories on globalization, democracy, and social spending hinge on the degree of insulation varying political institutions provide politicians. Compelled to win elections to survive politically, politicians who face electoral constraints may be more likely to compensate those segments of the population most vulnerable to fluctuations in the international market. If democracies do, in fact, make politicians more accountable to groups in society, we still know very little about who those groups are: Do they represent a sizeable proportion of the electorate, or do they own a relatively sizeable proportion of the economy? It is not quite clear whether politicians constrained by elections are more likely to respond to groups that are numerically or economically more influential. Although some preliminary evidence suggests that democracies perform a compensatory role, theory on democratization's intervening influence between globalization and social spending is relatively underdeveloped.

INCREASING VARIANCE THROUGH
INTERREGIONAL COMPARISONS

The developing world constitutes an interesting and relatively under-studied area for the analytical questions at hand. A number of factors set the developing world apart from the Western European countries. To illustrate those differences, we discuss the potential gains from comparing the Western European countries with Latin America (the region we know best). On the one hand, some of these factors encourage actions in accordance with the efficiency thesis. Others make Latin America more likely to adopt compensatory schemes.

The relative weakness of unions and the paucity of social democratic parties, a historical support base for universal social protection policies in Western Europe, deprives Latin American citizens of two key organizational means to defend social services against budgetary cuts. Thus, while Cameron (1978) finds that trade openness in Western Europe resulted in the provision of greater public resources for social protection, such an outcome may not hold in Latin America.

The degree to which the countries of Latin America opened up their economies in the last two decades also sets them apart from the Western European countries. The rapid and dramatic process of stabilization and structural adjustment in the wake of the Latin American debt crisis—and the active accompanying role played by the International Monetary Fund (IMF)—is without parallel. IMF prescriptions for attaining fiscal solvency depend on reducing government expenditure on social programs by introducing user fees in health and education, and by effecting a more efficient distribution of goods and services to the poor.[4] When faced with the increasing burdens associated with meeting IMF-prescribed goals—decreasing government spending—Latin American governments are tempted to cut social spending. IMF prescriptions usually represent policies that would make any popularly elected government hesitate. Governments reluctant to initiate such actions still acknowledge the importance of expressing to the IMF their intent to adopt these economic reforms largely because they depend so strongly on IMF approval to access future finance from abroad. Although Western European nations can not ignore international economic constraints, the precarious position of many Latin American countries—burdened by high levels of debt—provides an additional dimension to decisions governments take regarding social programs.

Finally, the comparative weakness of Latin American states exposes welfare programs to the risk of retrenchment. The state in most Latin American countries, while never as strong as in most Western European countries, was weakened further by the economic crisis of the 1980s and

1990s. Governments in the region are notorious for their inability to carry out some of the most essential tasks—for example, tax collection—necessary to maintain generous welfare support (Huber, 1999).

Other factors relevant to Latin America provide some reason to expect globalization might promote compensation. Greater trade volatility heightens the insecurity of citizens unless governments take active measures to provide for social protection. Most Latin American countries tend to have relatively specialized patterns of trade compared to their counterparts in the developed world: there is an overreliance by some economies on certain exports and imports. With higher levels of volatility, the effects of trade exposure may be more severe, inducing a higher level of government spending in relation to total trade. This hypothesized relationship leads us to examine the role played by new democratic regimes.

As we noted earlier, most of the previous empirical work focuses on the Western European nations, for which comparable and extensive data are available. Understandably, these studies take stable democratic institutions as given. In these models, democracy has only an indirect effect: it works as a channel through which the effects of other relevant variables (such as political parties and union strength) can be analyzed (Garrett & Lange, 1996). Although it is important to consider political factors other than regime type (i.e., whether a country is democratic or not), issues relating to regime type and regime transition still deserve attention. At least two reasons justify this claim.

First, we need to know more about the conditions that make democracies work, the conditions that enable them to achieve economic growth, material security, freedom from arbitrary violence, and other widely desirable objectives (Przeworski, 1995). Due to the endemic political instability that has characterized developing countries, much of the comparative work on democratization assumes that these new regimes are inherently fragile. These works were mostly concerned with constructing etiologies of regime change or of emerging democratic regimes. Virtually ignored in this literature is the impact of democracy on public policies.[5]

The recent political and economic transformations experienced by many developing countries offer a unique opportunity to explore questions about how different political regimes react to an inescapable feature of globalization: external economic shocks (Rodrik, 1999). Recent evidence suggests that the interaction between democracy and globalization has a strong positive effect on government expenditures (Adserá & Boix, 2002). When facing the same international economic constraints as their authoritarian counterparts, politicians in democracies are more likely to continue funding education, health, and social security as a strategy to build electoral support. Because Latin American countries comprise a significant part of the recent wave of democratization, they provide an excellent opportunity to analyze the effects of different political regimes on these policy options.

Second, despite the euphoria sparked by the widespread democratization among developing countries, current attitudes toward new democracies are mixed. Although they represent an important change in comparison to previous authoritarian institutions, Latin American democracies have been criticized for not having fulfilled many of the expectations they generated. This growing disenchantment with democracy is particularly acute in the provision of public services, an area in which democratization was expected to have a tangible impact on the plight of the poor.[6]

As the empirical literature makes clear, democracies alone are unlikely to reverse deeply entrenched patterns of poverty and social inequality. Nevertheless, the prevalence of new democracies headed by governments with a presumed interest in maintaining social stability and winning reelection would seem to auger well for social welfare programs. The social dislocations produced by restructuring an economy toward competition in the international marketplace affect the middle class as well as poorer segments of the population. The middle class is not only well represented at the ballot box, it is also crucial to public opinion formation. Rebellion among indigenous peasant producers in southern Mexico, food riots in Argentina, and strikes by public-sector workers in a number of countries are among the expressions of protest that have emerged in the last decade. The widespread institution of social emergency programs, such as PRONASOL in Mexico and FONCODES in Peru, suggests that governments in the region are not unaware of the need to secure support for themselves and for their economic reforms.

Some of the same factors that make for interesting comparisons and contrasts between Latin America and Western Europe, as discussed above, set Latin America apart from other developing regions. They are of central relevance to the question of what mediates the impact of economic globalization on social spending outcomes. Identifying and testing the importance of these factors is grist for further research. Three variables in particular could conceivably cause a significant divergence between Latin America's social spending response to economic globalization and that of other developing regions.

The first concerns the strength of organized labor. While organized labor is stronger in Western Europe than it is in Latin America, it is stronger in Latin America as a whole than it is in other developing world regions. The weaker bargaining power of workers in less developed countries (compared with Western European countries) may place welfare spending in a more precarious position in the face of increased trade and capital flows (Rudra, 2002). Yet, at the same time, where labor is more powerful, workers are better positioned to successfully mitigate these effects. The greater union strength of many Latin American countries compared with many of their developing world counterparts might help to buffer social spending in the former from the deleterious effects of economic globalization. When measured by such factors as union density (i.e., union membership as a proportion of the population), contract coverage, quantified bargaining level, and quantified convention ratification indicators (McGuire, 1999), union strength has been more marked in the newly developing countries of Latin America compared with those of East Asia (McGuire, 1995). Of late, unions have been active in protesting government cuts in services ranging from social security pensions to health care. As such, they may be an important determinant of keeping social spending at a higher level than it would otherwise be in the region. Interregional comparisons could help establish whether the levels of social spending found in the Western European context are the result of union strength. Understanding how Latin America contrasts with the rest of the developing world can indicate whether intermediate steps in liberalizing trade union activity can produce change.

A second factor that could plausibly influence Latin America's social spending patterns vis-à-vis other developing nations concerns the prominent role of the IMF. The number of IMF standby agreements that various Latin American countries have entered into in the last 20 years attests to the leading role played by the fund in the region (International Monetary Fund, 2003). Latin America's reliance on IMF lending is especially pronounced when compared to Asia. The public expenditure limitations often sought as a condition for loans might conceivably cause Latin America to respond to the pressures of market reform by shrinking social funding more than its developing world counterparts who work less with the IMF. Well known is the IMF's insistence that countries use their social budgets in cost-effective ways. This means shifting resources away from expensive items that are not well targeted to the poor (e.g., university education), or that do not yield strong economic returns (e.g., food subsidies). The IMF has also encouraged the introduction of user fees in health and education toward the goal of relieving pressure on public budgets. The advocacy of social safety nets as a way to protect the poorest from adversity during times of economic transition also conforms to the "more bang for the buck" outlook of the fund (Deacon, 1999). Social safety nets have served as a key complement in the social sphere to both the IMF and the World Bank's call for adjustment and restructuring in the economic sphere. By comparing Latin America with other developing regions, more could be learned about the IMF's impact on social spending.

Latin America's tendency (compared to other developing regions) to adopt social policies from the developed world, especially from the United States, constitutes a third factor that may influence Latin America as a region toward reaching a set of outcomes that differ from other parts of the developing world. Historic and cultural reasons, or perhaps the greater number and density of linkages in international organizations between the United States and Latin America, may account for this higher rate of diffusion. Without further research, it is difficult to say whether Latin America's tendency to adopt first world reforms would increase or decrease spending; nevertheless, diffusion effects surely generate systematically different regional responses to the wake of globalization. Two prominent examples of social policy diffusion between leading Western European countries and Latin America concern social security reform and the decentralization of administrative, fiscal, and political institutions. While the lion's share of Latin American countries have followed first world trends and moved to implement some kind of social security reform, such has not been the case in Asia or Africa (Orenstein, 2001, p. 24). Latin America has also moved more rapidly (especially relative to East Asia) toward the implementation of decentralizing reforms. Since many such "neoliberal" reforms were designed to reduce the "size of government" and public expenditures, it is possible that policy diffusion has reduced systematically certain categories of social spending in the region.

INTERREGIONAL COMPARISONS AND DISAGGREGATION

An important facet of the approach we advocate includes recognizing that important differences that exist between regions may be more in *kind* than in *quantity*. Only by looking beyond aggregate figures can those differences be observed. The most common measure used in the empirical literature is social spending as a percentage of GDP, an aggregate measure based on the sum of three categories: health, education, and social security. Although each category encompasses a wide range of funds, institutions, and programs in their own right, scholars have paid little attention to examining globalization's impact even at this relatively high level of aggregation.[7] Understanding what happens to social spending at less aggregate levels is particularly important when comparing the Western European countries with middle-income and least developed countries. Countries at different points along the spectrum of economic development are confronted by different economic and political challenges. Scholars are only now beginning to refine their analyses to account for important differences in the way governments allocate benefits to their citizens.

Discovering the empirical patterns that lie below aggregate figures has important practical implications as well. For example, different spending priorities in health, education, and social security can influence both economic development and the political landscape. From a purely economic perspective, allocating resources to health and education increases the stock of human capital, an important component of growth. Devoting an overwhelming share of resources to social security, while politically expedient, may be less efficient economically. From a purely political perspective, spending on health, education, and social security benefits very different segments of the electorate. When politicians allocate resources to one program and not the other, an explicit political calculation is being made. By disaggregating social spending, we can better understand the opportunities and constraints globalization places on politicians.

To the extent that it influences the distribution of resources among health, education, and social security, globalization may hold a number of important implications for economic growth. If globalization forces politicians to compensate the losers, we might expect a very different allocation of public resources than if government behavior more closely reflects the efficiency hypothesis. How that process plays out in each country will influence prospects for long-term growth.

Economists have long recognized the importance that education (the accumulation of human capital) holds for economic performance (Schultz, 1961). Despite earlier work (Denison, 1962; Schultz, 1961), research on human capital formation remained relatively dormant until the mid- to late 1980s when economists abandoned Solow's neoclassical growth model and began to embrace theories of endogenous growth. Unlike the neoclassical model that relied heavily on technology (something that simply developed over time), the new endogenous growth models gave government a prominent role in long-term economic performance. By providing the right incentives to both invest in physical capital (property rights) and to accumulate human capital (spending on education), government policy was brought back into the growth equation. Early work based on endogenous growth models seemed unanimous in its view that education influenced growth positively (Barro & Lee, 1993; Mankiw, Romer, & Weil, 1992; Psacharopoulos & Woodhall, 1985; Schultz, 1989, 1993). Over the last decade, however, some have questioned whether the strong positive relationship exists (Benhabib & Spiegel, 1994; Levine & Renelt, 1992; Levine & Zervos, 1993). The last few years have witnessed a growing controversy: microeconomic studies based largely on experimental data have found significant gains in income due to the accumulation of human capital, while at the same time economists have been unable to find a corresponding relationship at the aggregate level (Kalaitzidakis & Mamuneas, 2001). Although the debate will no doubt continue as to exactly how and under what conditions human capital effects economic growth, the decision to adopt an economic strategy based on the development of human capital can have a tremendous impact on human welfare.

Described in more detail elsewhere in this volume by Michael E. Shin (Chapter 10), government spending priorities on health hold consequences for long-term economic performance as well. A fairly large number of studies have documented an important empirical relationship between general health indicators—life expectancy, caloric intake, and physical size—and economic growth (Arora, 2001; Bloom, Canning, & Sevilla, 2001; Mayer, 2001b; Strauss & Thomas, 1998). As an important aspect of human capital, the overall health of a population can have an important influence on economic performance. The mechanisms associated with this relationship usually involve health's impact on worker productivity. Better health increases the probability that individuals will make more extensive investment in accumulating human capital (Mayer, 2001a). By investing in programs that raise levels of health, governments can affect fertility rates and, in turn, investments in human capital: parents faced with relatively high levels of infant mortality are likely to have more children, thereby investing less in each child's human capital (Becker, Murphy, & Tamura, 1990).

Decisions on which specific health programs to fund are important. In developing countries, programs designed to reduce the incidence of malaria have a much different impact on an economy's trajectory than programs designed to address heart disease. Every 40 seconds a child dies of malaria, resulting in a daily loss of more than 2,000 young lives worldwide (Sachs & Malaney, 2002, p. 680). Malaria not only puts children's survival at risk, it can hold long-term consequences for its survivors. Reports from Kenya suggest that up to 11% of school days per year are missed by Kenyan children because of the disease. This absenteeism affects failure rates, repetition, and dropout rates (Sachs & Malaney, 2002, p. 683). Especially in Africa, directing the bulk of health expenditures toward AIDS prevention represents an important decision for future growth. Some projections suggest that over the next 7 years the AIDS epidemic will cost Botswana nearly one-third of its economic output (*The Economist*, 2002).

Disaggregating social spending can also provide information on how different political institutions, traditions, or cultures attenuate the opportunities, pressures, and demands of globalization. Recent work suggests that political systems based on proportional representation (consensual polities) will more likely protect social welfare (Lijphart, 1999). In the face of growing competition in the international economy, different kinds of regimes or political institutions may respond with very different distributional decisions. Political institutions may affect the constraints politicians face under the rapidly increasing expansion of international trade and finance (Garrett & Lange, 1996). Even if political institutions, ideas, and cultures do not affect the overall level of public expenditures on social programs, they may determine what programs receive more than their fair share of resources.

What little work exists in this area indicates that political institutions have a significant impact on where politicians allocate public funds. Some evidence suggests that both civilian and military politicians in Latin America adopted different budgetary strategies to survive politically (Ames, 1987). Important differences exist between authoritarian and democratic regimes in the emphasis they place on primary, secondary, and tertiary education (Brown, 2002; Brown & Hunter, in press). Still, at the aggregate level, the evidence is not conclusive: some recent work suggests that with increased globalization, regime type has no overall impact on social spending (Kaufman & Segura, 2001). When disaggregated into different programs, however, a positive association between democracy and expenditures on health and education is revealed (Kaufman & Segura, 2001). Although democracy may matter in the details more than at the aggregate level, those details can be substantively important.

INTERREGIONAL COMPARISONS AND MEASUREMENT

As noted elsewhere in this volume by Ward and Gleditsch (Chapter 8), when measuring economic output it is important to use comparative statistics that represent real differences in the goods and services countries produce. Most recognize the importance of using purchasing power parities (PPPs) when making cross-national comparisons. Interestingly, however, in previous empirical work, trade openness (exports + imports/ GDP) measures have been constructed by using real exchange rates rather than PPPs to make cross-national comparisons. To our knowledge, all previous work employs trade openness figures based on real exchange rates. Real exchange rates, however, distort estimates of economic output and trade, calling into question previous estimates of trade integration's impact on social spending. Two issues are involved. First, measures based on exchange rates rely exclusively on the price of goods and services traded internationally. Ignored are nontradable goods and services, economic activities that constitute a sizeable proportion of the economy. Since we are ultimately interested in trade integration, an accurate measure of both trade and domestic output is necessary. Statistics based entirely on internationally traded goods will therefore prove inaccurate. More importantly, exchange rates are influenced by capital flows that can vary considerably on an annual basis. As a result, estimates of an economy's size can vary dramatically from year to year without any real change in output. The solution is to calculate the value of goods based on a certain amount of currency. Using PPPs accomplishes this, reducing the aforementioned distortions by calculating prices in a common unit of currency. To observe the important difference between measures based on real exchange rates and PPPs, consider Figure 11.1 which plots annual observations from the Latin American region between 1980 and 1999 (19 countries; 20 years).

Figure 11.1 demonstrates that there are significant discrepancies between the two measures. Although the correlation coefficient between the two variables is .66, the scatterplot reveals some important differences. Most notably, when exports and imports as a percentage of GDP

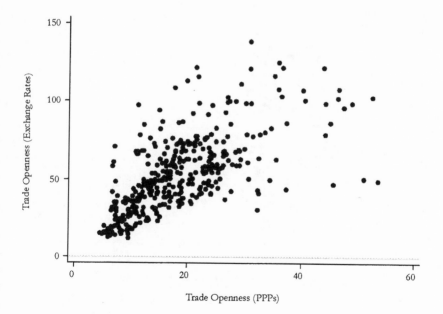

FIGURE 11.1. Scatterplot of trade openness based on real exchange rates and trade openness based on purchasing power parities.

are based on real exchange rates, there are a number of cases above 100%: the maximum value is 139% (Jamaica in 1992). When measured with PPPs, the maximum value is 53.6%. The corresponding value based on PPPs for Jamaica in 1992 is 30.5%.

The rather large discrepancies revealed in Figure 11.1 stem from the differences described above. Among the cases plotted in Figure 11.1, Honduras, Jamaica, Nicaragua, Paraguay, Panama, and Peru all have very different levels of trade openness depending on how it is measured: in some cases the two measures differ by a factor of 10! The seriousness of the problem becomes even more evident when we plot social spending as a percentage of GDP against the ratio of different trade measures (estimates of trade based on the real exchange divided by estimates of trade based on PPPs) for the same Latin American cases (Figure 11.2).[8] Figure 11.2 demonstrates that the biggest discrepancies between trade openness measures (the highest ratios) exist among the lowest spenders. In other words, measures based on real exchange rates assign the most distorted figures to countries with the lowest levels of spending. Moreover, those distorted figures are systematically biased in one direction: figures based

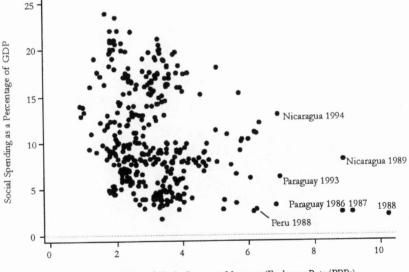

FIGURE 11.2. Scatterplot of trade openness measures (measure based on real exchange rates divided by measure based on purchasing power parities) and social spending as a percentage of GDP.

on real exchange rates exaggerate the amount of trade. The implication, at least for Latin America, is that low levels of social spending will be incorrectly matched with high levels of trade integration.

Another interesting feature of Figure 11.2 is the apparent nonlinear shape of the relationship. The nonlinear shape of the relationship suggests that there are relatively small differences in the trade openness measures for countries that spend a relatively large proportion of their GDP on social spending. The vast majority of the biggest discrepancies between the two measures do not occur until social spending dips below 10% of GDP. Countries whose figures differ dramatically (a ratio of 5 or greater) and who spend relatively little on social programs are Nicaragua, Paraguay, and Peru. Nicaragua, over the last two decades, has witnessed wild fluctuations in the inflow of foreign capital—both in aid and in loans—which could have significant effects on the price of its currency. The large difference in the measures for Peru, having at one time repudiated its debt, could also be the result of currency fluctuations responding to the inflow and outflow of capital. The discrepancy in Peru's numbers could also result from the large informal, or "black," market that exists as a result of Peru's large indigenous population: factor prices will more likely be distorted as a result. Wildly fluctuating prices in Brazil's currency during the 1980s and early 1990s could also have affected the large differences we see in the measures for Paraguay—Brazil's neighbor and trading partner.

The nonlinearity implies that comparisons among countries with similar structures of social spending (the Western European nations) will not be affected by one's choice of measure—trade openness based on real exchange rates or on PPPs. It does suggest, however, that when comparing countries with very different spending profiles—something more likely to occur when making interregional comparisons—the choice of measure is critical. As scholars proceed to incorporate all of the regions of the world in their analysis, more care will be required when measuring globalization.

CONCLUSION

Despite the recent flood of work on globalization and its consequences, we argue that there is much more to learn. For the reasons outlined above, more direct interregional comparisons can help shed light on the relationship between globalization and social spending. Interregional comparisons can give us leverage on a number of different questions that are raised but not quite answered by previous work on the Western European nations. For example, how important are unions for sustaining high levels of social spending? Does bureaucratic inefficiency—the inability to collect taxes—constrain social spending in the face of increased commercial and financial integration? By comparing the Western European nations with Latin America and then with the rest of the developing world, we can anticipate both the difficulties poor countries will face and the benefits they might realize as commercial and financial integration proceeds.

Interregional comparisons not only provide more variance on the independent variables, they obviate the need to disaggregate the dependent variable. Disaggregating social spending into its component parts can provide important information regarding the process of globalization and its potential winners and losers. Spending on education, health, and social security each affects a different segment of the population. Even within each category, different spending priorities can have an important effect on specific targeted groups. Health programs that emphasize new and sophisticated ways of treating cancer and heart disease will reach a different segment of society than will programs designed to prevent AIDS or malaria. Not only can disaggregating the dependent variable identify the constraints politicians face under a number of different institutional contexts, it can reveal how politics mitigates the effects of globalization, ultimately influencing the welfare of human beings.

Expanding the analysis to include countries with very different economies demands that we take a more careful approach to measuring globalization. How globalization is ultimately conceptualized and operationalized can have an important impact on the conclusions we draw from the resulting analysis. As we noted above, important differences exist among measures that purportedly record the same phenomenon. The difference between measures based on real exchange rates and PPPs is systematic, potentially generating biased estimates of globalization's impact on social spending.

Chances are that the efficiency/compensation debate will never be resolved definitively. Gaining important insight on how globalization and democracy affects social spending under a number of different conditions is more likely. We are also likely to find that not all forms of social spending are created equal, both in terms of who benefits and in terms of their economic consequences. Our call for greater geographical coverage, however, does not imply abandoning more focused case study analysis. Our call to disaggregate the dependent variable and to use more exacting measures of globalization implies that future work must combine cross-national analysis with context-specific inquiry. Interregional comparisons are not limited to regression analysis based on large multicountry data sets. Indeed, some of the most important questions yet to be addressed can only be analyzed by matching two cases from different regions, allowing scholars to delve into important relationships that remain obscured at the aggregate level. It will take a concerted effort by a broad range of scholars from different traditions and disciplines to fully understand globalization's impact on human welfare and the attempts by politicians to either mitigate or enhance its effects. Despite the considerable amount of work already completed, the quest to understand globalization, democracy, and their consequences for social spending is just beginning.

NOTES

1. The debate is often infused with ideological overtones since critics of globalization endeavor to show that globalization has harmful effects, reducing the safety nets previously provided by national governments before the onset of the latest round of increased trade and financial integration. Proponents of globalization are given additional ammunition to the extent that they can show globalization does not reduce spending on health, education, and social security.

2. While summary figures do not address the distributional impact of social spending, there is some evidence that they exert a positive impact on the poorer sectors of the population in Latin America (Mostajo, 2000; Petrei, 1996).

3. Although the different definitions of democracy scholars use in their empirical work could be responsible for producing contradictory results, we have seen no evidence that different measures of democracy are responsible for fueling the current debate. Instead, we show that different measures of globalization may be the more likely candidate for explaining the divergent results.

4. See Deacon (1999) for a list of the IMF's prescriptions vis-à-vis the social policies of borrowing member countries. See also Grosh (1996) for the theory behind targeting and its practice.

5. For a survey of different "types of democracy," mostly based on institutional characteristics, see Collier and Levistky (1997). As the ability of the poor to make effective demands depends on the institutional design of democratic regimes, a natural extension of the work done here is to test the impact of different types of democracy over public policies.

6. A recent poll by "Latinobarometro," published in The Economist (2001), attested to the decline of support for democracy in Latin America. This disenchantment, however, does not disregard the fact that changes in democracies are usually moderate and incremental, as claimed by many authors (Huntington, 1989; Schmitter & Karl, 1991). In most cases, the disenchantment stems from the perception that new democracies have not represented a shift in government priorities, even an incremental one, toward the interests of the poor.

7. Ames's early work (1987) is one exception. For more recent work, see Kaufman and Segura (2001).
8. The figures for social spending come from Comminetti and Ruiz (1998) and ECLAC/CEPAL (2001). Data on trade are from the World Bank (2000).

REFERENCES

Adserá, A., & Boix, C. (2002). Trade, democracy, and the size of the public sector: The political underpinnings of openness. *International Organization, 56*(2), 229–262.

Ames, B. (1987). *Political survival: Politicians and public policy in Latin America.* Berkeley and Los Angeles: University of California Press.

Arora, S. (2001). Health, human productivity, and long-term economic growth. *Journal of Economic History, 61*(3), 51.

Barro, R., & Lee, J.-W. (1993). International comparisons of educational attainment. *Journal of Monetary Economics, 32*(3), 363–394.

Becker, G. S., Murphy, K. M., & Tamura, R. (1990). Human capital, fertility, and economic growth. *Journal of Political Economy, 98*(5), 12–37.

Benhabib, J., & Spiegel, M. M. (1994). The role of human capital in economic development: Evidence from aggregate cross-country data. *Journal of Monetary Economics, 34*(2), 143–173.

Bloom, D. E., Canning, D., & Sevilla, J. (2001). The effect of health on economic growth: Theory and Evidence [Report #8587]. Cambridge, MA: National Bureau of Economic Research.

Brown, D. S. (2002). Democracy, authoritarianism and education finance in Brazil. *Journal of Latin American Studies, 34*(1), 115–141.

Brown, D. S., & Hunter, W. (in press). Democracy and spending on education in Latin America. *Comparative Political Studies.*

Cameron, D. (1978). The expansion of the public economy: A comparative analysis. *American Political Science Review, 72*(4), 1243–1261.

Collier, D., & Levitsky, S. (1997). Democracy with adjectives: Conceptual innovation in comparative research. *World Politics, 49*, 430–451.

Cominetti, R., & Gonzalo, R. (1998). *Evolución del gasto público social en America Latina: 1980–1995.* Santiago, Chile: Comisión Económica para América Latina y el Caribe.

Deacon, B. (1999). Social policy in a global context. In H. Andrew & W. Ngaire (Eds.), *Inequality globalization, and world politics.* Oxford, UK, and New York: Oxford University Press.

Denison, E. F. (1962). *The sources of economic growth in the United States.* New York: Committee for Economic Development.

Economist, The. (2001, July 28). The Latinobarometro Poll: An alarm call for Latin American democracies, pp. 37–38.

Economist, The. (2002, May 11). How to live with it, not die of it. p. 12

ECLAC/CEPAL (Division de Desarollo Social). (2001). *Base de datos de gasto social (Actualizada hasta fines de 1999).* Santiago, Chile: Author.

Garrett, G. (1998). *Partisan politics in the global economy.* Cambridge, UK, and New York: Cambridge University Press.

Garrett, G., & Lange, P. (1996). Internationalization, institutions and political change. In R. O. Keohane & H. Milner (Eds.), *Internationalization and domestic politics.* New York: Cambridge University Press.

Garrett, G., & Mitchell, D. (1999). *Globalization and the welfare state.* Unpublished manuscript, Department of Political Science, Yale University.

Garrett, G., & Nickerson, D. (2001). Globalization, democratization and government spending in middle income countries. Unpublished manuscript, Department of Political Science, Yale University.

Grosh, M. E. (1996). *Administering targeted social programs in Latin America: From platitudes to practice.* Aldershot, UK: Avebury.

Hicks, A., & Swank, D. (1992). Politics, institutions, and welfare spending in industrialized democracies. *American Political Science Review, 86,* 658–674.

Huber, E. (1999). *Globalization and social policy, developments in Latin America.* Paper presented at the Conference on Globalization and the Future of Welfare States: Interregional Comparisons, October 22–23, Brown University, Providence, RI.

Huber, E., & Stephens, J. D. (2001). *Development and crisis of the welfare state: Parties and policies in global markets.* Chicago: University of Chicago Press.

Huntington, S. P. (1989). The modest meaning of democracy. In R. A. Pastor (Ed.), *Democracy in the Americas: Stopping the pendulum.* New York: Holmes & Meyer.

International Monetary Fund. (2003). *IMF lending arrangements, 2003.* Available online at *http://www.imf.org/external/np/tre/tad/extarr1.cfm.* Accessed June 27, 2003.

Kalaitzidakis, P., & Mamuneas, T. P. (2001). Measures of human capital and nonlinearities in economic growth. *Journal of Economic Growth, 6,* 229–254.

Katzenstein, P. J. (1985). *Small states in world markets.* Ithaca, NY: Cornell University Press.

Kaufman, R., & Segura, A. (2001). Globalization, domestic politics and social spending in Latin America: A time-series cross-section analysis, 1973–1997. *World Politics, 53,* 553–587.

Levine, R., & Renelt, D. (1992). A sensitivity analysis of cross-country growth regressions. *American Economic Review, 82*(4), 942–963.

Levine, R., & Zervos, S. J. (1993). What we have learned about policy and growth from cross-country regressions? *American Economic Review, 83*(2), 426–430.

Lijphart, A. (1999). *Patterns of democracy: Government forms and performance in thirty-six countries.* New Haven, CT: Yale University Press.

Mankiw, N. G., Romer, D., & Weil, D. N. (1992). A contribution to the empirics of economic growth. *Quarterly Journal of Economics, 107,* 407–438.

Mayer, D. (2001a). The long-term impact of health on economic growth in Mexico, 1950–1995. *Journal of International Development, 13*(1), 123–126.

Mayer, D. (2001b). The long-term impact of health on economic growth in Latin America. *World Development, 29*(6), 1025–1033.

McGuire, J. W. (1995). Development policy and its determinants in East Asia and Latin America. *Journal of Public Policy, 14*(2), 205–242.

McGuire, J. W. (1999). Labor union strength and human development in East Asia and Latin America. *Studies in Comparative International Development*, *33*(4), 3–33.

Mostajo, R. (2000). *Gasto social y distribución del ingreso: Caracterización e impacto redistributivo en países seleccionados de América Latina y el Caribe.* Santiago, Chile: ECLAC.

Orenstein, M. A. (2001, October). *Mapping the diffusion of pension innovation.* Paper presented at the annual meeting of the American Political Science Association, San Francisco.

Petrei, H. (Ed.). (1996). *Distribuición del ingresso: El papel del gasto público social.* Santiago, Chile: ECLAC/CEPAL.

Pierson, P. (1996). The New Politics of the Welfare State. *World Politics*, *48*(2), 143–179.

Przeworski, A. (Ed.). (1995). *Sustainable democracy.* Cambridge, UK: Cambridge University Press.

Psacharopoulos, G., & Woodhall, M. (1985). *Education for development: An analysis of investment choices.* New York and London: Oxford University Press.

Rodrik, D. (1998). Globalisation, social conflict and economic growth. *World Economy*, *21*(2), 143–158.

Rodrik, D. (1999). *The new global economy and developing countries: Making openness work.* Washington, DC: Overseas Development Council.

Rodrik, D. (2001). ¿Por qué hay tanta inseguridad económica en América Latina? *Revista de la CEPAL*, *73*, 7–31.

Rudra, N. (2002). Globalization and the decline of the welfare state in less-developed countries. *International Organization*, *56*(2), 411–445.

Rudra, N., & Haggard, S. (2001). *Globalization, domestic politics, and welfare spending in the developing world.* Paper presented at the annual meeting of the American Political Science Association, San Francisco.

Sachs, J., & Malaney, P. (2002). Review article: The economic and social burden of malaria. *Nature*, *415*(6872), 680–686.

Schmitter, P. C., & Karl, T. L. (1991). What democracy is . . . and what is not? *Journal of Democracy*, *2*(1), 75–88.

Schultz, T. P. (1993). Investments in the schooling and health of women and men: Quantitites and returns. *Journal of Human Resources*, *28*(4), 694–734.

Schultz, T. W. (1961). Investment in human capital. *American Economic Review*, *61*, 1–17.

Schultz, T. W. (1989). Investing in people: Schooling in low income countries. *Economics of Education Review*, *8*(3), 219–223.

Stallings, B., & Wilson, P. (2000). *Growth, employment, and, equity: The impact of the economic reforms in Latin America, and the Caribbean.* Washington, DC: Brookings Institution Press.

Strauss, J., & Thomas, D. (1998). Health, nutrition, and economic development. *Journal of Economic Literature*, *36*(2), 766–817.

World Bank. (2000). *World development indicators* (CD-ROM). Washington, DC: Author.

Part IV

GLOBALIZATION'S OUTCOMES

Citizenship and Civil Society

12

Questioning Citizenship
in an "Age of Migration"

CAROLINE NAGEL

It has become a commonsense notion in "advanced capitalist" societies that this is an era of unprecedented human mobility, both in terms of the volume and the geographical diversity of migrants (Castles & Miller, 1998). While for some the societal diversity created by global migration is a cause for celebration, for many others the growing presence of "foreign" peoples gives rise to concerns about the ability of national societies and national citizenship to cope with and to accommodate cultural differences. It is ironic—but hardly surprising—that the regulation of migration has been placed so high on the political agenda of Western states at a time when boundaries to financial and trade flows are being systematically dismantled (Sassen, 1998). For many national governments, it seems, the enthusiasm for global economic interconnectedness and free trade is matched only by the dread of "floods" of foreigners.

Most public discourses about migration, of course, are surrounded by hyperbole. While some have termed the postwar period as the "age of migration" (Castles & Miller, 1998), recent historical analysis suggests that large-scale population mobility (in the form of displacement, dispossession, and dispersal) has been more the rule than the exception in the history of both Western and non-Western states (Held, McGrew,

Goldblatt, & Perraton, 1999; Sassen, 1999; Spellman, 2002). At the same time, what is almost as remarkable as the level of human mobility today is the level of nonmobility (Held et al., 1999). The sense of crisis that often surrounds the issue of migration belies the fact that only about 2–3% of the world's people are living outside of the country of their birth—the same proportion as the mid-1960s. In the United States, the proportion of foreign-born is smaller than it was in 1910; in Britain, the total foreign population stands at approximately 4%, and half of these are from the European Union or the EEA (Home Office, 2001). Still, it would be unfair to entirely dismiss concerns with migration as simple hyperbole. The effects of migration are unevenly felt and highly concentrated in a limited number of regions and cities (see, e.g., Waldinger, 2001). And regardless of the actual numbers involved, the *perception* of global migration as a problem for the modern nation-state has very real political consequences that cannot be ignored.

This chapter evaluates some of the key arguments coming from both academic and nonacademic literature about the nature of the contemporary migration "problem." In particular, this chapter focuses on claims about the role of migration and of migrants in transforming the nature of political membership and identity in the modern nation-state system. Such claims revolve around the concept of "citizenship"—a term that signifies both a legal relationship between individuals and the state and the more substantive integration of individuals into the political, economic, and cultural systems and institutions that comprise the modern nation-state (Brubaker, 1989). There are three major sets of arguments about citizenship and migration that I wish to evaluate. The first speaks of the unwillingness of the state to project a unifying vision of common nationhood on its inhabitants, and, as a consequence, the proliferation of identity-based movements among immigrants (Joppke, 1999a). The second and third arguments—which I will term "postnational" and "transnational," respectively[1]—perceive a more fundamental challenge to national citizenship posed by the forces of globalization. According to the postnational viewpoint, the importance of nationally based citizenship in the legal and substantive sense has diminished because of the formation of supranational human rights regimes (e.g., Jacobson, 1996; Soysal, 1994). The transnational viewpoint, meanwhile, contends that the transnational networks forged by migrants undermine the "traditional" model of citizenship (Mandaville, 1999). Both of these arguments suggest that the nation-state no longer serves as the main locus of political identity, community, or participation, and that we must look to other scales of analysis to understand political membership in a globalized world.

This chapter evaluates these three arguments and, in doing so, presents a more modest claim about the fate of contemporary citizenship.

Current analyses of multiculturalism, transnationalism, and postnational rights usefully point to the need to look beyond the nation-state to understand political identity and participation in an era of global migration. Yet it is difficult to speak in blanket terms about the inexorable transformation of nation-state-based citizenship when such transformations are partial, uneven, and experienced in very different ways. The incorporation of immigrants into the political community, this chapter argues, hinges on a multitude of shifting and ambivalent policies and political discourses that may operate unevenly even within a single nation-state. At the same time, experiences and understandings of citizenship may vary widely between and within immigrant groups, reflecting their variable positioning in the social, political, and economic systems of receiving societies. So while broader processes cannot be ignored, analyses of citizenship must account for the lack of uniformity and consistency in the way that these processes operate, and the different ways in which immigrants experience, practice, and assert their political and social membership.

THE CITIZENSHIP IDEAL

Turner (1990) notes that several distinct forms of citizenship have emerged historically in the West, reflecting varied understandings of the role of the state, civil society, and individuals in creating a political order. But while citizenship in practice has been very diverse, there is also a more unified, common understanding of what constitutes citizenship in Western liberal democracies. It is this highly idealized version of citizenship that many argue is being transformed, challenged, and/or undermined by forces related to mass migration.

This ideal of citizenship encompasses two competing political theories that have been instrumental in shaping the political order of the modern nation-state system. The first, liberalism, emphasizes the inherent rights of individuals and the role of the state in preserving and protecting these rights. The second, civic republicanism, while also holding the individual supreme, emphasizes the position of individuals within a collective. In contrast to liberalism, which envisions a minimalist state, this school of thought highlights the importance of the collective will, emotional attachment to the state, and the development of "civic virtue" and citizenship values through education and public service. Liberalism and civic republicanism historically have given rise to conflicting political agendas.[2] Nevertheless, commonsense understandings of citizenship in Western states encompass principles of both philosophies, and notions of individual liberty and collective will are reconciled with little difficulty in political discourse (Heater, 1999). In ideal terms, then, citizenship in the nation-state includes several key components: formal equality under the law, access to rights that derive from the state, a sense of duty and allegiance to the state and to the polity, and a common identity and purpose that serves as the basis for political participation.

In both modern political theory and in political discourse, citizenship has been regarded as a vehicle for greater inclusion in national societies (Joppke, 1999a). Highly influential in conceptualizing citizenship (especially in Great Britain and the United States) has been T. H. Marshall's (1950/1992) work, *Citizenship and Social Class*, which describes a gradual expansion of rights to include civil rights (i.e., basic legal rights and protections), political rights (i.e., the right to vote and access to political institutions), and by the mid-20th century, social rights (i.e., state-provided entitlements to ensure basic living standards). According to Marshall's model, this expansion of rights has been accompanied by the widening bounds of who is included in the polity, from the landowning classes to the middle classes and eventually to the working classes.

It is clear, however, that while political, civil, and social rights have been expanded and extended in contemporary liberal democracies, as described by Marshall, citizenship has always been and remains contested terrain. Indeed, Marshall has been criticized for ignoring the role of social struggles in expanding citizenship (Turner, 1990). Debates continuously erupt in liberal democracies about the ways in which the polity is defined and the nature of the rights available to its members. In the past few decades, many such debates have revolved around multiculturalism, a term that has been applied to and utilized by many groups, but that is most often discussed in relation to immigrants and their descendents. In the following section, I discuss multiculturalism as a political philosophy and as a tangible set of policies, and then explain the ways in which multiculturalism has been viewed as a challenge to citizenship and the modern nation-state.

IMMIGRATION, CITIZENSHIP, AND MULTICULTURALISM

In recent years, political theorists have been more inclined to think of citizenship not as a means of inclusion, à la Marshall, but as a means of persistent exclusion (Joppke, 1999a). Scholars of immigration have focused, in particular, on the processes of exclusion that emerge from the intersection between citizenship and nationalism. In so doing, they have drawn attention to the fundamental mismatch between the liberal ideals of equality and individualism that form the basis of citizenship, and notions of cultural exclusivity that form the basis of the nation (Baubock, 1992; Brubaker, 1989). Because of the historical merging of political community with national community, those who do not conform to dominant conceptions of cultural membership may be systematically excluded from political membership, including access to rights, political representation, and participation (Piper, 1998).

Scholarship on immigration to liberal democratic states is replete with examples of the ways in which members of immigrant groups are excluded from full, substantive participation in the polity. Germany, with its "ethnic" model of citizenship (which has recently been modified), has often been identified as an example of the conflation between cultural membership and political membership and the subsequent exclusion of "foreign" population from full rights and participation. But Germany is hardly a unique case, and even those countries with relatively undemanding naturalization policies (such as Great Britain) are often guilty of denying immigrants full social acceptance and access to rights (Kymlicka, 2001).

For some of those concerned with the exclusion of immigrants and other minorities, multiculturalism represents a means of severing the link between national ideologies and citizenship, and of allowing minorities to engage fully with the host society. Political theorists have proposed many different versions of multiculturalism, spelling out a variety of ways in which societies can respond to group-specific needs and redress group-specific forms of discrimination in a manner consistent with key tenets of liberal democracy (e.g., equal treatment under the law) (Carens, 2000; Kymlicka, 2001).[3] At the same time, multiculturalism has been embraced by political activists and translated into a multitude of distinctive policies that reflect particular social–political contexts. In some cases, multiculturalism involves an incremental, localized, and pragmatic approach to accommodating officially recognized minority groups, while in others cases it involves a more interventionist, top-down approach to protecting the cultural distinctiveness of groups (see Favell, 1997). But, in general, we can think of multiculturalism in theory and in practice as a rejection of the ideologies and processes of assimilation that have been integral to the creation of modern national societies. Multiculturalism, in other words, attempts to place minority groups on the same footing as members of the dominant cultural group by implementing policies to prevent or to redress discrimination and by tolerating, including, and/or actively promoting minority cultures (e.g., Parekh, 2000).

Multiculturalism and Identity Politics

Many criticisms have been leveled against multiculturalism as a philosophy and as a policy. Critics on the left have argued that multiculturalism represents a mindless celebration of diversity that essentializes group differences while it ignores broader structures of inequality and racism (Anthias & Yuval-Davis, 1992; see also Modood & May, 2001, for a review). Conservatives, meanwhile, contend that multiculturalism violates the "color-blind" principles of liberal democracy and foments cultural discord (D'Souza, 1998; Schlesinger, 1992). At the intersection of arguments from the left and the right are empirical claims about the role of multiculturalism in transforming national politics by encouraging immigrant groups and other minorities to adhere to particularistic identities and to press for recognition (rather than merely for inclusion). For better or for worse, multiculturalism, it seems, has made claims about "difference" an endemic feature of contemporary politics in immigrant-receiving contexts (Glazer, 1997; Isin & Wood, 1999). Joppke's (1999a) analysis of U.S. politics, for instance, suggests that multiculturalism has unleashed a rampant racial politics that counteracts the processes of integration. Joppke contends that while early waves of immigrants embraced "ethnic" identities, which facilitated assimilation, post-1965 immigrants accepted multicultural discourses and affirmative action policies and have embraced "racial" identities. The content of racial identities, Joppke argues, is "not a positive heritage transplanted into (and modified by) the receiving society. Its content is a negative one; its direction is not integration into a white majority but restitution for harm and guaranteed existence as a separate group" (p. 633). According to Joppke, multiculturalism, in this sense, provides an entirely new justification to maintain cultural differences, and, in fact, encourages these differences by increasing the political stakes involved in being recognized as a cultural minority.

Many of the concerns about multiculturalism as a political philosophy and as a state of being in contemporary nation-states reflect a common assumption about the nonassimilation and unassimilability of contemporary immigrants. Castles and Davidson (2000), for instance, contend that earlier waves of immigrants to core capitalist states were more ethnically similar to their host societies, and that the difficulties involved in communicating with their homelands meant that they had no choice but to assimilate to the dominant national culture. But while assimilation was an option for earlier generations of immigrants, Castles and Davidson assert, it has become untenable today because of the sheer volume of immigrants and the great cultural differences that, according to the authors, separate them from host societies. The claim that today's migrants are fundamentally different than past generations of migrants is echoed in many theoretical approaches, notably "segmented assimilation" scholarship (e.g., Portes & Zhou, 1993; Portes, 1996) and the literature on transnationalism.[4] Unlike generations past, today's immigrants, according to these perspectives, resist conformity and turn the political sphere into an arena for pressing group-based interests and claims. Thus, nonassimilating immigrants, in conjunction with multiculturalism, are said to disrupt the homogenizing project of the nation-state and, according to some accounts, threaten to undermine the social coherence necessary for the functioning of a political community.

Participants in these debates, regardless of their particular stance, consider multiculturalism a challenge to traditional conceptions of citizenship. But they generally regard the nation-state as the main "container" of citizenship, as the primary locus of political participation and rights, and as the focus of efforts to achieve social justice. Identity politics, in other words, represents a challenge to citizenship, not the end of nation-state sovereignty. In contrast, a number of recent theories of citizenship contend that the nation-state, in fact, is no longer the sole or even the main locus of political identity, participation, or community. These theories suggest the delinking of citizenship, identity, and political community from the territorial boundaries of nation-states due to forces operating "above" the nation-state—for example, new modes of global or "postnational" governance—and from "below"—for example, transnational migrant networks. These processes, as I will describe in the following section, are viewed as reconfiguring every dimension of citizenship, from political identity to rights and obligations to political participation.

THE END OF MODERN CITIZENSHIP?
Postnational Citizenship

Critical approaches to citizenship, as stated above, have long focused on the exclusionary nature of citizenship in the nation-state system. Of particular concern has been the presence of millions of noncitizens in immigrant-receiving societies, many of whom are socially and economically marginalized. Yet empirical evidence suggests that the willingness and capacity of a nation-state to exclude "outsiders" and to deny noncitizens access to social, political, and civil rights has been compromised by the development of a postnational regime of human rights (Jacobson, 1996; Sassen, 1998; Soysal, 1994, 2000). Even more than multiculturalism, human rights norms and discourses, as expressed in various supranational agreements, charters, and conventions, have had a profound impact on the way in which citizenship is structured and practiced in contemporary societies.

Membership statuses have proliferated in contemporary liberal democracies. "Society," it appears, is no longer coterminous with "the citizenry," as there are millions of individuals—many of them long-term residents or even second-generation settlers—who occupy positions between citizen and noncitizen. Yet, regardless of their particular status, individuals residing in contemporary nation-states have the capacity to participate in society and to access a multiplicity of social, civil, and political rights to an ever-greater degree (Miller, 1989; Baubock, 1992). In some instances, the rights of noncitizens can override the prerogative of states. For example, long-term noncitizen residents (i.e., "denizens") in some countries may exercise a right to reunite with members of their immediate family despite the prerogative of receiving societies to restrict immigration (Joppke, 1999b). With the extension of rights and entitlements to noncitizens and the relaxation of restrictions on the political participation of noncitizens in their country of settlement (as well as their country of origin), citizenship no longer appears to be predicated on legal membership in an individual nation-state, but on a broader system of human rights that operate supranationally. This process, many argue, renders the distinction between citizens and noncitizens less meaningful, and tests the entire basis of the nation-state and national citizenship (Jacobson, 1996; Schuck, 1998).

As with multiculturalism, normative evaluations of postnationalism vary widely. Some advocate a postnational system of citizenship as a way of overcoming the exclusionary tendencies of the nation-state system. More to the point, some conceptualize postnational membership as the only form of political membership that is tenable in a global era (for a review, see Turner, 2002; also see Castles, 2001). But some speak of postnational membership, and the decoupling of identity and rights that it implies, in more ambivalent terms. Soysal (2000), for instance, speaks of new forms of political claims making and the proliferation of sites in which people make and enact citizenship. Mirroring many of the arguments against multiculturalism, Soysal describes the increasing tendency of immigrants and minorities to press particularistic, group-based identity claims, which they then legitimate by appealing to universalistic discourses about the "right to be different." Providing examples of Muslim activists in Europe, she contends that the aims of contemporary immigrants are qualitatively different than the aims of civil rights movements of the past: while the latter was primarily interested in redefining individuals as part of the national collectivity, the former envisions participation in multiple civic spaces simultaneously. Similarly, Jacobson (1996) describes the "de-territorialization" of immigrants' political identity and participation, and, prefiguring arguments about immigrant transnationalism, asserts that immigrant loyalties are no longer necessarily situated in the nation-state in which they are resident.

Immigrant Transnationalism

Transnationalism, like postnationalism, suggests a diffusion of sovereignty away from the nation-state and a weakening of the link between political identity and participation and the territorial state. But the mechanism for this process, according to this perspective, lies not in postnational human rights discourses but in the political, economic, and cultural linkages that migrant groups maintain with their places of origin. Similar to arguments about nonassimilation described earlier, the idea of transnationalism suggests that in an era of rapid telecommunications and inexpensive long-distance travel, immigrants no longer make a definitive break with their homelands, but instead, construct "transnational social fields" through which they continue to participate in the social, economic, and political life of their place of origin (Basch, Glick, Schiller, & Szanton-Blanc, 1994; Vertovec, 2001). From this perspective, developed largely by anthropologists and some sociologists, issues of rights and entitlements take a back seat to concerns about the ways in which people organize their everyday lives across nation-state boundaries.

Transnational theorists describe a world in which identities are "multifocal" and political activities are "multilocal." Mandaville (1999), for instance, argues that individuals operate in a "new global market for political loyalties" and that people's loyalties and affiliations no longer correspond to the nation-state. Mandaville provides a variety of examples—"diasporic communities," global spiritual movements, and "border zones"—to illustrate the ways in which the politics of modernity have been radically transformed by transnational processes. Other transnational theorists are less insistent on the irrelevance of the nation-state. Itzigsohn (2000), for instance, shows that the nation-state itself is instrumental in generating transnationalism, a point he illustrates by explaining the Salvadoran government's efforts to maintain connections with Salvadoran immigrants in the United States (also see Guarnizo & Smith, 1998). But like Mandaville, Itzigsohn speaks of a growing disjuncture between one's place of residence and one's political identity and contends that the boundaries of political participation have expanded and multiplied.

The ever-expanding literature on transnationalism is replete with examples of the border-defying actions and identities of contemporary immigrants. Some scholars, such as Portes (1999) are cognizant of nativist fears about immigrant transnationalism. He contends that transnationalism does not pose a barrier to incorporation, but, in fact, gives immigrants the opportunities and sense of cultural pride necessary to integrate successfully into their host societies. But at the heart of this literature is a sense that contemporary migrants, unlike their forbears, defy the homogenizing project of the nation-state and disrupt the traditional trajectory from migrant to citizen (Clifford, 1994). Engaged in cultural, political, and economic exchanges that link together different sites, these individuals imagine themselves not as members of a single national community, as described by Benedict Anderson (1983/1991), but as members of multiple communities. Thus, while the nation-state is still with us, it no longer has the capacity to confine participation and identity within its borders.

DISCUSSION AND ANALYSIS

Multiculturalism, postnationalism, and transnationalism describe very different challenges to citizenship: the first signifies a decoupling of political membership from cultural belonging; the second, a decoupling of rights and entitlements from formal citizenship in the nation-state; and the third, a decoupling of political membership and cultural belonging from the territorial boundaries of the nation-state. Together, these three concepts render citizenship an increasingly perplexing idea. While most accounts reject the view that the nation-state has become irrelevant, proponents of these approaches suggest in different ways that the ability of states to exercise exclusive power over the terms of membership, access to rights, political participation, and identity has significantly diminished. At the same time, they suggest that the sites and scales of political membership and identity have multiplied, as have the types of claims being pressed in public arenas. The overriding sense is that the exclusive relationship between citizenship and territory—the "national order of things," as Soysal (2000) puts it—has broken down.

In this section, I wish to consider these major arguments in more detail, not so much to refute them as to qualify them and to indicate their limitations. To begin, I return to the "citizenship ideal"—the benchmark against which these various approaches measure transformations to modern citizenship. The three arguments I have reviewed here suggest that a variety of forces challenge or even undermine the modern citizenship ideal. But it should be emphasized that the "modern citizenship ideal" is just that—an ideal. Joppke's (1999b, p. 270) assessment of the postnational citizenship argument—that it "is premised on a colossus of 'national citizenship' that never was"—applies equally to arguments about transnationalism and multiculturalism.

The critical approaches to citizenship described above have long noted that the modern citizenship ideal has rarely been realized in practice, and that citizenship has always been a contested terrain. All three of these approaches, in this sense, exaggerate the ability of the nation-state historically to contain and to structure political identity, participation, loyalty, and rights. Claims about immigrant nonassimilation, dual loyalties, and predilections toward particularistic identities found in each of these theoretical approaches therefore should be approached cautiously, as they rest upon an assumption that the nation-building projects of liberal democracies previously exercised complete control over political identities and that they subsumed all modes of political engagement. Political participation and political identity—with respect to both immigrants and nonimmigrants—have always been enacted at multiple scales and in multiple sites. This point is made indirectly by critics who question the historical novelty of immigrant transnationalism.[5] It is also a point made by feminist political theorists, who have long argued that mainstream conceptions of politics privilege the nation-state at the expense of understanding the different scales at which power is imposed and contested (e.g., Staeheli, 1994).

Still the "National Order of Things"?

At the same time that we question whether citizenship, with its myriad meanings, has ever been subsumed entirely by "the national order of things," we must also consider the ways in which the nation-state remains a locus of political identity, participation, and rights. Recent accounts suggest that the neither the sovereignty of nation-states nor their capacity to structure the terms of citizenship has diminished significantly.

To begin, those concerned with immigrant rights have been quick to point out the limitations of postnational human rights regimes. Citizenship as a bundle of rights, for the most part, continues to be structured at the level of the nation-state. States in some instances do indeed defer to international law in treating individuals within their borders, and many states have incorporated international law into national law (Sassen, 1998; Joppke, 1999b). But it is states, rather than international bodies, that interpret and implement rights and decide the extent to which certain rights apply to certain groups. States also retain the power to designate legal categories and to parcel out rights and entitlements accordingly.[6] So while it is true that many immigrants lacking in formal citizenship are able to participate in the political, economic, and social life of the host society, it is also true that access to rights and privileges of membership is highly uneven and, in some cases, quite tenuous (Kofman, 2002; Morris, 2002).

The case of the United States since the mid-1990s illustrates the unevenness and reversibility of rights. "Legal resident aliens" (or "green-

card holders) in the past have enjoyed most of the legal rights and privileges of U.S. citizens with the main exception of voting rights. But since
the 1990s the U.S. government has moved to restrict the rights of legal
aliens, as well as their access to social welfare. In 1996, for instance, the
government adopted welfare reform measures denying resident aliens
certain forms of social security benefits to which they had previously
been entitled.[7] In that same year, the government also signed into law the
Anti-Terrorism and Effective Death Penalty Act, which widely expanded
the government's ability to deport immigrants thought to be linked to
terrorist groups, even if these individuals have not participated in any
criminal activity, and to prosecute permanent residents and citizens
supporting alleged "terrorist organizations." One of the most notorious
elements of the legislation concerns "secret evidence hearings," which
allow the government to "present evidence against suspected alien terrorists in secret, non-adversarial sessions in which the defendant is not
present and does not have a chance to respond to evidence presented by
the government" (Moore, 1999, p. 84). Such provisions, by explicitly
limiting the rights of both documented and undocumented aliens to appeal to constitutional protections, sharpen distinctions between citizens
and noncitizens and narrow conceptions of "civil community."

The narrow reading of the U.S. Constitution's due process provisions has had the most pronounced impact on Arab and Muslim immigrants, who, even before September 11, 2001, were the primary targets
of domestic antiterrorism activity. Arab immigrants have been named as
defendants in the major court cases that have emerged from the 1996
antiterrorism legislation—a notable example being the seven Palestinians
and one Kenyan national (often referred to as the "LA 8") who were initially arrested in 1987 for their activities on behalf of the Front for the
Liberation of Palestine. Not surprisingly, some of the important legal
battles surrounding the 1996 antiterrorism act have been initiated by the
American-Arab Anti-Discrimination Committee, a national Arab American civil rights organization.

The government's efforts to restrict the legal rights of immigrants,
and the targeting of Arab and Muslim immigrants, have become all the
more pronounced since the terrorist attacks of September 11. The U.S.
Patriot Act, legislation signed into law following the September 11 attacks, gives the attorney general wide powers to detain documented and
undocumented immigrants suspected of terrorist activities. It also mandates the tighter monitoring of foreign students and widens the power of
intelligence and law enforcement agencies to gather information and to
carry out surveillance on suspected terrorists. Interestingly, the bill begins with a lengthy section condemning discrimination against Arab or
Muslim Americans and reaffirming the positive contributions made by

Muslims and Arab immigrants to American life. But Arab American and Muslim civil rights groups, while in many instances eager to cooperate with the Justice Department, have not failed to notice that domestic antiterrorism measures, including airline profiling and the deportation of visa overstayers, have been directed primarily against their communities.[8] Attorney General Ashcroft's efforts to gain support for a new and even farther reaching antiterrorism bill, dubbed "Patriot II," are being actively opposed by Arab American and Muslim activists, as well as by civil libertarians.[9]

The Limits to Difference

The recent welfare and antiterrorism legislation in the United States, the world's largest recipient of immigrants and supposedly its beacon of freedom and democracy, indicates that a system of "postnational rights" has not been instituted uniformly or universally in immigrant-receiving countries. This case also indicates that restrictions on immigrant rights are driven by and reinforce hierarchical conceptions of social differences. In this respect, just as we must recognize the limits to postnational rights, we must also recognize the limits of multiculturalism and the celebration of "diversity" in contemporary liberal democracies. For all of the evidence of the increasing flexibility of national ideologies, as indicated by the widespread embrace of antidiscrimination policies and multiculturalist discourses, there is also a clear unwillingness to accommodate differences perceived to be fundamentally at odds with "national values" (Saggar, 1999).

The uneasiness with which governments have approached group differences has been most clear with respect to Muslim minorities, who, as intimated above, have been at the heart of ongoing debates in United States and elsewhere about the desirability of societal diversity (e.g., Dwyer & Meyer, 1995; Modood, 1994). Both a fear of Islam and a more condescending attitude reflecting the belief that Islamic values and practices are inferior to Western standards have been evident in numerous high-profile events over the past two decades, including the "Headscarves Affair" in France and the "Rushdie Affair" in Great Britain. Such events, along with the numerous physical and verbal attacks against Muslims in post-9/11 America, point to a current of sentiment within immigrant-receiving societies that Muslims are irreconcilably different from the "mainstream" and threatening to "French" or "British" or "American" values of tolerance and freedom (see Asad, 1990; de Wenden, 1991; Favell, 1997; Tlatli, 1996). Commenting on the negative sentiments toward Muslim immigrants, Brubaker (2001) notes the growing challenge to discourses of difference and diversity and the resurgence of discourses of assimilation in Western states.[10] The perception that multiculturalism, both as a political program and as an empirical reality, generates societal divisions and cultural extremism has led to calls for a renewed emphasis on national solidarity and citizenship responsibilities rather than group-based rights and entitlements.[11]

Understanding Immigrant Politics

The political participation of immigrant groups therefore takes place in an ideological context that is, at best, hesitant about the benefits of cultural diversity. As stated previously, the political identities and claims of contemporary immigrants have been described increasingly in terms of cultural particularism and/or transnationalism—both of which suggest the partial or nonassimilation and limited incorporation of immigrants into host society political and social systems. The view that immigrants cling to far-flung allegiances and remain wedded to homelands, however, seems as much a stereotype as an empirical observation. With rights and political inclusion in many instances hinging on full legal citizenship and on notions of cultural proximity, it is important to understand not only how immigrants preserve and express cultural differences but also how they position themselves as the "same" within the host society.

The case of Arab immigrants in the United States is again instructive. "Arab American" is a term that has come into wide use mainly since the early 1980s, though Arabic-speaking communities have existed in the United States since the late 19th century (Orfalea, 1988). This identity has been cultivated by many Arab-origin activists, some the second- and third-generation descendents of early, primarily Christian, settlers, and others part of the more recent and more heavily Muslim waves of Arab immigrants. Many Arab American organizations, such as the Palestine Solidarity Committee and the activist network Al-'Awda, have increasingly asserted themselves politically in an effort to gain support for Palestinian statehood, the Palestinian "right of return," and other "Arab causes." Some groups, as well, provide direct philanthropic assistance (particularly medical equipment, medicines, and scholarships) to their war-torn homelands. At the same time, many Arab American organizations have pressed for greater representation of Arab-origin communities in university ethnic studies departments and in other forums of "diversity" (see Saliba, 1999). But neither the assertive stance on Arab issues nor the cultivation of Arab American identities in the public sphere represents an unambiguous case of transnationalism or communal insularity. Arab Americans—whether activists or nonactivists— almost by necessity must respond to commonsense perceptions of Arabs and Muslims as prone to terrorism and fundamentally anti-American.

Reflecting this, political activities within Arab American communities tend to be accompanied by an insistence on *sameness* and on the compatibility and even equivalence between Arab/Muslim culture, identity, and interests and U.S. citizenship and values. Arab Americans, in other words, are engaged in a politics of assimilation as much as a politics of difference, and in a politics of "home" as much as a politics of "homeland."

To elaborate, an ongoing study of Arab American communities[12] reveals a striking pattern whereby activists describe themselves as multiple communities—referring to ethnicity, to countries or villages of birth, to religion, and/or to actual locale of residence—yet insist upon the primacy of their U.S. citizenship in terms of their sense of loyalty, duty, and allegiance. Further, assertions of various identities—Arab American, Muslim American, Palestinian American, and so on—tend to be couched within wider narratives of immigrant integration and assimilation. In a sense, the activists interviewed for this study appear eager to make "Arab American" a mainstream, assimilable identity, similar to more innocuous ethnic identities such as "Irish American" and "Italian American." The issuance of an Eid postal stamp (celebrating the end of Ramadan) and the recent decision by Hallmark to create "Eid Mubarak" greeting cards have important symbolic value for some Arab and Muslim American activists, not so much because they affirm cultural difference, but because they are seen to normalize the status of Arabs and Muslims in U.S. society. The activities of Arab American organizations, while very diverse, should be read at least in part as a play for *equivalence* and *inclusion* in national narratives and "mainstream" politics, rather than as an effort to remain separate, distinctive, and attached to homeland. Indeed, given the political climate in the United States and in the Arab world, this is a group that does not have the luxury of being either "different" or transnational.

Arab Americans, of course, face a distinctive set of opportunities, constraints, pressures, and motivations and their experiences do not match those of other immigrant groups. But while it may not be a representative case, it does indicate the following two points. First, transnationalism may be best understood as an uneven process both *between* and *within* immigrant groups, and one that is highly contingent on a variety of conditions in sending and receiving countries (a point made in some recent scholarship on transnationalism; see, e.g., Al-Ali, Black, & Koser, 2001). It should not be assumed that all contemporary immigrants are equally attached to homeland, or that questions of assimilation are not important to them. This leads to a second point that assertions of group difference may coexist with—and indeed, feed into—assertions of full citizenship within a national community (e.g., Karpathakis, 1999).

In this respect, it is useful to understand "assimilation" not as an obsolete model of immigrant adaptation, but as a conception of acceptance and belonging that is constantly negotiated between immigrants and nonimmigrants (Nagel, 2002).

CONCLUSIONS

Citizenship—along with the nation-state—has been regarded as one of the defining features of the modern world, and the prospect that it is being transformed and possibly diminished has led to a flurry of theoretical claims about the changing nature of political membership, participation, and identity. Current analyses suggest that the "traditional" model of citizenship today faces a postnational human rights regime that circumvents and devalues state-based citizenship, multicultural ideologies that encourage minorities to assert their cultural differences, and transnational linkages that facilitate multiple political memberships and that disrupt immigrant assimilation. It seems that we can no longer regard citizenship as a one-to-one relationship between individuals and the territorial state, a mode of community held together by common national identity, or as a territorially bound system of rights and obligations.

Claims about the transformation of citizenship rest on the notion that this is an era of unprecedented global migration in which territorial borders and national societies are radically transformed. But, contrary to popular belief, today's migrations and their demographic, economic, and political impacts are not unprecedented; nor are the ways in which various political actors (and academics) have construed migrations as crises to national sovereignty, identity, and citizenship. It has long been recognized that "bounded national societies" have never been truly bounded or static or self-enclosed, despite the best efforts of political leaders to militarize borders and to exclude undesirables from national space (Hall, 1990; Hobsbawm, 1990). It is therefore problematic to measure transformations of citizenship against a model that has never existed in practice. While we cannot ignore important changes taking place today with respect to citizenship—including the development of postnational rights, multicultural ideologies, and dual citizenship—we need to look beyond arguments about what citizenship was like before and what it is like now in order to understand the contemporary politics of citizenship.

The politics of citizenship (and, more generally, of immigration) reflect ongoing contradictions and tensions between inclusion and exclusion, fluidity and closure. For instance, the growing availability of rights for immigrants at the supranational level has been tempered by the revocation of rights at the national level, and the integration of multicultural principles in policies toward immigrants has been counteracted by the reassertion of exclusionary national ideologies. The purported transformations of citizenship, in this way, should be regarded as partial and uneven and far from straightforward, whether viewed from the perspective of nation-states or that of immigrants.

The contradictions and tensions within discourses and practices of citizenship, in part, reflect the fact that various components of citizenship are structured and practiced at different geographical scales by different social actors. Current theoretical frameworks, despite their problematic elements, have usefully drawn attention to the multiple scales at which rights may be allocated, political community created, and political membership enacted. The way in which citizenship politics unfold is inevitably contingent on factors operating nationally, transnationally, and within localities and municipalities. Many accounts of citizenship and immigration have considered national variations in citizenship and integration policies and how these impact immigrants (Favell, 1997; Joppke, 1999b; Piper, 1998; Soysal, 1994). Fewer studies, however, have considered the ways in which the politics of citizenship (e.g., the allocation of rights, the articulation of ideologies of membership and belonging, the formulation of policies of integration, etc.) emerge at different levels and intersect within local institutions and forums of participation (Staeheli, 1999). By examining such contingencies and recognizing the multiple modes of identity, belonging, and participation created therein, we can understand citizenship not as an ideal challenged by globalization, but as a contentious mode of articulating and defining political community.

NOTES

1. I use these terms with the recognition that they have different meanings and definitions in different disciplines—indeed, even in the same discipline. What I describe as "postnational," for instance, is often referred to as "transnational" in the literature (e.g., Baubock, 1992; Sassen, 1998). And what political theorists call "transnational" is often very different from what anthropologists call "transnational." I endeavor to define these terms carefully and to use them consistently throughout this chapter to avoid any confusion.

2. These conflicting agendas can be seen in the recent political history of the United States and Great Britain, both of which have experienced swings between neoliberal thinking (e.g., Thatcher and Reagan) and communitarian influences (e.g., Blair and Clinton) (Heater, 1999).

3. Some political theorists, it should be noted, have urged a more radical departure from liberalism, and have advocated different forms of group recognition as a means to undermine societal inequalities and oppression (e.g.,

Benhabib, 1996; Isin & Wood, 1999; Young, 1990). While these radical arguments do feed into public debates (especially with respect to their rejection of a hegemonic national identity), they do not have the same level of public salience as multiculturalism.

4. For rebuttals to the nonassimilation argument, see Waldinger and Perlmann (1998) and Alba (1995).

5. These critics have documented the existence of "transnational" (i.e., homeland-oriented) politics among previous generations of immigrants and refute the idea that present-day transnationalism is qualitatively different in form or content from that of the past (e.g., Kivisto, 2001).

6. Geddes (2000), for instance, shows that the interests of member states' decision making have been paramount with respect to immigration policy and with respect to the allocation of rights to third-country nationals in the European Union. There has been little political will until recently to break the link between formal citizenship and entitlements.

7. The 1996 Welfare Reform Act denied legal immigrants access to food stamps, supplemental security income and other assistance programs. It should also be noted that the 1996 welfare reforms were targeted not only against noncitizen immigrants, but also against citizens, whose use of welfare was limited to 5 years over one's lifetime. The impact of this legislation has been mitigated, in part, by increased state-level provision of welfare services to immigrants and by the reinstatement of some food stamp benefits. Nonetheless, this legislation, along with California's Proposition 187 and other anti-immigrant measures, succeeded in making clearer distinctions—however symbolic—between citizens and noncitizens.

8. The Bush administration, it should be noted, in September 2003 confirmed its intention to continue deportation proceedings against two of the "LA 8" under the provisions of the Patriot Act (Smith, 2003).

9. Opponents of Patriot II are concerned with proposed measures to significantly enhance the government's power to monitor the private communications of citizens and noncitizens and to detain individuals (again, citizens and noncitizens) suspected of "terrorist activities" in the absence of judicial review. Unlike the Patriot Act, Patriot II does not contain a sunset clause.

10. Favell's (1997) description of citizenship and integration policies in France provides an especially vivid example of the reaction against the politics of difference and the reassertion of assimilatory policies and rhetoric.

11. In Britain, traditionally lax naturalization requirements have given way to new policies to promote a more robust conception of citizenship. New naturalization procedures proposed by the home secretary will require that immigrants learn a British language (English, or, less realistically, Welsh or Scottish Gaelic), British history, and British governmental procedure. Also signaling a shift from its race-conscious (albeit ad hoc and incremental) brand of multiculturalism, the British government, in the wake of the 2001 "race riots" in the northern mill towns of Bradford, Burnley, and Oldham, has advocated a more active sense of citizenship in British society, with policy documents emphasizing concepts such as "civil renewal" and "social cohesion" (see Home Office, 2003).

12. This account is based on preliminary findings from research conducted with
Lynn Staeheli (University of Colorado), entitled "Community, Immigration,
and the Construction of Citizenship." The project, which is funded jointly
by the National Science Foundation and the Economic and Social Research
Council (UK), involves interviews and focus groups with Arab American
and British Arab activists and nonactivists in four U.S. cities (Los Angeles;
Washington, DC; San Francisco; and Detroit) and two British cities (London
and Sheffield).

REFERENCES

Al-Ali, N., Black, R. & Koser, K. (2001). The limits to "transnationalism":
Bosnian and Eritrean refugees in Europe as emerging transnational communities. *Ethnic and Racial Studies, 24*(4), 578–600.

Alba, R. (1995). Assimilation's quiet tide. *The Public Interest, 119,* 3–18.

Anderson, B. (1991). *Imagined communities* (2nd ed.). London: Verso. (Original
work published 1983)

Anthias, F., & Yuval-Davis, N. (1992). *Racialised boundaries: Race, nation, gender, colour, and class and the anti-racist struggle.* London and New York:
Routledge.

Asad, T. (1990). Multiculturalism and British identity in the wake of the Rushdie
affair. *Politics and Society, 18*(4), 454–480.

Basch, L., Glick Schiller, N., & Szanton-Blanc, C. (1994). *Nations unbound.* Am-
sterdam: Gordon & Breach.

Baubock, R. (1992). *Immigration and the boundaries of citizenship.* Coventry,
UK: Centre for Research in Ethnic Relations.

Benhabib, S. (Ed.). (1996). *Democracy and difference: Contesting the boundaries
of the political.* Princeton, NJ: Princeton University Press.

Brubaker, W. R. (1989). Introduction. In W. R. Brubaker (Ed.), *Immigration and
the politics of citizenship in Europe and North America* (pp. 1–27). Lanham,
MD, and London: University Press of America.

Brubaker, W. R. (2001). The return of assimilation?: Changing perspectives on immigration and its sequels in France, Germany, and the United States. *Ethnic
and Racial Studies, 24*(4), 531–548.

Carens, J. (2000). *Culture, citizenship and community: A contextual explanation
of justice as evenhandedness.* Oxford, UK: Oxford University Press.

Castles, S. (2001). Globalization and citizenship: An Australian dilemma. *Patterns
of Prejudice, 35*(1), 91–109.

Castles, S., & Davidson, A. (2000). *Citizenship and migration: Globalization and
the politics of belonging.* Basingstoke, UK: Macmillan.

Castles, S., & Miller, M. (1998). *The age of migration* (2nd ed.). Basingstoke, UK:
Macmillan.

Clifford, J. (1994). Diasporas. *Cultural Anthropology, 9*(3), 302–338.

de Wenden, C. W. (1991). Immigration policy and the issue of nationality. *Ethnic
and Racial Studies, 14*(3), 319–331.

D'Souza, D. (1998). *Illiberal education: The politics of race and sex on campus.* New York: Free Press.

Dwyer, C. & Meyer, A. (1995). The institutionalization of Islam in the Netherlands and in the UK: The case of Islamic Schools. *New Community, 21*(1), 37–54.

Favell, A. (1997). *Philosophies of integration: Immigration and the idea of citizenship in France and Britain.* Basingstoke, UK: Macmillan.

Geddes, A. (2000). *Immigration and European integration: Toward a fortress Europe?* Manchester, UK: Manchester University Press.

Glazer, N. (1997). *We are all multiculturalists now.* Cambridge, MA: Harvard University Press.

Guarnizo, L. E., & Smith, M. P. (1998). The locations of transnationalism. *Comparative Urban and Community Research, 6,* 3–34.

Hall, S. (1990). Cultural identity and diaspora. In J. Rutherford (Ed.), *Identity, community, culture, difference* (pp. 222–237). London: Lawrence & Wishart.

Heater, D. (1999). *What is citizenship?* Malden, MA: Polity Press/Blackwell.

Held, D., McGrew, A., Goldblatt, D. & Perraton, J. (1999). *Global transformations: Politics, economics and culture.* Cambridge, UK: Polity Press.

Hobsbawm, E. J. (1990). *Nations and nationalism since 1780.* Cambridge, UK: Cambridge University Press.

Home Office. (2003). Community and race website. Available online at *www.homeoffice.gov.uk/comrace/index.html.* Accessed September 15, 2003.

Home Office. (2001). *International migration and the UK: Recent patterns and trends* (Final Report to the Home Office, Research and Development Statistics). Available online at *www.homeoffice.gov.uk/rds/occ75sub.html.* Accessed February 2003.

Isin, E. F., & Wood, P. K. (1999). *Citizenship and identity.* London: Sage.

Itzigsohn, J. (2000). Immigration and the boundaries of citizenship: The institutions of immigrants" political transnationalism. *International Migration Review, 34*(4), 1126–1154.

Jacobson, D. (1996). *Rights across borders: Immigration and the decline of citizenship.* Baltimore: Johns Hopkins University Press.

Jacobson, M. F. (1998). *Whiteness of a different color: European immigrants and the alchemy of race.* Cambridge, MA: Harvard University Press

Joppke, C. (1999a). How immigration is changing citizenship: A comparative view. *Ethnic and Racial Studies, 22*(4), 629–652.

Joppke, C. (1999b). *Immigration and the nation-state.* Oxford, UK, and New York: Oxford University Press.

Karpathakis, A. (1999). Home society politics and immigrant political incorporation: The case of Greek immigrants in New York City. *International Migration Review, 33*(1), 55–78.

Kivisto, P. (2001). Theorizing transnational immigration: A critical review of current efforts. *Ethnic and Racial Studies, 24*(4), 549–577.

Kofman, E. (2002). Contemporary European migrations, civic stratification and citizenship. *Political Geography, 21*(8), 1035–1054.

Kymlicka, W. (2001). *Politics in the vernacular: Nationalism, multiculturalism, and citizenship.* Oxford, UK: Oxford University Press.

Mandaville, P. (1999). Territory and translocality: Discrepant idioms of political identity. *Millennium: Journal of International Studies, 28*(3), 653–673.

Marshall, T. H. (1992). *Citizenship and social class.* London: Pluto. (Original work published 1950)

Miller, M. (1989). Political participation and representation of noncitizens. In W. R. Brubaker (Ed.), *Immigration and the politics of citizenship in Europe and North America* (pp. 129–143). Lanham, MD, and London: University Press of America.

Modood, T. (1994). The end of the hegemony: The concept of "black" and British Asians. In J. Rex & B. Drury (Eds.), *Ethnic mobilization in a multicultural Europe* (pp. 87–96). Aldershot: Avebury.

Modood, T., & May, S. (2001). Multicultural education in Britain: An internally contested debate. *International Journal of Educational Research, 35*(ER3), 305–317.

Moore, K. M. (1999). Arabs and the American legal system: Cultural and political ramifications. In M. W. Suleiman (Ed.), *Arabs in America: Building a new future* (pp. 84–99). Philadelphia: Temple University Press.

Morris, L. (2002). Britain's asylum and immigration regime: The shifting contours of rights. *Journal of Ethnic and Racial Studies, 28,* 409–425.

Nagel, C. (2002). Constructing difference and sameness: The politics of assimilation in London's Arab communities. *Ethnic and Racial Studies, 25*(2), 258–287.

Orfalea, G. (1988). *Before the flames: A quest for the history of Arab Americans.* Austin: University of Texas Press.

Portes, A. (Ed.). (1996). *The new second generation.* New York: Russell Sage Foundation.

Portes, A. (1999). Conclusion: Towards a new world—The origins and effects of transnational activity. *Ethnic and Racial Studies, 22*(2), 463–480.

Portes, A., & Zhou, M. (1993). The new second generation: segmented assimilation and its variants. *Annals of the American Academy of Political and Social Sciences, 530,* 74–96.

Parekh, B. (2000). *The Parekh report* (Report of the Commission of the Future of Multiethnic Britain). London: Profile.

Piper, N. (1998). *Racism, nationalism and citizenship: Ethnic minorities in Britain and Germany.* Aldershot, UK: Ashgate.

Saggar, S. (1999). Immigration and minority policy debate in Britain: multicultural narratives contested. In A. Geddes & A. Favell (Eds.), *The politics of belonging: Migrants and minorities in contemporary Europe* (pp. 42–59). Aldershot, UK: Ashgate.

Saliba, T. (1999). Resisting invisibility: Arab Americans in academia and activism. In M. W. Suleiman (Ed.), *Arabs in America: Building a new future* (pp. 304–319). Philadelphia: Temple University Press.

Sassen, S. (1998). *Globalization and its discontents: Essays on the new mobility of people and money.* New York: New Press.

Sassen, S. (1999). *Guests and aliens.* New York: New Press.

Schlesinger, A. M. (1992). *The disuniting of America.* New York: Norton.

Schuck, P. (1998). *Citizens, strangers and in-betweens.* Boulder, CO: Westview Press.

Smith, R. J. (2003, September 23). Patriot Act used in 16–year-old deportation case. *Washington Post*, p. A03.

Soysal, Y. N. (1994). *The limits of citizenship: Migrants and postnational membership in Europe.* Chicago: University of Chicago Press.

Soysal, Y. N. (2000). Citizenship and identity: Living in diasporas in post-war Europe? *Ethnic and Racial Studies, 23*(1), 1–15.

Spellman, W. M. (2002). *The global community: Migration and the making of the modern world.* Stroud, UK: Sutton.

Staeheli, L. (1994). Restructuring citizenship in Pueblo, Colorado. *Environment and Planning A, 26,* 849–871.

Staeheli, L. (1999). Globalization and the scales of citizenship. *Geography Research Forum, 19,* 60–77.

Tlatli, S. (1996). French nationalism and the issue of North African immigration. In L. C. Brown & M. S. Gordon (Eds.), *Franco–Arab Encounters* (pp. 392–414). Beirut: American University of Beirut Press.

Turner, B. S. (1990). Outline of a theory of citizenship. *Sociology, 24*(2), 189–217.

Turner, B. S. (2002). Cosmopolitan virtue, globalization and patriotism. *Theory, Culture, and Society, 19*(1), 45–63.

Vertovec, S. (2001). Transnationalism and identity. *Journal of Ethnic and Migration Studies, 27*(4), 573–582.

Waldinger, R. (Ed.). (2001). *Strangers at the gates: New immigrants in urban America.* Berkeley and Los Angeles: University of California Press.

Waldinger, R., & Perlmann, J. (1998). Second generations: Past, present, future. *Journal of Ethnic and Migration Studies, 24*(1), 5–24.

Young, I. M. (1990). *Justice and the politics of difference.* Princeton, NJ: Princeton University Press.

13

Global Governmentality and Graduated Sovereignty
National Belonging among Migrants in Ecuador

VICTORIA A. LAWSON

Questions about the transformation of governance and national identity are being reexamined in the context of contemporary economic globalization (Castells, 2000; Ohmae, 1990; Herod, O'Tuathail, & Roberts, 1998; Sparke & Lawson, 2003; Sassen, 1998; Ong, 1999). Scholars are debating the ways in which globalization is reworking national identities through a shifting of economic governance away from "the territorially defined boundaries of the nation-state . . . [and into] "unbundled" space for which there is not yet a name" (Gupta, 1998, p. 321). Sassen (1998) argues that the rise of global governance without government has strengthened the claims of powerful actors (e.g., global corporations, global financial institutions, global treaties, etc.), opening vulnerable economies and pushing large numbers of people into casualized work and poverty. These processes raise unanswered questions about how the globalization of production and finance are reworking state sovereignty and political identities, as argued by Peterson (1996, p. 12): "identities conventionally "grounded" in state territoriality are losing ground to a politics of new, or even non space(s)."

253

In this chapter I examine the ways in which Ecuadorian state sovereignty is being reworked in the context of economic globalization and the adoption of neoliberal policies. There are two aspects to this reworking of sovereignty. First, I examine the rise of global governmentality— the increasing control of supranational institutions over macroeconomic governance of the country in recent decades. Second, in the context of the neoliberal policies enforced by these institutions, I explore changing state control over (and support for) particular people and places within Ecuador. I focus specifically on how supranational governance of the Ecuadorian economy has led to a deregulation of work, as the state has withdrawn from protecting wages, benefits, union rights, and so on. Instead, under economic globalization, international corporations, investors, and state managers have set the terms and conditions of work by demanding low-wage, nonunion workers. As Ong (1999) has argued in other contexts, the neoliberal state has abandoned its own poorer sectors to global circuits of capital and policy implementation.

I then link this examination of shifting scales and sites of sovereignty to an examination of national identity and belonging for poor migrants in Ecuador. As the consequence of neoliberal policies and a deep economic crisis, the Ecuadorian state has been unable to provide social rights and protections to all its citizens. In this context I pose questions about the spatial scales at which migrants locate the causes of their own economic distress and about the scales with which they most identify. In other words, do these migrants see themselves as workers in a global economy, as national citizens engaged with neoliberal modernization, and/or as ethnic- or rural-identified groups? Much work examining questions of national identity and belonging under globalization has emphasized questions of mobility, memory, and identity in diasporic communities (Eschle, 2001; Rouse, 1995; Cheah, 1998; Clifford, 1988). By contrast, I work with rural-to-urban migrants *within* Ecuador to emphasize how economic globalization reaches inside national territories and reconstitutes and reinvigorates preexisting social hierarchies and spatial identities (Alexander & Mohanty, 1997).[1]

I begin by arguing that there has been a shift of economic governance away from a territorially defined Ecuadorian state and into the unbundled spaces and institutions of globalization (such as the International Monetary Fund [IMF], the World Bank [WB], the Inter-American Development Bank [IDB], the Corporacion Andina de Fomento [CAF], and the Paris Club). I trace how diminished national control over the economy, but not outright erasure of the state, has reworked the relations between elite Ecuadorian state actors, supranational institutions and regimes of governance, and poor migrants within Ecuador.

Next, I argue that cross-border movements of capital have prompted large-scale migrations inside "adjusting" countries like Ecuador. Indeed, from the 1960s to the 1980s, investments in urban/industrial modernization and associated transformations of rural livelihoods have fomented substantial rural-to-urban migration (Delaunay, 1989). During the 1980s, when neoliberal openings to global markets began in earnest, some 1,120,000 persons migrated in Ecuador (approximately 10% of the total population) and 59% of these migrants went to the two largest cities, Guayaquil and Quito (International Research and Training Institute for the Advancement of Women, 1994).[2] Between 1985 and 1990, all of the highland provinces (except Pichincha, where Quito is located) had out-migration rates of between 20% and 39%. For example, two of every five people born in the province of Bolivar (located in the central Andes) migrated away (Luz Borrero & Vega Ugalde, 1995, p. 28).

While these poor rural-to-urban migrants do not have access to globalized circuits of capital and culture, they are nonetheless deeply affected by economic globalization (Cheah, 1998). As Ong (1999) argues, globalization is not really about deregulation of the economy and society, but rather about reregulation in favor of certain groups. Under neoliberal restructuring, weaker social groups—such as poor economic migrants—are essentially given over to deregulation by supranational entities that relegate many to unregulated work with little security and few rights (Ong, 1999; Mittelman, 2000). In Ecuador, state managers are reworking the discourses of national development that historically sought to incorporate popular classes. I draw on survey data and interviews to illustrate how this shift of governance has expanded flexible workforces, which deeply affects economic in-migrants to Quito.

Third, I look at how global governance is refracted through the economy and social policy in ways that reformulate the daily lives of migrants. I draw on migrants" experiences and interpretations of their lives in the city to pose questions about how globalization and insertion into supranational economic and political systems has reworked migrants" framing of their national (or regional, ethnic) identities, and how these reworkings intersect with preexisting social hierarchies of difference. The broad questions framing this study are (1) how has state sovereignty been reworked in the context of neoliberalism and economic globalization and (2) how do migrants frame their spatial and political identities in the context of state withdrawal from social reproduction?

METHODOLOGICAL CONTEXT

In making these points, I draw on interviews with economic migrants themselves to open an analytical space for critical interpretations/analyses of globalization which emanate from those people who are experiencing these dislocations firsthand.[3] I start from the standpoints of people who are impoverished under globalization to argue that spatial identities are being reshaped through profoundly material processes of migration and work. Interpretations from migrants themselves are often neglected in research, and yet they can counter and complicate Western, advanced economy analyses of economic globalization. Weiss's (1997) work on Ecuador clearly demonstrates that poorer sectors are sophisticated in their analyses of the causes of their poverty. Weiss found that people linked their poverty to the role of international organizations and the capitulation of successive governments to the global regime. Similarly, the migrants I spoke with analyzed neoliberal modernization as an unstable system, in which discrimination, uneven access to resources, and erasure of indigenous identities persists despite discourses of individual freedom (this view is elaborated in Lawson, 1999, 2000).

This insistence on bringing attention to the too-often neglected subjects, scales, and places of globalization at the "periphery" rather than at the "center" resonates with Escobar's (2001) call for a reconceptualization of globalization that pays close attention to the role of place and local knowledges. Escobar argues that other stories can be told, stories from marginalized places that emphasize differences rather than similarities, diversity rather than homogeneity. As Escobar notes,

> The erasure of place is a reflection of the asymmetry that exist[s] between the global and the local in much contemporary literature on globalization, in which the global is associated with space, capital, history and agency while the local, conversely, is linked to place, labor, and tradition—as well as with women, minorities, the poor, and one might add, local cultures. (2001, pp. 155–156)

This challenge to reconceptualize the places of globalization and to work against the erasure of other readings from women, minorities and those from the South is central to a feminist analysis of globalization (Nagar, Lawson, McDowell, & Hanson, 2002). It is crucial to challenge the claims of the center, of capital-centrism and Eurocentrism, that are the privileged, authoritative terrain of global-speak, as well as to insert into analyses of globalization a focus on the *connections between* migration, employment, reproduction, bodies, households, and communities.

GLOBAL GOVERNANCE

There is much debate over shifts in nation-state power in the context of globalization (Slater, 1992; Ong, 1999; Sassen, 1998; Gupta, 1998).

This has led to questions about "the significance of the national as a site of collective identity and the state as a force that can function to serve the collective interests of those who reside within the nation" (Bergeron, 2001, p. 3). Some argue that the power of global corporations and supranational institutions of economic regulation has "hollowed out" the state (Jessop, 1995; Scholte, 1996; Korten, 2001). This "hollowing" refers to a decline of national sovereignty in terms of managing the national economy and the abandonment of commitments to vulnerable sectors of the population. Other theorists argue that economic globalization has shifted the role of the state rather than erased it. They advocate analysis of the integrated nature of supranational and national governance emerging in the context of globalization (Ong, 1999; Glassman, 1999; Bergeron, 2001). Of course, these broad questions can only be answered in the analysis of specific places. My analysis traces the reworking of economic governance and the rise of both global governmentality (Gupta, 1998; Herod et al., 1998) and "graduated sovereignty" (Ong, 1999, p. 217) in Ecuador. Here I argue that global governmentality "refers to that ensemble of institutions, procedures and tactics that allow the exercise of a certain kind of power" (Gupta, 1998, p. 320; Foucault, 1991; Brown & Boyle, 2000) and that it is emerging in the unbundled spaces of globalization.[4] This has given rise to new forms of governance in vulnerable states like Ecuador that rework economic sovereignty as well as the relations between elites and poor people. In the following section, I elaborate on the idea of graduated sovereignty by tracing the ways in which state actors embrace neoliberal agendas and deregulate the workforce to illustrate how different sectors of the Ecuadorian population are differentially protected and valued in the context of globalization.

The Ecuadorian state has been steadily internationalized since at least the Alliance for Progress (Kofas, 2001). However, this process has intensified over the last 20 years in concert with a deepening crisis of indebtedness and a growing influence of the IMF and the WB in economic policy. State sovereignty over economic and social policy has been steadily eroded such that Ecuador is now in a vulnerable position within global networks of corporate and financial power. Today Ecuador carries one of the highest burdens of public external debt in Latin America (total debt is approximately $17 billion). This has dramatically reduced fiscal budgets for the state, leaving a tiny proportion for social sectors— in 1995, social security and welfare 1.9%; health 11.2%, and housing 0.6%. In 1999 the crisis deepened with several weeks of severe currency devaluation and a national banking crisis, followed by a steep rise in inflation and drastic erosion of real wages.[5] The combination of massive debt service payments and capital flight from the devastated economy means that the country *must have new loans to stay afloat.* Thus situation ensures that the cycle of debt and inescapable ties to the supranational sphere continue.

And yet the role of the state in neoliberal adjustment is complex. The state's role is not erased, but it is transformed into a regulator of networks and flows in the global economy. Harvey (2000, p. 65) observes that "to make the contemporary wave of neoliberalism work, the state has to penetrate even more deeply into certain segments of political–economic life and become in some ways even more interventionist than before." While Ecuador is clearly vulnerable and impoverished in the global context, the state is self-disciplining and governing itself within the material and discursive frame of neoliberalism. To understand these contradictions, I draw on the concept of governmentality to argue that Ecuador is enmeshed in a neoliberal regime of global governmentality which operates through transnational institutions and networks of power including free trade treaties such as GATT and NAFTA and institutions like the WTO, the WB, and the IMF. These treaties and institutions combine their considerable ideological and material force to rework the relations between states and the global economy. Vulnerable states such as Ecuador become enmeshed in supranational networks of power with which they actively engage, even as they lose sovereignty through that engagement.[6]

During the 1980s and 1990s Ecuadorian state actors have increasingly governed the country in the mentality of neoliberal institutions through their engagements with the IMF, the WB, the IDB, and the CAF. These institutions have constructed analyses of the causes of economic crisis across Southern nations and so have defined both the "problem" and the "solution" (Escobar, 1995). Ecuadorian state managers self-discipline the state, and so the country, in the context of this analysis. Specifically, in order to compete in global financial markets for the loans necessary to keep the economy afloat, Ecuadorian state actors themselves take the lead in reforming the national economy, in ways clearly prescribed by these institutions. Prior to receiving emergency loans, state managers commit to, and are often already implementing, a suite of reforms that bring the economy into line with the needs of global capital (reforms include opening their markets, privatizing state functions, further reducing fiscal expenditures, devaluing currency, etc.).

Processes of global governmentality help explain why, despite a succession of governments (seven presidents in the last 10 years) with different party and ideological positions, the process of internationalization has proceeded steadily. However, a full understanding of this inexorable reorientation toward global capitalist interests must also consider the class position of elite state actors. Specifically, within Ecuador, the architects of neoliberalization are Western-educated elites—bankers, merchants, and industrialists—whose interests align with global circuits of investment and finance rather than with poorer sectors (Conaghan, 1988; Corkill & Cubitt, 1988; Kofas, 2001; Glassman, 1999). Key state actors have espoused neoliberal discourse and enacted austerity policies in order to access resources from the international financial and development communities that are firmly committed to neoliberal restructuring. State actors have aligned themselves with global elites and have acted strategically to gain resources and to maintain legitimacy. As Dirlik (1998, p. 11) argues, globalization and developmentalism go forward with

> the complicity of Third World states, corporations, intellectuals and experts who are allowed increasingly to participate in the discourse and processes of development. The condition of their participation . . . is the *internalization of the knowledge and norms of the system.* (emphasis added)

The actions of Presidents Febres-Cordero (1984–1988) and Duran-Ballen (1992–1996) illustrate this complex mix of governmentality and elite agency. Among the series of presidents noted above, they were among the most aggressive architects of neoliberalization.[7]

The election of Leon Febres-Cordero in 1984 brought the onset of "Andean Reaganomics," and exemplifies the gulf between the executive branch and the majority of Ecuadorians. As Conaghan (1988, p. 120) argues, "from the perspective of the bourgoisie, Febres-Cordero was truly one of their own." A millionaire industrialist, Febres-Cordero came from a powerful family. His presidency was a key moment for industrialists to reestablish control over the state after the military reformist period of the 1970s. At that time Ecuador was heavily dependent upon oil exports, which represented 63% of merchandise exports and over 60% of central government budget revenue (*Business America*, 1986). Febres-Cordero ran on a platform of populism that was coupled with "market-friendly" reforms in response to declining export revenues, increasing economic instability, and international pressures.

> Mr. Cordero . . . raised hopes with slogans that included "bread, roofs and jobs." However, the collapse of world oil prices in 1986 seriously hurt the country's finances, prompting the government to drastically reduce its budget, freeze wages and eliminate price controls on many basic goods. The government says its policies have brought inflation down to 20% from 40% two years ago. . . . Mr. Cordero's political foes accused the government of *putting the people at the mercy of the marketplace.* (Latin American Index, 1986, p. 67; emphasis added)

As oil prices steadily declined, the government took on additional debt to keep the economy moving. Febres-Cordero free-floated the *sucre*, rolled back tariffs for domestic industries, and complied with U.S. government demands for liberalized investment rules (Corkill & Cubitt, 1988). A key element of these adjustments was sucretization of privately held debt. The state had originally assumed these loans as public debt. As Acosta (1990, p. 306) noted, this was *the gift of the century for private industrialists*. The result of this "gift" to private elites has been a fiscal crisis for the state with widespread punishing ramifications for popular classes as social subsidies, health care, and education spending were drastically reduced. This move of private debt into the public sphere was one of the most explicit moments of an ongoing process of reorienting the Ecuadorian state away from developmentalism and toward domestic and international capitalist interests. By the end of Febres-Cordero's presidency, Ecuador had embraced liberalization, coupled with extensive new loans. The country's total external debt had risen to $8.6 billion and debt service was 31% of export revenues.

By 1992 Ecuador was deeply embroiled in the internationalization of finance capital through loans and debt service payments. The newly elected Sixto Duran-Ballen has been a right-wing player in the Conservative and the Social Christian Parties since the 1970s and was a key architect of a united electoral front for the Right (Frente de Reconstruccion Nacional), which was a strategic response to the defeat of rightist candidates in 1979. Duran-Ballen, working closely with the IMF, moved rapidly to implement drastic structural reforms to the Ecuadorian economy (Lind, 2000). This shock package was designed to "cool" the economy and to further liberalize trade and finance to bring in new revenues. It resulted in a series of policies that favored elite financial and commercial interests and had dramatic and immediate effects on daily life for poor sectors.[8] This aggressive structural adjustment program (SAP) led to over 50% reduction in real wages and purchasing power, a dramatic decline in standards of living for many, and increased domestic burdens as women and families took on various tasks of social reproduction (such as community organizing, schooling, and medical care) previously provided by the state. At the end of this two-decade period, SAPRIN (Structural Adjustment Participatory Review International Network) released its *Report on Ecuador*, noting that the social impacts of neoliberalism were severe.

> Research on the impact of adjustment in Ecuador since 1982 concluded that trade and financial-sector liberalization in Ecuador have led to marked contraction in the national productive apparatus . . . as well as a greater concentration of productive resources. This, in turn, has in-

creased unemployment and underemployment while, along with labor market "flexibilization" policies, reducing job security. The lack of adequate, stable employment and the further concentration of wealth have generated an increase in poverty and a deterioration the living conditions of a majority of the Ecuadorian population. (SAPRIN, 2001, p. 3)

The internationalization of the Ecuadorian state under the influence of global governance focuses elite state actors toward global capital. This process of neoliberalization embeds poor communities in global regimes that emphasize deregulation of labor markets and a radically reduced state role in social policy.

GRADUATED SOVEREIGNTY AND FLEXIBLE WORK

> . . . the poor are excluded, they are not citizens.
> —SHORRIS (2001, p. 5)

These two intertwined processes of (1) deregulation of the workforce to compete for foreign investment and (2) social disinvestment due to severe fiscal crisis of the state impact poor migrants in particular ways. In this section, I look at the ways in which domestic economic migrants are embedded in global regimes of governance through the restructuring of labor regulations and economic crisis. As Ong argues,

> Globalization has induced a situation of *graduated sovereignty,* whereby even as the state maintains control over its territory, it . . . let[s] corporate entities set the terms for constituting and regulating some domains. . . . *Weaker and less desirable groups* are given over to the regulation of supranational entities. (1999, p. 217; emphasis added)

I focus on migrants because they are economically vulnerable under neoliberalization. I examine the terms of their insertion into the globalized economy as reflected in their labor force participation.

I start from migrants' experiences in work, because this reveals the ways in which poor migrants are directly incorporated into the globalized economy. This is important, because we know very little about whether those subjects working in globalized and yet devalorized sectors connect the global regimes of regulation that define their conditions of work with the scales at which they frame their identities. I draw this approach from Sassen (1998) and Hartsock (2001), who argue that understanding the situation of immigrant women is a strategic move in understanding the dynamics of global capitalism. Sassen argues that highly specialized finance and trading activities of the internationalized sector of the economy impose new pricing criteria (e.g., on rents, on business services, on certain workers) such that low value-added sectors of economic activity cannot compete with the extremely high profit-making capacities of globalized firms. This results in a crisis of profitability for domestic manufacturing and service activities, resulting in diverse cost-cutting strategies. One such strategy is to shift to part-time, unregulated jobs carried out by undervalued populations such as immigrant women. What is important here is that this casualized work is not excluded from global circuits, but rather *underwrites and constitutes globalization.*

Poor domestic migrants within Ecuador are a similarly vulnerable group. The deregulation of the labor market in Quito is both a direct outgrowth of explicit state policies and of broader processes that devalorize domestic manufacturing and service activities in the context of globalization. Through flexible work, the Ecuadorian state cheapens labor costs for business and reduces costly bureaucracies involved in regulating and protecting workers. I argue here that deregulation of the labor market by state actors explicitly links poor migrants to globalized circuits. I make the case here that elite state actors have devalorized many sectors and jobs in which migrants work. In the next section I trace out how migrants experiences in the job market shape their sense of belonging, alienation, and discrimination.

From the 1950s to the early 1980s a series of nationalist, developmentalist governments enacted protectionist economics and socially redistributive policies to incorporate *both* elites and popular sectors into a national modernization program (Lawson, 1995; Corkill & Cubitt, 1988; Schodt, 1987; Conaghan, 1988). With the rise of global governmentality, we see state managers break with nationalist development agendas. State managers themselves have internalized the discourses of neoliberalism that lead to reduced social spending and that marginalized workers. I saw this shift in the early 1990s in a series of interviews with ministers in the Duran-Ballen government (1992–1996). As noted in the previous section, Duran-Ballen implemented an economic shock package, involving drastic structural reforms to the Ecuadorian economy. It was in this context that the minister of industry told me:

> "I am convinced that foreign capital helps national industrial development. *The nationalistic mentality doesn't make sense* . . . [nor does] the idea that we have to take a great deal of care with foreign capital, that they do us harm, these foreign companies. . . . We have to give them guarantees so that their investments here are profitable. They are bringing employment. . . . *We have to open the country up more to foreign capital. We have to be more competitive with other*

countries in order to attract foreign capital." (personal interview, 1992; emphasis added)

This official explicitly discredits a nationalist development strategy and argues for a global perspective. *A central element of attracting foreign investment and competing against other countries in global markets is deregulation of the workforce.* This is underscored during my interview with the minister of labor.

"All of the actions we have taken lately . . . are specifically directed at lowering the temperature [conflict between employers, unions, and workers] *to liberate us from the fossilized traditional arrangements and open ourselves to new perspectives* in accordance with the current world situation." (personal interview, 1992; emphasis added)

Similarly, the vice-president of the Chamber of Industry explained:

"Our priorities, which we are discussing with the government are first *flexibilization of labor laws* . . . another closely related issue is the treatment of foreign capital. We have to facilitate the entry of foreign capital to a much greater extent." (personal interview, 1992; emphasis added)

"Flexible work" translates into rolling back union power, reducing the cost of labor by holding down real wages and eliminating benefits, and opening the market to foreign competition, thereby destroying domestic production—which cannot compete (Lawson, 1995, 1999). This withdrawal of the internationalized state from regulating the labor market essentially "results i[n] a system of variegated citizenship in which populations subjected to different regimes of value enjoy different kinds of rights, discipline, caring and security" (Ong, 1999, p. 217). As the state withdraws from protecting family wage jobs, international corporations and investors, and state managers, set the terms and conditions of work by demanding low-wage, nonunion workers.

This struggle for global financial solvency and competitiveness through deregulation of the Ecuadorian economy simultaneously dislocated thousands of economic migrants. These are primarily economic migrations, in the context of severe poverty rates, exacerbated by the faltering economy and withdrawal of the state from social programs. In 1994, 35% of the population (4 million people) lived in poverty (measured as the inability to purchase a minimum basket of food and nonfood goods), with 60% of the total poor living in rural areas (World

Bank, 1996). Poverty has continued to increase dramatically throughout the 1990s. At the end of the 1990s, 69% of the rural population was living in poverty (compared to 60% in 1995). The number of people living in extreme poverty—meaning that they cannot afford their basic food needs—jumped from 17% of the population in 1995 to over 34% of the population in 1999 (World Bank, 2000). These dynamics ensure a continuing flow of poor migrants into the cities in search of paid work.

The combination of "flexibilization" of work, high poverty rates, and substantial rural-to-urban migration flows find their expression in rising levels of informal work. Although notoriously difficult to measure precisely, estimates suggest that informal work is on the rise in Ecuador since the 1980s. Teltscher (1993, p. 64) estimates that the informal sector constituted 28.6% of the labor force in 1980 and that this number rose to 57% by 1990. Guasch's (1999) research produces similar numbers, estimating that 51% of the Ecuadorian labor force worked informally in 1990 and 54% by 1999. In our 1996 survey of migrants and long-term residents in Quito (see note 1) we also found strong evidence of widespread casualization of work, with 78% of our respondents working without a written contract and 89% without union representation. All the migrants we spoke with work in devalorized economic activities, with the majority working in either construction (12.9%), vending (24%), or services (44%—including domestics). Employment volatility also emerged as a serious problem in the early 1990s, coincident with the drastic neoliberal austerity measures of the Duran-Ballen government. Respondents reported relative job stability during the 1980s, whereas job loss/change increased dramatically in the 1990s.

Table 13.1 illustrates departures of migrants and long-term residents from their first job in Quito; losses increased from close to zero in the 1980s, to 19% of migrants reporting job loss in 1992, to 25% reporting job loss in 1995. When asked about the reasons for leaving their first job in Quito, 47.5% of migrants experienced job loss or incomes too low to support their families (34% of long-term residents reported these same reasons). These trends, combined with continuing in-migration, intensify competition for jobs as the ranks of the underemployed swell and real incomes sharply decline. Poor migrants embody this flexible, devalorized workforce enacted by the state in the context of global governance.

MIGRATION, BELONGING, AND EXCLUSION

In this concluding section, I begin from the standpoints of those marginalized and deregulated by global governance and the internation-

TABLE 13.1 Year in Which Migrants and Long-Term Residents Left Their First Jobs

Year	Migrants (n = 226)	Long-term residents (n = 706)
1984	0%	0.3% (1)
1985	0%	0%
1986	0%	0%
1987	0%	0%
1988	0%	0.3% (1)
1989	1.7% (1)	0%
1990	8.0% (8)	8.5% (29)
1991	10.0% (10)	9.0% (31)
1992	19.0% (19)	18.4% (63)
1993	16.0% (16)	14.3% (49)
1994	10.0% (10)	15.2% (52)
1995	25.0% (25)	18.7% (64)
1996	11.0% (11)	15.5% (53)
Total	100% (100)	100% (343)

Note. 1996 Migrant Survey Data compiled by V. Lawson and K. Van Eyck. Migrants moved to Quito after 1980. Both groups started their first jobs in or after 1980.

alized Ecuadorian state. I draw on interviews with recent migrants to Quito to explore the scales at which they locate the causes of economic crisis, the scale and spatial reach of their own networks, and the places with which they most identify. By starting from the insights of those excluded from much globalization research, I work against the silencing and erasure of poorer subjects. Their interpretations capture the hybridity and complexity of their discourses of accommodation and resistance, of ambivalence and dissent. While in some senses these migrants engage the discourses and promises of modernity and have embraced the national project of economic modernization, at the same time some express a strong sense of vulnerability and exclusion from the national project. Their words bring visibility to neglected subjects of economic globalization and reveal the often obscured social costs of neoliberal modernization (Nagar et al., 2002).

In 1997, one year after the initial survey, I returned to Quito and identified a subsample (n = 50) of migrants from our 1996 survey with whom we attempted to renew contact. A great many of these migrants no longer lived in the same place, suggesting that migrants move a great deal within the city after their arrival. I did locate and interview 20 individuals who had completed the original survey (described in note 3). These migrants are working-class and poor migrants of mestizo and indigenous ethnicities. The following quotes are drawn from five of those interviews. I draw on these interviews not as a basis for generalization, but rather to suggest some of the ways that economic migrants frame a sense of belonging in the context of crisis.

Spaces and Scales of Economic Crisis

Despite severe hardships, overwhelmingly migrants thought that Ecuador should continue on its path of economic modernization. Several people said that modernization would bring more well-paid jobs, especially industrial jobs. Migrants located their economic difficulties at the national scale: they repeatedly identified the problem of corrupt politicians who never do anything for the people, but always help themselves, always put resources in their own pockets. Very few people linked their increasing poverty to the global scale; issues such as debt, austerity policies, or foreign influence in the economy were rarely mentioned. Indeed, Luis argued that more privatization would solve the problem of corruption, which he argues is holding the country back.

Luis is a 25-year-old who migrated to Quito with his wife and young daughter. He works long hours as a security guard in a wealthy neighborhood, a 2-hours' commute from his cinder-block, two-room home. We asked Luis whether he would like to see Ecuador continue to modernize, and he replied:

> "The government must create more jobs to help the migrants, in this sense they must not abandon us. There are many in-migrants, but the government doesn't help them, and because of this there is a great deal of poverty in Quito.
>
> "Privatization is good for the country, because at the moment it is the same rich people who are making themselves even richer. . . . By contrast, if we had privatization of companies, then there would be more jobs for more people. Then we could share the wealth with the rest."

He argues that the poor are excluded and that elites are enriching themselves. His focus is on the national dimensions of economic crisis, with no sense of the global context.

Manuel is 35-years-old, married with six children. He came to Quito from a small rural village in the highlands. Manuel's first language is Quechua and he is of indigenous descent. When the family first came to Quito, Manuel had a good job in construction, but his job was eliminated and he is now selling vegetables in a market for a very low income. I asked what he thought about the process of economic modernization. He responded:

"There is no way to go even a single day without a job. When you have no work, when you go a week without work, then you have nothing to give your kids, nothing, and there is no help for you, nothing. Simply put, between the two of us, we struggle every day, you can't be without work here."

He eloquently describes the family's economic vulnerability, his inability to feed his family and the lack of any support for the poor. As we continued, I asked Manuel "Should the country continue to modernize?" He said:

"In Quito, everybody thinks only about themselves, including the politicians. . . . Everything is getting more expensive, *the politicians are working for themselves*, and they don't even want to know about the poor people."

The problems are with the national government and with corrupt politicians—corrupt government is holding back economic development for everyone.

"For those who need a job, it should be like this. Come on in, here is a job. But when a poor person presents an application or asks about a job, then the authorities should say come on in. They don't say this, the politicians never say this."

Economic crisis has left many people without work, and so the competition for jobs is intense in the city. In this context, crisis intensifies employment discrimination. Manuel, a dark-complected indigenous man, has been rejected for a series of jobs and he feels unprotected and abandoned by the state. He refers to his experiences of discrimination:

"[The government] should help us here, and that would improve our situation. . . . The government should help us, should think about us, should make an effort to help, then it would be great if the people here [in Quito] would treat us with understanding and kindness, it would be wonderful if they would help their citizens!"

Manuel consistently speaks to a sense of exclusion and a sense that the poor are being abandoned. Despite their poverty, many of the migrants I spoke with continue to embrace the imaginary of national economic modernization, while simultaneously locating their problems in the failures of the state to incorporate them. Nonetheless, they express a desire to belong to the national modernization project. It is notable, if unsurprising, that few people connect the national crisis of governance and economic decline to global or external processes.

However, there are exceptions. Some migrants were aware of the global dimensions to the crisis, and in that context they expressed a strong sense of national identity. Gloria is a young mestizo woman with two children. She and her husband came to Quito in search of steady work. Her husband has a family-wage factory job in Quito and this was the reason they migrated—they only had irregular work where they lived before. We discussed the lack of education and health care, and she commented:

> "It isn't the fault of the country, it is our [the people's] fault. We all want to earn more money and have more services, but *the country has foreign debts.* The people keep demanding more and more from the country. People keep demanding more from the country, more from the country, but I say that we should all try to help the country so that things can improve more and more."

José also holds an industrial job. Gloria and José represented the only households, among those we interviewed in 1997, where a member still held a formal job with a written contract and a family wage. Jose also located the economic crisis in the context of capital flight and a lack of investment in Ecuador:

> "We could modernize the country if we had the credit and investments to do so. Then there would be better employment opportunities. What I hear is that there is capital, but that it leaves the country [literally: flies away]. I don't know what happens, why the investors don't have confidence, but they could, they could invest here because there are possibilities, there are opportunities. This is why we keep getting further behind, we should believe in ourselves, invest, and work, and then we would get ahead."

These quotes suggest that incorporation into the urban economy, and experiences of poverty, may be key components of how spatial identities are reworked under globalization. Those who identified strongly with national modernization and located the crisis in the global scale— as problems of debt and capital flight—are migrants who have stable, relatively well-paid work. In contrast, those who are very poor, barely surviving in unregulated work, and who face discrimination in Quito express a sense of exclusion and abandonment. These migrants focused primarily on failings of the government and their sense of exclusion from the national modernization project (there were no differences in levels of education reported by these various respondents). It is ironic that those who have faced the brunt of deregulation, and who are especially vulnerable with the rise of flexible work, focus their critique on the national scale, while the global dimensions of the crisis remain obscured. Those whose unregulated labor underwrites and constitutes economic globalization simultaneously recognize the problem and situate it only at the national level. If those most vulnerable feel excluded from the nation as citizens, and do not articulate a sense of the global forces shaping the crisis, then what is the spatial register with which they identify?

Spaces/Places of Refuge and Identification

> There are no unmediated national identities—any identifications with the national are mediated profoundly by local/regional affiliations (anything from neighborhood, village, district to region, although *less at the supranational level*).
> —RADCLIFFE AND WESTWOOD (1996, p. 132; emphasis added)

A persistent theme in the interviews was a desire/nostalgia for origin places, which were frequently rural villages. This nostalgia springs from a combination of economic hardship and discrimination, and, for women, a sense of constrained movement, isolation, and lack of social supports. These hardships, which are intensified by globalization, seem to have exaggerated regional differences and identifications. The larger context of economic crisis has deepened poverty and the scarcity of paid work, and has simultaneously intensified discrimination in the city. Feelings of exclusion and difference expressed by the migrants translate into strong regional and ethnic identifications, which works against any sense of national belonging or a coherent critique of structural adjustment. This phenomenon of rural nostalgia is important in at least two ways. I focus on the ways in which migrants articulate strong regional identifications as a positive source of identity and personal valuation in the face of material hardship and alienation. It may also be that migrants are actually returning to their places of origin, that there is a process of reruralization in response to crisis in Quito. However, my research does not capture this second possibility because the research was conducted with migrants still resident in Quito.

Maria lives in Quito with her husband and young child. She came from small rural community on the coast. She is a dark-complected mestiza woman. Maria's husband is in the military. They live in a one-room rental in a boarding house. I asked Maria how her life in Quito had turned out and about the place where she would prefer to live. She said that it hasn't worked out so well and that at times she feels very lonely. She explained their poverty:

> "The economy is very difficult, and here [in Quito] you have to buy everything. Absolutely everything. Back on the coast, if you wanted an orange, you could just go and get one from a tree. By contrast here, if you don't work, you don't have anything.
>
> "Here, as much as my husband works, it isn't the same [as back home], I mean, both of us would have to work for us to live better, as much as he is working, we would both have to work."

Maria has a young child and so it would be very difficult for her to work, but she has been looking for something. Her husband's salary is very low, and they are barely surviving in Quito. Economic liberalization, overseen by the IMF and the WB, has rolled back social subsidies (on cooking gas, food, transportation, etc.), which in turn has increased the cost of urban living dramatically. Without access to work and a living wage, the poor cannot participate in neoliberal modernization. I asked Maria where she would prefer to live. She replied:

> "Well, I prefer the countryside, you see, when I was single I used to go an help my uncle harvest rice and beans and he would give me food in return. You see, I prefer the countryside, but my husband works here. So we can't live back there, but I prefer the countryside."

Maria links this preference for the countryside with their economic difficulties first. But she also talks about her lack of social networks that once helped her with domestic work, and also painful experiences of discrimination as a dark woman of coastal origins.

> "I was together with my sisters and if one of us didn't do the chores, another did it. . . . We also used to go out to dances. Here that just stopped. I can't go out alone at night. You would have to organize a special group to go out, you would have to have a car, because you can't walk [in Quito].
>
> "Here in Quito, in order to get a job you must be Quitena . . . you have to be from here. . . . Yes, there are people who call us "monkeys', that's what they call people from the coast, yes, that's what they call us, they call us 'monkeys,' it is an insult. . . . When I go for a job they say, we have already found a 'girl,' someone from Quito. This is ugly and it hurts."

Maria expressed her sense of painful loneliness, isolation, and discrimination in Quito. She longs to return to her coastal community. She reveals the ways in which a "flexible" workforce, enacted by the state to compete for global capital, plays out on the bodies of migrants, reinscribing preexisting social hierarchies of difference. Gender, ethnic difference, and poverty mark who is excluded under "neoliberalization."

Manuel also remains very identified with his origin community. He came from a small village in the highlands of the Sierra, where his family has a small farm. He has worked in a number of cities since the age of 12, but he still goes home every month to see his mother and to spend time in his home community. He made it clear to us that he has not abandoned his home. Even though he expects his children to live in Quito, he is teaching them about indigenous culture and to speak Quechua. In the previous section, Manuel expressed dismay at the economic crisis and his sense of exclusion from jobs, lack of government support, or services. Manuel was very explicit about his dual identities and dual lives. I asked where he would prefer to live. He responded, "Well, well, at the moment I am young and I can work here [in Quito], but, well, my life is really much more back there [in Riobamba]." He explained that he is in Quito to build a life for his children and that

" . . . my children are baptized and their godparents are here [in Quito] as well as everything else, but for myself, I can't say that I am going to live back there or that I will live here, rather I will live in both places.

"I am here because my children are here, until they are grown up. But for me, I am losing my spiritual identity here in Quito, so for me it is back there. . . . When I die, my family cannot leave me here, my body belongs back there on my land. This is how it is. I would simply lose myself here."

In conclusion, it was common among the migrants I interviewed to hold onto the possibility of return, and to insist on a rural/regional identity even in the face of discrimination and poverty in Quito. It is notable that these are asserted identities as much as they are material practices. Most of the migrants in this research do not receive any sort of material support (income, food, etc.) from their origin places, nor do they typically express an intent to return permanently. Rather, they retreat into a rural/regional identity as a positive force in their urban-based lives. This resonates with Mills's (1999) work in Bangkok where migrants responded with ambivalence to the city. Mills found that migrants frequently asserted

> the moral and emotional superiority of rural life [and] an attachment to the home village as a positive source of identity and sense of self-worth as well as a sense of dislocation and moral incoherence in the city. . . . Migrants" desires for greater participation in the modern nation . . . clash with their lived experiences of urban wage labor. . . . The bonds of compassion, trust and mutual help that (ideally) link fellow villagers stand in opposition to the alienating quality of many women's (and men's) experiences as low status laborers in Bangkok. (Mills, 1999, pp. 136–137)

In both cases, one coping strategy for dealing with the hardships of the city was the maintenance of strong regional identities and connections to rural kin and communities. This strategy has several dimensions. First, rural regional and ethnic identities that are ridiculed or even erased in the city are nonetheless a source of affirmation and strength in the context of urban poverty, exclusion, and outright discrimination. Related to this, rural nostalgia and the possibility of return may suppress migrants' critique of their situation, making the city more bearable, such that they remain in the city despite hardships. One of the most striking findings of this project was the stunningly low level of organizing, or incorporation of migrants into community organizations, urban civil movements, NGO networks, and the like. These migrants were surprisingly disconnected from various forms of "people-level globalization" (Mittelman, 2000) or from transnational networks of engagement (Bebbington, 2001; Bebbington & Batterbury, 2001) or resistance (Radcliffe, 2001). The relative recency of their migration, coupled with their experiences of discrimination in Quito and strong identifications with their places of origin, may also help us understand why migrants are not more networked in Quito. In this chapter, I suggest that migrancy, gender and ethnic discrimination, and economic vulnerability play key roles in the constitution of their spatial identities and in their isolation from organized responses to globalization.

EPILOGUE

I conducted the research reported here in 1996–1998. In January 2000 the contradictions of Ecuador's development trajectory exploded into a popular uprising and the ouster of President Mahuad in the wake of his decision to dollarize the Ecuadorian economy. Thousands of indigenous people, students, workers, and members of women's groups were supported by elements of the military as they brought the concerns of the newly formed National People's Parliament onto the international scene. Despite the success of this uprising in ousting President Mahuad, the very next day Vice-President Noboa, of the same government, was installed as president. Noboa continued the very same program of structural adjustment and dollarization which had been the rallying call for protest (because it symbolized the loss of national control over the economy). At the same time, popular protests intensified between 2000 and 2003, with the vast majority of organized resistance emanating from indigenous communities in rural areas. Indigenous leaders who lead the uprising, under the umbrella organization Confederation of Indigenous Groups in Ecuador (CONAIE), are exerting continuing pressure on the government. For the first time in 500 years, indigenous people in Ecuador are represented by their own political party and they have also won a historic (albeit still small) level of representation in congress (six members in the 123-member Ecuadorian congress in 2002). In the fall of 2000 members of CONAIE were involved in Ecuadorian negotiations with the Paris Club. In February 2001 another march of thousands on Quito renewed the call for a permanent national dialogue on reforms. In 2003, Lucio Guttierez, a military man of humble origins and a leader of the coup that ousted Mahuad, was elected President, strongly supported by Ecuador's indigenous people. Suddenly, the "invisible" (indigenous peoples) became "visible," and these developments suggest that "marginalized," rural, regional identities may yet be a progressive force.

ACKNOWLEDGMENT

An earlier version of this chapter appeared in *Scottish Geographical Journal*, 2003, vol. *188*(3). I appreciate their permission to update and reprint this paper.

NOTES

1. Recent work emerging on the rescaling or resiting of identities under globalization has looked to the transnational sphere (Radcliffe, 2001; Sen, 1997; Moghadam, 1998). This work has focused on various forms of transnational engagement (with NGOs) or activism (environment, debt relief, etc.) to argue that reworked spatial identities can be strategic and political in relation to globalization. By contrast, I explore the implications of economic globalization for migrants who are not globally networked, even as they are deeply affected by global flows. The migrants I spoke with were surprisingly disconnected from various forms of "people-level globalization" (Mittelman, 2000) or from transnational networks of engagement (Bebbington, 2001; Bebbington & Batterbury, 2001) or resistance (Radcliffe, 2001).

2. There have also been substantial international migration flows from Ecuador, again indicating the deep insertion of the economy into global flows. The country witnessed a dramatic growth in flows of international migrants to the United States and particularly to Spain and Italy in the 1980s and 1990s (Jokisch, 2002). Indeed, the estimated 1 million overseas Ecuadorians sent home remittances of more than U.S. $1.3 billion in 2000, which amounts to more than the total export earnings for bananas, shrimp, coffee, cocoa, and tuna (Acosta, 2002).

3. I draw on a larger project (collaborative with Kim Van Eyck, Ana Maria Albuja, and Lastenia Rumbo) in which we compare migrants' and long-term residents' experiences of Quito's labor markets to understand how migrant status, gender, and severe economic shocks in the last two decades impact daily life and work in Quito. In 1996 and 1997, I was part of a team of researchers who developed a survey of in-migrants to Quito (632 households, 1,406 individuals). We then followed up a year later with 50 of the respondents to the in-migrant survey and completed 20 in-depth interviews in Quito. Our sample was limited, primarily mestizo, few indigenous, no Afro-Ecuadorians. We focused on households and so missed the very poor who don't have a place of residence. These interviews were transcribed and analyzed through close readings of the transcripts and listening to the tapes. The interviews involved a series of open-ended questions probing their experiences in Quito as compared with other places they had lived. Interviews were conducted in people's homes, where both the author and Ecuadorian colleagues were involved in the conversation. Each interview was recorded in one visit, which lasted approximately 1 hour. There was little opportunity for the respondents to develop a substantial level of familiarity with us and our project, and so the interviews were quite formal and circumscribed by these conditions. All of the quotes employed here were drawn from these interviews and they have been translated from Spanish by the author. I use pseudonyms for respondents. I collaborated with the Center for Population Studies in Ecuador (CEPAR) and a team of Ecuadorian researchers.

4. Governmentality has most commonly been theorized and studied at the national scale. According to Brown and Boyle (2000, p. 89), "Governmentality calls attention to the simple point that the state's administrative apparatuses play a key role in the way nationals come to know themselves as a coherent nation. . . . Specifically, a state's own knowledge of its population powerfully frames the conditions and terms through which its citizens can see themselves as a nation. In this way, they come to "govern" themselves through the state's "mentality.' " Gupta (1998) has argued for the consideration of global governmentality in this contemporary era and I explore this idea here.

5. See the Epilogue section of this chapter for an update on Ecuador's political and social context. This analysis focuses on the period for which the survey data were collected, the 1980s to the mid-1990s.

6. As Webber (1998) argued for Australia, this is not simply a case of structural determinism, because state managers themselves internalize the transnational ideologies of neoliberalism and globalize their own economies (Herod, O'Tuathail, & Roberts, 1998).

7. Since the democratic transition in 1980 there have been 10 presidents, beginning with the transition government of Roldos (1980–1981). He was followed by Hurtado (1981–1984), Febres-Cordero (1984–1988), Borja (1988–1992), Duran-Ballen (1992–1996), Bucaram (1996–1997), Alarcon (1997–1998), Mahuad (1998–2000), Noboa (2000–2003), and Gutierrez (2003–present). Gutierrez may entail a break with the elite political establishment represented by many of his predecessors. Thus this is an interesting time for tracking these relationships.

8. These included a 100% increase in the cost of gasoline and cooking fuel; a 40% increase in the cost of utilities, transportation, and government services; and a 35% devaluation of the sucre and liberalization of interest rates.

REFERENCES

Acosta, A. (1990). *La deuda eterna*. Quito: Ediduende Cia. Ltda.

Acosta, A. (2002). Ecuador: Un modelo para America Latina? Available online at *www.portoalegre2002.org/publique/cgi/public.*

Alexander, M. J., & Mohanty, C. T. (1997). Introduction: Genealogies, legacies, movements. In M. J. Alexander & C. T. Mohanty (Eds.), *Feminist genealogies, colonial legacies, democratic futures* (pp. xiii–xlii). New York: Routledge.

Bebbington, A. (2001). Globalized Andes?: Livelihoods, landscapes and development. *Ecumene, 8,* 414–436.

Bebbington, A., & Batterbury, S. (2001). Transnational livelihoods and landscapes: Political ecologies of globalization. *Ecumene, 8,* 369–380.

Bergeron, S. (2001). Political economy discourses of globalization and feminist politics. *Signs, 26*(4), 983–1006.

Brown, M., & Boyle, P. (2000). National closets: Governmentality, sexuality and the census. In M. Brown, *Closet space: Geographies of metaphor from the body to the globe* (pp. 88–115). London: Routledge.

Business America. (1986, November). Ecuador acts to adjust its economy in the wake of falling oil prices. pp. 10–11.

Castells, M. (2000). *End of millennium.* Oxford, UK: Blackwell.

Cheah, P. (1998). Given culture: Rethinking cosmopolitical freedom in transnationalism. In P. Cheah & B. Robbins (Eds.), *Cosmopolitics: Thinking and feeling beyond the nation* (pp. 290–328). Minneapolis: University of Minnesota Press.

Clifford, J. (1988). *The predicament of culture: Twentieth-century ethnography, literature and art.* Cambridge, MA: Harvard University Press.

Conaghan, C. (1988). *Restructuring domination: Industrialists and the state in Ecuador.* Pittsburgh, PA: University of Pittsburgh Press.

Corkill, D., & Cubitt, D. (1988). *Ecuador: Fragile democracy.* London: Latin America Bureau.

Delaunay, D. (1989). Espacios demograficos y redes migratorias. *Estudios de Geografia, 1,* 71–98.

Dirlik, A. (1998). Globalism and the politics of place. *Development, 41,* 7–13.

Eschle, C. (2001). *Global democracy, social movements and feminism*. Boulder, CO: Westview Press.

Escobar, A. (1995). *Encountering development*. Princeton, NJ: Princeton University Press.

Escobar, A. (2001). Culture sits in places: Reflections on globalism and subaltern strategies of localization. *Political Geography, 20*, 139–174.

Foucault, M. (1991). Governmentality. In G. Burchell, C. Gordon, & P. Miller (Eds.), *The Foucault effect: Studies in governmentality* (pp. 87–104). Chicago: University of Chicago Press.

Gausch, J. L. (1999). *Labor market reform and job creation: The unfinished agenda in Latin American and Caribbean countries*. Washington, DC: World Bank. (See Table 3.3, pp. 17–20).

Glassman, J. (1999). State power beyond the "territorial trap": The internationalization of the state. *Political Geography, 18*, 669–696.

Gupta, A. (1998). *Postcolonial developments: Agriculture in the making of modern India*. Durham, NC: Duke University Press.

Hartsock, N. (2001, May). *Domination, globalization: Towards a feminist analytic*. Paper presented at the Berlin Institute for Critical Theory, Berlin.

Harvey, D. (2000). *Space of hope*. Berkeley and Los Angeles: University of California Press.

Herod, A., O'Tuathail, G., & Roberts, S. (1998). *Unruly world: Globalization, governance and geography*. London: Routledge.

International Research and Training Institute for the Advancement of Women. (1994). *La mujer migrante en el Ecuador*. Quito: Instituto Ecuatoriano de Investigaciones y Capacitacion de la Mujer, Punto Focal del INSTRAW.

Jessop, R. (1995). Towards a Schumpeterian workfare regime in Britain?: Reflections on regulation, governance and the welfare state. *Environment and Planning, 27*, 1613–1626.

Jokisch, B. (2002). Migration and agricultural change: The case of smallholder agriculture in highland Ecuador. *Human Ecology, 30*, 523–550.

Kofas, J. (2001). The IMF, the World Bank, and U.S. foreign policy in Ecuador, 1956–1966. *Latin American Perspectives, 28*, 50–83.

Korten, D. (2001). *When corporations rule the world* (2nd ed.). San Francisco: Kumarian Press.

Latin American Index (1986, September 15). Washington, DC: Latin Research Group.

Lawson, V. (1995). Beyond the firm: Restructuring gender divisions of labor in Quito's garment industry under austerity. *Environment and Planning D: Society and Space, 13*, 415–444.

Lawson, V. (1999). Questions of migration and belonging: Understandings of migration under neoliberalism in Ecuador. *International Journal of Population Geography, 5*, 1–16.

Lawson, V. (2000). Arguments within geographies of movement: The theoretical potential of migrants" stories. *Progress in Human Geography, 24*(2), 173–189.

Lind, A. (2000). Negotiating boundaries: Women's organizations and the politics of restructuring in Ecuador. In M. Marchand & A. Runyan (Eds.), *Gender and global restructuring: Sightings, sites and resistances* (pp. 161–175). London: Routledge.

Luz Borrero, A., & Vega Ugalde, S. (1995). *Mujer y migracion: Alcance de un fenomeno nacional y regional.* Quito: ILDIS.

Mills, M. B. (1999). *Thai women in the global labor force: Consuming desires, contested selves.* New Brunswick, NJ: Rutgers University Press.

Mittelman, J. (2000). *The globalization syndrome.* Princeton, NJ: Princeton University Press.

Moghadam, V. (1998). Gender and the global economy. In M. Feree, J. Lorber, & B. Hess (Eds.), *Revisioning gender* (pp. 128–160). London: Sage.

Nagar, R., Lawson, V., McDowell, L., & Hanson, S. (2002). Locating globalization: Feminist (re)readings of the subjects and spaces of globalization. *Economic Geography, 78,* 257–284.

Ohmae, K. (1990). *The borderless world: Power and strategy in the interlinked economy.* New York: Harper & Row.

Ong, A. (1999). *Flexible citizenship.* Durham, NC: Duke University Press.

Peterson, V. S. (1996). Shifting ground(s): Epistemological and territorial remapping in the context of globalization(s). In E. Kofman & G. Youngs (Eds.), *Globalization: Theory and practice* (pp. 11–28). London: Pinter.

Radcliffe, S. (2001). Development, the state, and transnational political connections: State and subject formation in Latin America. *Global Networks, 1*(1), 19–36.

Radcliffe, S., & Westwood, S. (1996). *Remaking the nation: Place, identity, and politics in Latin America.* London: Routledge.

Rouse, R. (1995). Thinking through transnationalism: Notes on the cultural politics of class relations in the contemporary United States. *Public Culture, 7,* 353–402.

Structural Adjustment Participatory Review Imternational Network. (2001). Fact sheet, February 6, 2001. Available from *www.saprin.org/ecuador/ecuador6feb. htm.*

Sassen, S. (1998). *Globalization and its discontents.* New York: New Press.

Schodt, D. (1987). *Ecuador: An Andean enigma.* Boulder, CO: Westview Press.

Scholte, J. A. (1996). Beyond the buzzword: Towards a critical theory of globalization. In E. Kofman & G. Youngs (Eds.), *Globalization: Theory and practice* (pp. 43–57). London: Pinter.

Sen, G. (1997). Globalization, justice and equity: A gender perspective. *Development, 40,* 21–26.

Shorris, E. (2001, September 20–October 3). The light of power: An interview with Adam Holdorf. *Real Change,* p. 5.

Slater, D. (1992). Theories of development and the politics of the post-modern: Exploring a border zone. *Development and Change, 23,* 283–319.

Sparke, M., & Lawson, V. (2003). Entrepreneurial geographies of global–local governance. In J. Agnew, G. O'Tuathail, & K. Mitchell (Eds.), *A companion to political geography* (pp. 315–34). Oxford, UK: Blackwell.

Teltscher, S. (1993). *Informal trading in Quito, Ecuador.* Saarbrucken, Germany: Verlag Breitenbach.

Webber, M. (1998). Producing globalization: Apparel and the Australian state. In A. Herod, G. Tuathail, & S. Roberts (Eds.), *Unruly world. Globalization, governance and geography* (pp. 135–161). London: Routledge.

Weiss, W. (1997). Debt and devaluation: The burden on Ecuador's popular class. *Latin American Perspectives, 24*(4), 9–33.

World Bank. (1996). *Ecuador poverty report: A World Bank country study.* Washington, DC: Author.

World Bank. (2000). *Ecuador country brief.* Available online at *http://wbln0018. worldbank.org/external/.*

14

Discourses of Globalization and Islamist Politics

Beyond Global–Local

Anna J. Secor

The globalization literature of the past 15 years has inserted "Islamist politics," that is, movements for the moral, social, and political Islamization of society, into discussions of the role of the local within the global condition. Within many globalist theories, Islamism becomes a central, cohesive instrument of antihegemonic forces unleashed by globalization processes. Islamism and other "fundamentalisms" are frequently represented as particularistic, and resistant both to the so-called universal values of the Enlightenment and to the economic and cultural processes of globalization. For many theorists of globalization, "Islamism" thus comes to bear the critical mark of (local) fragmentation and (global) homogenization as two constitutive trends of globalization (Barber, 1996).

Despite this attention, the relationship between globalization and the rise of political mobilizations that aim to enact particular visions of Islamic life has more often been asserted than interrogated. Since the Iranian revolution of 1979, the proliferation of Islamist movements across the world has been cast in terms of an outcome of globalization processes that are assumed to have spurred new forms of identity politics

(Ahmed and Donnan, 1994; Bauman, 1998). But what are the everyday processes through which globalization produces Islamism as a (the?) plausible site of identification and political mobilization in Muslim societies? Islamist politics is too often represented as a "natural" outcome of processes such as neoliberal restructuring and the intensification of cultural flows through new media technologies (see, e.g., Stone, 2002). Globalization narratives thus seem to take for granted the association between economic globalization processes and particular modes of identity formation.

The idea that Islamism represents a response to the uncertainties of globalization draws strength from theories of postmodernism as "the cultural dominant of the logic of late capitalism" (Jameson, 1984, p. 81; see also Anderson, 1998; Harvey, 1990). Such analyses often interpret religious movements as responses to the agony of postmodern ephemerality and expressions of authoritarian longing under conditions of spatial and temporal reorganization. For Giddens (1991), the interconnectivity and risk that characterize "high modernity" catalyze religious mobilization: "For reasons that are to do precisely with the connections between modernity and doubt, religion not only refuses to disappear but undergoes a resurgence" (1991, p. 195). Islamist political mobilization is thus inserted into a globalization discourse that projects an evolutionary idea of progressive interconnectivity accompanied by increasing doubt, uncertainty, and risk.

As globalization is argued to erode the "local" cultural practices of Muslim culture, questions of gender and women's roles in society have been critical to both Islamists and their foes. Within Islamist movements, family law, gender relations, and sexual norms have been prominent targets for social reform.[1] At the same time, the image of the veiled, secluded, and oppressed Muslim woman has been used to justify colonial and imperial incursions into Muslim societies (Ahmed, 1992; Yeğenoğlu, 1998). Further, women have played active roles in the mobilization and administration of Islamist movements in urban areas. Because Islamist women, like Islamist men, have been found to be most often lower-middle-class, rural–urban migrants, women's involvement in Islamist politics has been understood as a product of globalization processes that have brought young women into the orbit of identity politics in the city (Ong, 1990).

The purpose of this study is both to evaluate the claims of globalization theory vis-à-vis Islamism through a case study of Islamist political support in Istanbul and to interrogate "globalization" as a discourse that strategically positions Islamism in various ways in relation to the local and the global. There are two questions that guide this investigation: First, how does globalization theory operate in relation to Islamism as a regulatory discourse of scale, power, and resistance? Second, does globalization theory help us to interpret the correlates of Islamist support among women and men in Istanbul? By drawing out gender differences in the bases of Islamist political support, this chapter shows the unevenness and variability of globalization processes and their effects. I begin by tracing the multiple positions of Islamism within globalization discourses that produce "the local" and "the global" as polysemous sites for the practice of politics. This general discussion is followed by an overview of how the relationship between globalization and Islamism has been represented in the case of Turkey. Finally, the question of how globalization theory applies to Islamist support in Istanbul is addressed through a statistical analysis of survey data collected using a proportionate random sample design in four lower- to lower-middle-class districts of Istanbul. As the title of this chapter suggests, neither the processes invoked in discourses of global or of local can single-handedly explain support for Islamist parties, particularly among women. Rather, the sources of support reflect a tension between the processes that is expressed in complicated, and perhaps contradictory, ways.

STRATEGIES OF LOCALISM AND GLOBALISM

> Cultural imperialism rests on the power to universalize particularisms linked to a singular historical tradition by causing them to be misrecognized as such.
> —BOURDIEU AND WACQUANT (1999, p. 41)

Globalization is a discourse that creates, colonizes, and naturalizes "local" and "global" scales as sites of power. This operation is rendered invisible as these scales become codified, fixed, and filled—that is, when certain movements are understood to be local and others global, and when these scalar identities become welded to ideas of authenticity, foreignness, and universality. As Marston (2000) has put it, scale is not an "ontologically given category," but a "contingent outcome of the tensions that exist between structural forces and the practices of human agents" (p. 220). Globalization is not the only scaling discourse in circulation, but today it has become a formidable constitutive language through which political action is staged and countered. That this matters, that the political role of globalization as discourse is critical in the contemporary world, becomes apparent when one considers the ways in which power, threat, and response have been mobilized through the discourse of globalization as it situates Islamist politics in the world today.

The operation of this discourse as a rehearsal of power is also made manifest in its gendered language; as J. K. Gibson-Graham (2002) has argued, "Where globalization is seen in terms of penetration, the parallels with rape are obvious" (p. 121). The language of domination and irreversible spatial invasion are part of a "scripted narrative of power" that naturalizes globalization as an inevitable process and unstoppable force (Gibson-Graham, 2002, p. 125). In a recent piece of geopolitical writing, Naval historian Thomas Barnett (2003) describes those areas which, in his view, have been "resistant" to globalization as "the Gap." A region of the world that includes all majority Muslim areas, the Gap is represented as a feminized and dangerous space where "delinking" from the global economy inevitable provokes U.S. military intervention (Barrnett, 2003; see also Roberts, Secor, & Sparke, 2003).

In this section I argue that globalization discourse, as a narrative of power, works to situate Islamism within the fields of geostrategy and geoculture. Further, Islamists, orientalists, and others find in globalization discourse a resource through which claims to authority and legitimacy can be mobilized. Globalization is thereby both produced through these struggles over power and meaning and is itself productive of new staging grounds for political contest, the negotiation of power, and even violent intervention. In order to show how this takes place, I present a number of ways in which Islamism has been scripted as "global" or "local" within globalization discourse.

"Local" Islamism

Both "the local" and "the global" are signifiers that have multiple meanings, variously privileged and debased within the multisited discourse of scale and power that is foundational to globalization theories. Social theoretical productions of "the global" and "the local" provide critical links between constructions of locality, authenticity, and political legitimacy within the practices of Islamist mobilization. In her discussion of the ascendance of social scientific interest in local knowledges, Abaza (2000) writes: "All these endeavors in the sociology of knowledge, through the search for alternative paradigms, together with the Post-Orientalism debate paved the way, indirectly or directly to self-representation of the voices of the South as 'indigenous' and 'authentic,' local voices" (Abaza, 2000, p. 55). Within the revived anti-imperialist discourse that Abaza identifies, globalization appears as a vessel carrying the foreign and impure into the sanctified realms of home and homeland. As Nelson and Rouse (2000) write in the introduction to their edited volume on Egyptian perspectives on globalization, "The global was *represented as 'foreign,'* external (to the local) [in earlier work on globalization]. The struggle therefore was posited as consisting of a dialectic encounter between hegemonic forces of globalization and 'indigenous' forces of resistance" (pp. 9–10, original emphasis). Through this set of analytical moves, Islamism is scripted as a local reaction against globalization, and simultaneously the local becomes a site of identity and power from which political claims may be launched. Further, as in anticolonial nationalisms, this realm of the authentic and local is frequently mapped onto the "private" realm of the family, gender relations, and women's honor (see Chatterjee, 1993).

"Global" Islamism

In another register of globalization theory, Islamism appears as an "alternative globalism," a contrapuntal strand within a polyphonic global system. Pieterse (2001) summarizes this perspective:

> What might be considered an alternative globalism is political Islam. Unlike other movements this represents a holistic alternative, it encompasses a worldview, a way of life, a historical formation as well as a geographical space, stretching from Morocco to Southeast Asia. Its scope includes Islamic politics and law (sharia), Islamic geopolitics, Islamic economics and social policy, Islamic science, and Islamic identity and culture. (p. 32)

Despite these apparent affinities between Islamist politics and a particular idea of the global, Pieterse (2001) goes on to suggest that the internal fragmentation of Islamism renders "Islam as a global alternative . . . a political fiction that is being held up for propagandistic reasons on both sides of the fence, in a game of mirrors, a game of mutual stereotyping" (p. 35). Indeed, the claim that Islamism presents an alternative globalism not only projects an image of a unified, integrated, and identified "Islamism," but also roughly cleaves Islamist mobilization from the particular ways in which it is articulated in everyday life across a range of contexts.

Advocates for a globalist view of Islamism occupy a range of positions vis-à-vis Islamist politics. Not only is the idea of Islamism as an alternative globalism deployed from within a threat-based Western discourse that builds in part upon Huntington's (1993) "clash of civilizations" edifice, but it is also present within Islamist agendas. Through his study of Egyptian Islamist texts, Marfleet (2000) traces the ways in which the global is seen as strategically important for Islamization movements. He suggests that while globalization is associated with Western capitalism and imperialism, the global context is seen as providing opportunities: "The notion of globalization appears repeatedly in strategic documents of the Islamization movement, where it is associated with a growing sense that the new state of affairs offers possibilities of realization or fulfillment of foundational Islamic values" (p. 25). Likewise, in the case of the Islamist movement in Istanbul, discourses of localism and authenticity were paired with claims to an alternative vision of globalism, in which the city was to (re)occupy the position of an Islamic capital under the new administration (Bora, 1999). What is at stake here is municipal government, but the language of globalization infuses both the everyday politics of electoral campaigns and the ways in which the contest and its outcome are framed in Turkey and around the world.

Globalization is a discourse that fixes "the global" and "the local"—or the globalized and the "delinked"—as geopolitical sites of power and domination. This becomes evident not only in the overwrought language of American militarism, but also when Islamist movements present themselves as authentic and local alternatives to a predatory globalization emanating from the imperial impulses of the West, or when social theorists represent Islamist movements as place-based movements. It is also made manifest when Islamist movements position themselves as "global" or call for a greater globalism in their agendas, and when others represent Islamism as a threatening or libratory "alternative globalism." The relationship between globalization and Islamism can no longer be understood as causal or dialectical; rather, it is a relationship comprised of strategic maneuvers, of a continual jockeying for position and power. It is a relationship characterized by an ongoing contest over the meaning and operation of "the global" and "the local" as the scalar categories of globalization.

REPRESENTATIONS OF GLOBALIZATION AND ISLAMISM IN TURKEY

Globalist accounts situate the rise of Islamist politics in Turkey within the context of the political, economic, and cultural restructuring that the country has undergone in the past two decades. Globalization in Turkey is usually dated from the 1980 coup d'etat, which enabled the 1980–1983 military regime to launch the unpopular International Monetary Fund (IMF) reforms required of Turkey by one of the earliest structural adjustment programs (SAPs). Through these reforms, the Turkish economy shifted course from the import substitution, statist policies that had dominated the republican era to the pursuit of an export-oriented and increasingly privatized economy. Despite the corruption and profiteering that have accompanied this still incomplete transition, the neoliberal reorientation of the state following the 1980 coup and the new constitution of 1983 have come to constitute the meaning of the term "globalization" in the Turkish context (see Gülalp, 2001; Isin, 2001; Önis, 2000).

At the same time, beyond the observation of a new culture of conspicuous consumption in urban areas and widening income gaps, there is no consensus as to what this economic restructuring of the Turkish economy has meant for various sectors of society such as manufacturing and small businesses (see Birtek & Toprak, 1993; Tokatli & Eldener, 2002). The mixed outcomes of neoliberal restructuring in Turkey can be seen as part of what Keyder (1999) calls "really existing globalization." Informal and uneven, really existing globalization in Istanbul consists mainly of illegal flows associated with money laundering, the drug trade, and the "suitcase trade" that has made Istanbul a marketplace for Russian, Eastern European, and Central Asian merchants (Keyder, 1999). As merchants leave Turkey with suitcases full of unrecorded and untaxed textiles and leather goods, the demand for these products has further spawned the growth of flexible and informal production strategies. Through these processes, differences in income, consumption patterns, and lifestyles have created what Keyder calls a "divided city," a city that manifests the effects of structural adjustment in disparate ways in the *gecekondu* (squatter) areas of rural–urban migrant communities and the new gated subdivisions of the rich.

The rise of Islamist politics in Turkey and, more specifically, in Istanbul has taken place within the context of these economic and cultural shifts. In 1994, the Islamist party won the Istanbul municipal elections on a platform of moral regeneration and populist policies. The Islamists were able to position their political vision as counter to the unpopular "global city" project of the 1980s, through which the center–right municipal administration aimed to transform Istanbul from a prime national city to a newly imagined global metropolis marked by "urban renewal," tourism, and international business (Bora, 1999). In the 1990s, backlash against these policies made clear that the architects of the global city project could not gain reelection in a city dominated by migrant voters who not only did not benefit from these investments, but desperately needed city funds to be redirected toward infrastructure and services for the urban periphery. In 1999, despite slipping from first (in 1995) to third place in the national elections, the Islamist party (then Fazilet Partisi [FP]) continued its success at the local and urban levels, once again winning control of Istanbul's city government. In 2002, a new post-Islamist party (Adalet ve Kalkinma Partisi [AKP]) comprised of some of the younger and more popular members of the previous parties, won both a plurality of seats in Parliament and maintained control of

Istanbul municipality. This new party explicitly rejects the label of Islamist but has managed to maintain much of the same leadership and support base as its predecessors.

Some have viewed Islamist politics in Turkey as a movement for the "recovery" of lost stability (e.g., Önis, 2000), either in terms of a reaction to the fragmentation of previously unified identities, or as the "return of the repressed" through which moral questions disallowed by modern institutions return to the public sphere (Giddens, 1991, p. 207; see also Aksoy & Robins, 1997). Other analyses, however, have rejected the idea of Islamist politics in Turkey as representative of an "authentic" resurgent identity, viewing it instead as a constitutive element of globalization processes. One of the most compelling approaches to globalization and Islamism in Turkey situates the rise of Islamist politics within the context of the retreat of the state under conditions of neoliberal restructuring and the rise of civil society as an alternative avenue of communal identity and social security (Gülalp, 2001; see also White, 2002). Pasha (2002) suggests that Islamism has become the political expression of a wide range of "Islamic cultural areas" across the world as a result of "the growing disconnect between an already fractured political community and an increasingly illegitimate state [that] provides Islamicists the opening to capture key institutions in civil society" (p. 121). In the case of Turkey, Islamism appears to glean its strength from its ability to appeal to both the "winners" and the "losers" of these globalization processes, papering over class cleavages with Islamic symbolism, imagery, and rhetoric (Gülalp, 2001; White, 2002).

In the mid-1990s, Islamist politics in Turkey came to be represented as a kind of postmodern moment. Focusing on the roles of university students, young professionals, and the new middle class, studies of Islamism in Turkey suggested that these communities were actively forging new Islamist vocabularies and lifestyles, practices that became associated with upward mobility and the rise of an Islamist "counterelite" in Istanbul (Göle, 1996, 2000; Ismail, 2001; White, 1999). With a focus on consumption and self-presentation, this literature highlighted the practices of young Islamist women who had become increasingly visible, not only in urban spaces but in political and social activism. In scholarly and journalistic accounts alike, the new "Islamist woman" was typically represented as an educated, upwardly mobile, first- or second-generation migrant to the city. Through her choice of veiling style and clothing, she would achieve or aspire to what White (1999) calls an "Islamic chic" associated both with urban Islamist identities and "the road to success" in Istanbul (p. 80). Islamism in Turkey could thus be inserted into broader narratives of globalization and postmodernization as an "identity politics" born of neoliberal restructuring, the associated construction of a new individualist consumer culture, and the search for fixity and identity within and against currents of change and flux.

Despite the salience of these articulations of Islamism in Turkey, they leave open the critical question that Sayyid (1997) raises in his conceptual narrative of the emergence of Islamism: "Why is it the name of Islam, rather than another name, that has become so central to Muslim politics?" (p. 40). While analyses of Islamism within civil society may well point to a crucial opening within Muslim polities in the late 20th century, why has the redefinition of state and society given rise to Islamist politics rather than to some other political expression? Likewise, while Islamism may function at multiple levels as an identity politics and even a consumer lifestyle, what makes it a viable site for postmodern identity-formation processes? This chapter does not pretend to answer these questions, but bearing them in mind, it does provide one clue to the role globalization processes and discourses play in the constitution of Islamist political mobilization.

ISLAMIST SUPPORT IN ISTANBUL

This section establishes a dialogue between findings from fieldwork conducted in Istanbul in 1998–1999 and the recent literature on postmodernity and Islamist politics in Turkey. It begins with an attempt to unravel a puzzle: How is it that, while studies of Islamist politics in Turkey have emphasized the role of globalization—as a set of economic processes that has both created "winners" and "losers" and produced new modes of identity formation through consumption and lifestyle—my own survey research, which explored the bases of electoral support for the Islamist party in Istanbul, revealed a different sort of dynamic of Islamist support? The analysis presented in the following pages suggests that the answer is more complicated than either the discourses of globalization or of a reactionary localism would suggest. This study is based on a 1999 survey of 735[2] Istanbul residents conducted door-to-door in four relatively low-land-value districts of Istanbul: Fatih, Üsküdar, Ümraniye, and Gaziosmanpaşa.[3] Within the four districts, neighborhoods were chosen randomly and sampled using proportional to population size (PPS) sampling techniques.

Using a statistical analysis of the factors associated with the Islamist vote provides evidence of the unevenness and contingency of globalization and its outcomes. While it is difficult to specify the experience of "globalization" or "postmodernity," such things as urban mobility, engagement with media, or perceptions of personal transformation can serve as imperfect indicators of individual engagement with the flows and processes associated with identity formation under conditions of globalization. Likewise, support for Islamist politics is not perfectly encapsulated by the vote for the Islamist party, but for this analysis the vote is taken to be a reasonable measure of Islamism among this population. Evidence from this survey suggests that "globalization" fails to explain or describe the everyday dynamics of Islamist support in lower income areas of Istanbul. Furthermore, the globalization thesis becomes especially problematic for understanding women's voting choices.

In attempting to identify the factors associated with support for the Islamist party, this study produced two main findings (Secor, 2000). First, with the sampled population, neither class nor economic attitudes were significantly associated with the Islamist vote. In other words, voting for the Islamist party was not associated with income, occupation, or with feelings or expectations regarding household well-being in general.[4] Support for the Islamist party was, however, associated with religious practice and self-identity, as well as with migrant status and migrant self-identification. These relationships are illustrated by Figure 14.1, in which the probability of certain identity choices is shown to be significantly higher among supporters of the Islamist party than others. Respondents were asked to list, in order of importance, up to three ways in

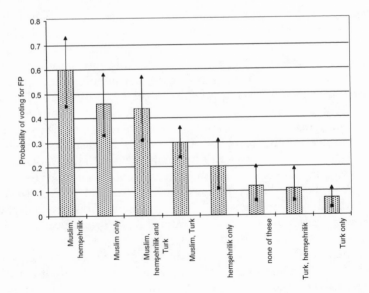

FIGURE 14.1. Probability of voting for Islamist party (FP) in 1999 elections by combinations of identity choices. Error bars show confidence intervals; $n = 436$. Made using King et al.'s (1999) "Clarify" statistical software.

which they identified themselves in their everyday lives. While Muslim self-identification is clearly associated with support for the Islamist party, it is also notable that *hemşehrilik*, a term that refers to identification with those who migrated from the same village or province, was an important indicator of Islamist support in conjunction with Muslim identity. Among Islamist supporters, the national/ethnic identity category Turkish was less prevalent than this regional self-identification, which may in part have masked other ethnic identifications (namely, Kurdishness). Given that *hemşehrilik* is associated both with chain-migration neighborhoods and with networks of support and community in the urban environment (Erder, 1999), Islamism thus seems to fold into itself multiple routes of self-identification, some of which embody practices and discourses of locality while others (namely identification with the Muslim *umma*) are distinctly "global," or at least transnational.

The general finding of this study is that religious practice and self-identity, along with migrant status and self-identification, were the only significant predictors of Islamist support. A multinomial logistic model, which expresses the logarithm of the odds of voting for the Islamist party as a linear combination of parameters of categorical explanatory variables (i.e., religious practice, Muslim self-identity, and migrant status) fit the data moderately well ($R^2 < .20$) for both men and women (see Table 14.1). The degree to which individuals are affected by the economic processes of globalization does not itself serve to differentiate supporters and non-supporters within the population of these four districts. In other words, among lower and lower middle class urban residents, while self-identification is closely linked to Islamist support, these identities and political choices do not appear to be determined by relative economic well being, or by particular economic aspirations or frustrations. The Islamist supporters do not appear any more downtrodden or any more upwardly mobile than their neighbors. What we are left with here is an "identity politics" among the lower and lower-middle classes that is more easily explained in terms of value systems and communities than any socioeconomic indicators. The black box within which economic globalization is thought to give rise to particular modes of identity formation seems to dissolve into an insubstantial conjuncture.

The picture appears more complicated, however, when gender is considered. For women respondents, there were a number of other factors besides self-identification and religious practice that were associated with support for the Islamist party. Many of these can be seen as related to engagement with the wider urban environment. For example, although men's reported mobility in the city is not significantly related to their support for the Islamist party, female Islamist supporters had a significantly lower probability of reporting frequent travel outside of their

TABLE 14.1. Multinomial Logistic Regression for Binomial Islamist (FP) Vote in Istanbul City Elections: Migration, Religion, and Religious Identity

Variable	Women		Men	
	Parameter est.	SE est.	Parameter est.	SE est.
Intercept	0.952**	0.380	0.602	0.420
Migration				
Migrant	−0.631*	0.378	0.326	0.384
Istanbul parents	−2.638**	1.112	−0.610	0.780
Migrant parents	ref.		ref.	
Religion				
Never	−1.668***	0.485	−1.208**	0.561
Holidays	−1.931***	0.596	−2.051*	1.132
Fridays	−1.127**	0.441	−0.646*	0.340
Always	ref.		ref.	
Muslim				
No	−1.189**	0.397	−1.790***	0.381
Yes	ref.		ref.	
	Test statistics			
Valid cases	223		223	
Log likelihood	51.101		48.772	
McFadden pseudo R^2	0.175		0.194	

Note. "SE est." refers to standard error estimate for parameter. Categories marked "ref." are those that have been set to 0.0 as redundant parameters. "Other" is the reference category for the FP vote, and so does not appear.

$*p < 0.10;$ $**p < .05;$ $***p < .001.$

own districts in the city (1:5). Further, 27% of Islamist party voters, compared to 16% of voters for other parties, reported *never* leaving the districts where they live. Another example of women being differently positioned in relation to Islamist politics in Istanbul arises from media engagement. Both men and women who voted for the Islamist party (FP) were more likely than other voters to report reading religiously oriented or nationalist newspapers and watching Islamist television channels as their primary news sources. Gender differences in media engagement related to the Islamist vote emerge around the question of television viewing choices. Women who voted for the FP (though not men) were less likely than others to report watching foreign films on television (only 15% of them do, compared to 30% of those who voted for other parties), and were significantly more likely to claim that they never watch any movies at all (21%, compared to 5% of other voters). Film watching, like women's travel in the city, may be an issue of exposure to diversity and integration into wider "imagined communities" beyond one's family or neighborhood, but at the same time may also be a moral issue. The relatively high probability of Islamist women's absence from the wider urban environment or from certain sorts of media points toward gender differences not only in "urbanism" as a way of life, but also in the ways men and women interact and engage with trends associated with cultural globalization.

Finally, while no socioeconomic factors were associated with Islamist support among men, Islamist support is also tied to educational achievement for women. In this survey, the higher the degree of education women achieved, the lower their probability of voting for the Islamist party.[5] Seven percent of female FP supporters finished high school compared to 18% of supporters for other parties. The proportions are reversed for illiteracy, with 17% of female FP supporters being illiterate compared to 6% of women who voted for other parties. In light of women's overall lower levels of education, the finding that only women's support for Islamist politics is associated with educational achievement is suggestive of men's and women's different positions in relation to both globalization and the construction of Islamist politics.

As the relationships between mobility, media, education, and Islamist politics crystallize for women, the image of the "new Islamist woman" seems to recede. Finally, the lack of evidence for the globalization thesis in this study, and the gender differences in the dynamics of Islamist support that it reveals, suggest that support for Islamist politics cannot be interpreted through a totalizing lens.

CONCLUSION

Globalization is a theory that itself is mobilized by academic and political agents for particular ends. Globalization analyses take the form of historicizing and epochal tales that make epistemological and ontological claims about identity and society. In doing so, these globalization stories script Islamist political mobilization in a way that privileges particular understandings of "the local" and "the global" as sites of political action, power and resistance. I argue that this discourse works to fix Islamism within constellations of power, space, and time, and that it is in fact this contest over meaning that has defined both Islamism and globalization. In this way my argument is sympathetic to Marfleet's (2000) point when he writes: "Both currents [Islamism and globalization] impose an imagined unicity upon a volatile and disordered world: neither is appropriate as a means of understanding contemporary conditions, including the condition of the mass of Muslims" (p. 26).

Certainly, "globalization" denotes a set of material practices with important consequences for everyday life. Changes in production and employment processes, neoliberal structural adjustment policies implemented in accordance with IMF protocols, the intensification of flows of information and culture, and the proliferation of critical perspectives on modernization projects are all processes that have had enormous effects on everyday lives. But bundling them together into the package of "globalization" not only forecloses important questions about the relationships between these processes, but also imposes particular narratives of space and time (Low & Barnett, 2000). It is therefore critical that we trace the ways in which "Islamism" bears the mark of its multiple articulations within globalization discourse.

The discourse of globalization sifts power through the categories of "the global" and "the local," naturalizes a given set of associations, and asserts a particular spatial and temporal story. My analysis of the role of globalization in political support for Islamist politics in Istanbul calls into question the dominant narratives of globalization and Islamism and the scripted relationship between economic restructuring and "identity politics." Further, it shows that men and women reveal different relations to both Islamist politics and the processes associated with globalization in their everyday urban engagements. Not only were identity and religious practice the strongest predictors of the Islamist vote, while class and economic satisfaction were immaterial, but Islamist support appeared to be constituted through different sets of practices and meanings for men and women. Thus we can conclude that the lower- to lower-middle-class locations within which this survey was conducted represent a particular urban milieu in Istanbul, one in which relative economic position does not correlate with Islamist support.

Finally, understanding how globalization processes operate in particular contexts may help us to understand Islamist movements, but it is just as likely that case studies will show that the ways in which these practices play out are neither predictable nor scriptable within the parameters of the globalization story. Islamism cannot be "explained" by globalization processes, but we may be better able to understand the operation of globalization, as a practice and as a discourse, by understanding the variable and contingent constitution of Islamist politics.

NOTES

1. For examples of the gendered aspects of Islamist movements across the globe, see Haddad and Esposito (1998).
2. Of the 735 respondents, 452 both voted and answered the question of for

whom they voted in the municipal elections (see Secor, 2000, for more detail).

3. The four study sites were identified based on migration rates, low land value, relatively high support for the Islamist party, and geographic location on the European or Asian side of the city. This study was funded by a National Science Foundation Dissertation Grant.

4. Respondents were asked to characterize their households in terms of economic situation, to rate their household's economic situation relative to 5 years ago, and to project what they expected their situation to be in 5 years time. They were also asked whether they expected their children's economic situation to be better, worse, or the same as theirs.

5. Only five women respondents who reported their vote had a university degree. One of the five voted for the Islamist party.

REFERENCES

Abaza, M. (2000). The Islamization of knowledge between particularism and globalization: Malaysia and Egypt. In C. Nelson & S. Rouse (Eds.), *Situating globalization: Views from Egypt* (pp. 53–95). Bielefeld, Germany: Transcript Verlag.

Ahmed, A. S., & Donnan, H. (Eds.). (1994). *Islam, globalization and postmodernity.* London: Routledge.

Ahmed, L. (1992). *Women and gender in Islam.* New Haven, CT: Yale University Press.

Aksoy, A., & Robins, K. (1997). Peripheral vision: Cultural industries and cultural identities in Turkey. *Environment and Planning A, 29,* 1937–1952.

Anderson, P. (1998). *The origins of postmodernity.* London: Verso.

Barber, B. (1996). *Jihad vs. McWorld.* New York: Ballantine Books.

Barnett, T. P. M. (2003, March). The Pentagon's new map. *Esquire.* Available online at *http://www.nwc.navy.mil/newrulesets/ThePentagonsNewMap.htm.* Accessed April 5, 2003.

Bauman, Z. (1998). Postmodern religion? In P. Heelas (Ed.), *Religion, modernity and postmodernity* (pp. 55–78). Oxford, UK: Blackwell.

Birtek, F., & Toprak, B. (1993). The conflictual agendas of neo-liberal restructuring and the rise of Islamic politics in Turkey: The hazards of rewriting modernity. *Praxis International, 13,* 192–212.

Bora, T. (1999). Istanbul of the conquerer: The "alternative global city" dreams of political Islam. In Ç. Keyder (Ed.), *Istanbul between the global and the local* (pp. 47–58). Lanham, MD: Rowman & Littlefield.

Bourdieu, P., & Wacquant, L. (1999). On the cunning of imperialist reason. *Theory, Culture and Society, 16,* 41–58.

Chatterjee, P. (1993). *The nation and its fragments: Colonial and postcolonial histories.* Princeton, NJ: Princeton University Press.

Erder, S. (1999). Where do you hail from?: Localism and networks in Istanbul. In Ç. Keyder (Ed.), *Istanbul between the global and the local* (pp. 161–172). Lanham, MD: Rowman & Littlefield.

Gibson-Graham, J. K. (2002). *The end of capitalism (as we knew it): A feminist critique of political economy*. Cambridge, MA: Blackwell.

Giddens, A. (1991). *Modernity and self-identity: Self and society in the late modern age*. Stanford, CA: Stanford University Press.

Göle, N. (1996). *The forbidden modern: Civilization and veiling*. Ann Arbor: University of Michigan Press.

Göle, N. (2000). Snapshots of Islamic modernity. *Daedalus, 129*, 91–118.

Gülalp, H. (2001). Globalization and political Islam: The social bases of Turkey's Welfare Party. *International Journal of Middle East Studies, 33*, 433–448.

Haddad, Y. Y., & Esposito, J. L. (Eds.). (1998). *Islam, gender, and social change*. Oxford, UK: Oxford University Press.

Harvey, D. (1990). *The condition of postmodernity*. Cambridge, MA: Blackwell.

Huntington, S. P. (1993). The clash of civilizations? *Foreign Affairs, 72*, 22–49.

Isin, E. (2001). Istanbul's conflicting paths to citizenship: Islamicization and globalization. In A. J. Scott (Ed.), *Global city regions: Trends, theory, policy* (pp. 349–368). Oxford, UK: Oxford University Press.

Ismail, S. (2001). The paradox of Islamist politics. *Middle East Report, 221*, 34–39.

Jameson, F. (1984). Postmodernism; or, The cultural logic of late capitalism. *New Left Review, 146*, 53–92.

Keyder, Ç. (1999). The setting. In Ç. Keyder (Ed.), *Istanbul between the global and the local* (pp. 3–30). Lanham, MD: Rowman & Littlefield.

Low, M., & Barnett, C. (2000). After globalization. *Environment and Planning D: Society and Space, 18*, 53–61.

Marfleet, P. (2000). Globalization, Islam and the indigenization of knowledge. In C. Nelson & S. Rouse (Eds.), *Situating globalization: Views from Egypt* (pp. 15–52). Bielefeld, Germany: Transcript Verlag.

Marston, S. (2000). The social construction of scale. *Progress in Human Geography, 24*, 219–242.

Nelson, C., & Rouse, S. (2000). Prologue. In C. Nelson & S. Rouse (Eds.), *Situating globalization: Views from Egypt* (pp. 9–14). Bielefeld, Germany: Transcript Verlag.

Ong, A. (1990). State versus Islam: Malay families, women's bodies, and the body politic in Malaysia. *American Ethnologist, 17*, 258–276.

Önis, Z. (2000). Neoliberal globalization and the democracy paradox: The Turkish general elections of 1999. *Journal of International Affairs, 54*, 283–306.

Pasha, M. K. (2002). Predatory globalization and democracy in the Islamic world. *Annals of the American Academy of Political and Social Science, 581*, 121–132.

Pieterse, J. N. (2001). Globalization and collective action. In P. Hamel, H. Lustiger-Thaler, J. N. Pieterse, & S. Roseneil (Eds.), *Globalization and social movements* (pp. 21–40). New York: Palgrave.

Roberts, S., Secor, A. J., & Sparke, M. (2003). Neoliberal geopolitics. *Antipode, 35*, 886–897.

Sayyid, B. S. (1997). *The fundamental fear: Eurocentrism and the emergence of Islamism*. London: Zed.

Secor, A. J. (2000). *Islamism in Istanbul*. Unpublished doctoral dissertation, University of Colorado–Boulder.

Stone, L. A. (2002). The Islamic crescent: Islam, culture and globalization. *Innovation, 15*, 121–131.

Tokatli, N., & Eldener, Y. B. (2002). Globalization and the changing political economy of distribution channels in Turkey. *Environment and Planning A, 34*, 217–238.

White, J. (1999). Islamic chic. In Ç. Keyder (Ed.), *Istanbul between the global and the local* (pp. 77–94). Lanham, MD: Rowman & Littlefield.

White, J. B. (2002). *Islamist political mobilization in Turkey: A study in vernacular politics*. Seattle: University of Washington Press.

Yeğenoğlu, M. (1998). *Colonial fantasies: Towards a feminist reading of orientalism*. Cambridge: Cambridge University Press.

15

Globalizing Democracy?

Autonomous Public Spheres and the Construction of Postnational Democracy in Mexico

PATRICIA M. MARTIN

On July 2, 2000, landmark presidential elections were held in Mexico ushering in a winning candidate from an opposition party for the first time in 71 years. The election of Vicente Fox, the candidate from the center-right party, the Partido Acción Nacional (PAN), brought to an end the longest standing one-party political regime of the 20th century. While presidential elections seem to affirm the national scope of democratic politics, there were many ways in which the 2000 elections in Mexico also represented a political event that extended beyond national borders. Thus the elections bore marks of the contemporary age. Several distinct facets demonstrate the transnational, even global, contours of the Mexican elections. First, the nature of the electoral campaign in the months prior to the elections was noted for reflecting U.S. influence (Frohling, Gallaher, & Jones, 2001). Continuous polling, televised debates, and televised political advertising mimicked U.S.-style electoral politics (Frohling et al., 2001). As a symptom of this trend, the English word "marketing" gained prominence in the Spanish lexicon to describe the process of image making employed by the presidential candidates (see, e.g., Cabildo, Montes, & Vera, 2000). To combat the success of Vicente Fox in this arena, Partido Revolucionario Institucional (PRI) candidate Fransisco Labastida hired James Carville (a former advisor of Bill Clinton) to serve as a political campaign consultant.

The presence of 860 "international visitors" acting as electoral observers provided a second clear dimension of the transnational nature of the Mexican elections. These visitors hailed from 57 countries and represented diverse groups such as Global Exchange, the Carter Center (with former president Carter also attending), the International Republican Institute, Common Borders of Canada, the European Parliament, and various political parties from countries in Latin America (Ibarra & Elizalde, 2000). The U.S. government reportedly spent close to $1 million in support of electoral observation programs in Mexico (Cason & Brooks, 2000). A third dimension that clearly demonstrated the globalized nature of the Mexican elections was provided by the lower house of the Italian parliament. They postponed ratification of the "Global Agreement" (formally, the Economic Partnership, Political Coordination and Cooperation Agreement) between the European Union (EU) and Mexico until after the presidential elections, emphasizing that ratification hinged upon evidence of transparent and clean elections. The political chapter of the agreement contained a "democracy clause," which articulated the general importance of free and fair elections, as well as respect for human rights (Ross, 2000). Finally, as another sign of the potential global dimensions of the electoral process in Mexico, the U.S. information firm ChoicePoint recently acquired Mexican voter registration data, along with that of seven other Latin American countries. This information was sold to the U.S. Justice Department in a move linked to the current antiterrorist campaign. A political uproar in Mexico over this series of events led ChoicePoint to purge and return the information to Mexican authorities (Ross, 2003).

As the brief discussion above indicates, there are many ways in which contemporary democratization in one state has become a transnational, if not a global, affair. Yet the implications of this trend remain uncertain. While some argue that globalization has contributed to the spread of democracy around the world (see Ward & Gleditsch, Chapter 8, this volume), others suggest that globalization has transformed national sovereignty in a way that leads to deficits in democratic governance (Held, 1991; McGrew, 1997; Sassen, 1996). Indeed, the political trends described above might lend support to both these characterizations. This chapter, based on research conducted in Mexico in the year 2000, offers an examination of some of the paradoxes and ambiguities of democracy in contemporary globalized times. I move away, nonetheless, from an analysis of democracy based on elections and instead follow the work of certain feminist and critical political theorists to examine the intersection of globalization and democracy from the lens of autonomous public spheres. Such a perspective emanates from an understanding of democratization as a struggle for the redistribution of power in multiple spheres of life (e.g., political, economic, and cultural) in a manner that effectively augments the autonomy and power of citizens and communities. Such a perspective resonates with popular and radical notions of democracy that emphasize the role of social movements within the terrain of civil society in constructing an active understanding of citizenship within an expanding horizon of rights (Dagnino, 1998; Laclau & Mouffe, 1985; Mouffe, 1992).

To develop this argument, this chapter has the following outline. In the next section I provide an introduction to the concept of autonomous public spheres within civil society. I argue, moreover, that under conditions of globalization, the geography of both civil society and autonomous public spheres may be shifting in a manner that is indicative of a "postnational" concept of democracy. The chapter then shifts onto Mexican terrain, with an analysis of the construction of autonomous public spheres. This provides a politicized account of the emergence of an autonomous, organized civil society with multiple historical and geographical origins. Such histories are imbricated, furthermore, with the reformulation of social, public identities. One of the primary goals of these organizations is the production of new kinds of information. Finally, I turn to the imagined territorial identities and networks within which these autonomous public spheres are emerging. My research indicates that the emergence of autonomous public spheres in Mexico is discursively linked to both local and global imagined communities. In light of this, I conclude the chapter by arguing that such a postnational engagement may be at odds with the nationally inscribed form of democracy that has emerged at a formal and institutional level. This examination leads me to the conclusion that the process of democratization in Mexico is caught between national and postnational political orders. This, I contend, raises complex questions about the potential for the consolidation of democracy.

AUTONOMOUS PUBLIC SPHERES

The concept of the autonomous public sphere, first articulated by Jürgen Habermas, has reshaped debates about participatory forms of democracy (Ferree, Gamson, Gerhards, & Rucht, 2002). In his conception, the autonomous public sphere is a discursive space that is public, yet autonomous from the state, comprised of citizens coming together to discuss issues of general concern or interest (Fraser, 1995; Garcelon, 1997). The autonomous public sphere therefore represents a mode of public political participation that moves beyond procedures such as voting. It provides a site through which citizens can engage in public dialogue and produce ideas about politics in a way that shapes governance and state action. The Habermasian notion of the public sphere locates the public within civil society (Devetak & Higgott, 1999; Garcelon, 1997; Habermas, 1989; Uribe-Urán, 2000). This location correlates, furthermore, with the emphasis on processes of democratization linked to social movements and the construction of citizenship from below. In this light, the term *public sphere* refers to the explicitly political, nonmarket, dimension of civil society (Habermas, 1989; Fraser, 1995; Garcelon, 1997).

Feminist and critical political theorists have both critiqued and reformulated Habermas's conception of the public sphere in important ways (Ferree et al., 2002). Some have argued that Habermas presented an idealized view of the public sphere that made claims to universality and openness, but that in fact it was shaped by exclusions based on class, race, and gender (Brooks-Higgenbotham, 1993; Ryan, 1998). Others have argued that the bourgeois public sphere played a role in constructing new forms of political domination. The public sphere could serve as a site, in other words, through which consent is manufactured (Eley, 1992; Urla, 1997). Nonetheless, there are aspects of the public sphere that some feminists and critical scholars have found useful (Landes, 1998). The notion of a public sphere resonates with the conditions of politics under modernity in which broad and diverse kinds of participation arise within the multiple, socially differentiated spaces of politics, economy, society, and culture (Benhabib, 1998; Cohen & Arato, 1992). Inasmuch as these differentiated realms are shaped by power, then "politics" has multiple dimensions; similarly there are multiple forms of participation.

In order to construct a more inclusionary model, the concept of the public sphere has been subject to important reformulations. For example, Ferree et al. (2002) highlight a concern for recognizing and valuing multiple modes of speaking or communicating. Others have interrogated both the issues and the spaces that define the boundaries of what is considered part of the "public" in order to demonstrate that these boundaries are not natural, but rather socially constructed (Benhabib, 1998; Collins, 1998; Fraser, 1995). Mouffe (1992) questions the degree to which consensus is a desired outcome within public deliberation (see also Ferree et al., 2002). Fraser (1995) provides a particularly important reinterpretation of the Habermasian autonomous public sphere that speaks to some of these reformulations. She suggests that rather than thinking of a homogenous and unified sphere, a metaphor of plural, multiple, and overlapping autonomous public spheres is more apt. Moreover, for societies traversed by systemic inequalities, Fraser provides the notion of "subaltern counterpublics" as vital spaces from which marginalized groups formulate public critique and discourse. Subaltern counterpublics provide locations in which marginalized groups can transform and rearticulate public, social identities. They can also serve as locations in which decision-making capacities can be developed in a manner that supports the construction of citizen governance.

The notion of subaltern counterpublics is important in the context of democratic transition in Latin America. As Alvarez, Dagnino, and Escobar (1998) point out, even in those Latin American societies that have undergone formal democratization, frequently only a small elite has access to the institutions of government. In such a circumstance, subaltern counterpublics that produce counterdiscourses, identities, and rights are "integral to an expansion and deepening of democratization" (Alvarez et al., 1998, p. 19). Thus subaltern counterpublics are sites in which new public identities can be formed and where the lines of what is considered public yet *autonomous from the state* can be shifted in ways that may deepen the process of democratization.

In Mexico the effective construction of autonomous political spaces has served as a particularly politicized aspect of the process of democratic transition. Olvera (2001) contends that agents within civil society have succeeded in constructing sites autonomous from the government. In so doing, "the common intention was to open new political spaces, achieve respect for social and political rights of citizens; and defend material interests through reform and negotiation with the government" (p. 8; my translation). Yet political struggle has marked this process (Fox, 1994; Robles Gil, 1998). In Mexico the right of associational autonomy, which is integrally related to the development of autonomous public spheres, is recognized in a de jure sense. Yet the recognition of the right of associational autonomy in a de facto sense has lagged behind (Fox, 1994). Movement toward the de facto right of associational autonomy has been constructed in a nonlinear manner through cycles of political conflict and negotiation (Fox, 1994). Robles Gil (1998) describes a similar trajectory, noting continued attempts by the Mexican state to bring organized civil society under political control, even as the transition to formal democracy has proceeded. Given this politicized and uneven history, recovering a perspective on the process of democratic transition through the lens of autonomous public spheres continues to be an important political and analytical task.

In assessing the emergence of autonomous public spheres in Mexico, it is also important to consider that globalization may also be qualitatively transforming the nature of political life. The processes of "globalization" in a Latin American context can certainly be traced back to the time of the Spanish Conquest (Klak, 1998; Saenz, 2002). Nonetheless, 1982, the year the Mexican government declared a moratorium on its international debt payments, serves as a pivotal moment in Mexico's most recent encounter with the contemporary round of global capitalist restructuring. Since that time the Mexican economy has been radically reorganized toward an export-oriented model of development, becoming fully integrated with the global economy (Gwynne, 1999). The heightened importance of the *maquiladora* (export processing) industry, Mexico's entrance into the Global Agreement on Tariffs and Trade in 1986, the privatization of communal land in 1991, and the final passage of the North American Free Trade Agreement (NAFTA) in 1993 are only a few indications of the scope of this process. This economic reorientation has been accompanied by a crisis in Mexican nationalism (Lomintz, 2001; Monsiváis, 1996) linked to the globalization and commodification of culture (Canclini, 1999). Additionally, increased transnational flows expressed through patterns of international migration and international tourism have also profoundly reshaped the Mexican social and cultural landscape (see, e.g., Brenner & Aguilar, 2002; Smith, 1998). It remains important to emphasize that this encounter with globalization has also been marked by profound crises in Mexico. Evidence for this lies not only in increased socioeconomic polarization (Dussel Peters, 2000), but also in repeated economic and political crises. While these have been periodic since 1982, the year 1994 is particularly notable in this regard. In January of that year, the Zapatista uprising occurred in the southern state of Chiapas; this was followed by a string of high-level political assassinations; and the year ended with the peso crisis in December (Castañeda, 1996). As I discuss below, the processes of globalization may impact the spaces of politics (Radcliffe, 1999) in a way that complicates a reading of the emergence of autonomous public spheres in Mexico.

In its original conception, Habermas's notion of the public sphere was linked to the rise of national societies; likewise modern democracy became institutionalized in the nation-state. As Habermas (2001) writes, "The phenomena of the territorial state, the nation, and a popular economy formed a historical constellation in which the democratic process assumed a more or less convincing institutional form" (p. 60). As such, there was a presumed spatial congruence between the state, the autonomous public sphere, and a common national identity. In his estimation, democracy included the process through which one part of society could reflexively intervene into this "historical constellation" in a manner that demonstrated that the concept of national society was politically meaningful. Yet, even if this congruence ever fully existed, it has been destabilized by the contemporary processes of globalization (Habermas, 2001; see also Devetak & Higgott, 1999; and Gupta & Ferguson, 1992). In light of these changes Habermas (and others) have argued that one contemporary transnational political task is the construction of a "postnational constellation," in which global economic processes can be brought under political control. Such a task implies not simply constructing more democratic transnational governance structures (Commission on Global Governance, 1995; Held, 1995), but the production of meaningful political and cultural identities that would foster democratic forms of political intervention (Habermas, 2001). Research in the arenas of transnational communities and identities (e.g., Smith & Guarnizo, 1998), cosmopolitanism (e.g., Cheah & Robbins, 1998), new forms of urban citizenship in cities (e.g., Holston, 1999), and global social movements (e.g., Cohen & Rai, 2000; Eschle, 2001) have begun to map the possible contours of a "postnational democratic constellation." Likewise, this chapter also pays attention to the imagined territorial communities that have supported the construction of autonomous public spheres in Mexico. Since Mexico is a country where a formal democratic transition—measured in terms of the victory and current tenure of a president from an opposition party—arrived well after the processes of globalization commenced (Lomintz, 2001; Rodríguez, 1997), the country provides fertile terrain for exploring the changing geography of contemporary democratic politics, particularly within the realm of civil society.

GEOGRAPHIES OF THE PUBLIC SPHERE IN MEXICO

During the year 2000 I spent 11 months in two urban locations in Mexico: Oaxaca, Oaxaca, and Monterrey, Nuevo León. These two places are situated in the south and the north of Mexico, respectively. In both sites I conducted a total of 72 open-ended interviews with individuals located in diverse institutions such as political parties, the government, the media, nongovernmental organizations (NGOs), unions, and the Catholic Church. The research I conducted aimed to capture a range of experiences with, and perspectives on, the process of democratic transition. One of the elements that I have investigated through these interviews is the emergence of autonomous public spheres in Mexico.

Taken together—as a set of social documents—the interviews I conducted form a web of interconnected and relational narratives shaped by social and spatial location (Nagar, 1997). As such, the interviews serve as transcripts at the interstices of discursive political change. Ideally, that is, they provide a view into the struggles over democracy in Mexico and the actors, institutions, and spaces involved in such a process. As a set of interconnected and relational narratives, the interviews are also embedded within the landscape of politics. I could examine the interviews I conducted in each location as a whole, looking for patterns and ruptures in the ways in which individuals describe their experience in politics. The kinds of patterns I examined, for example, included the mention of local issues of concern, the production and flow of information, descriptions of political change, and the political and social actors that were highlighted in such narratives.

An underlying assumption within this analysis is that politics is produced in places (Agnew, 2002). As such, the individuals whom I interviewed were embedded in a range of local networks. Taken together, their narratives provide a discursive topography of political space. In certain instances the interviews also describe the ways in which ideas and identities moved from informal and submerged local networks into the public arena. Still, evaluating the existence of multiple autonomous public spheres remains a difficult task. In part, this is because the notion of a "public sphere" is itself highly abstract. In order to approach such an examination, I have drawn inferences from both the landscape of political relationships and the nature of political practice in each place.

In what follows, I present a narrative of the emergence of autonomous public spheres that works through the voices and perspectives of the individuals I interviewed. I begin by placing the emergence of autonomous public spheres within the realm of civil society in Mexico. Next, I link the emergence of autonomous public spheres to the process of identity formation, drawing on examples from the feminist and indigenous movements. I then move to an overview of the kinds of political practices that shape these public spheres. Finally, I link these autonomous public spheres to a postnational set of imagined communities. All of the perspectives that I present in this narrative are drawn from individuals who worked outside of the government arena; they represent perspectives that emanate, in other words, from organized civil society in Mexico.

The Emergence of Autonomous Civil Society

In the long transition from a populist, semiauthoritarian, nationalist society, to a globalized, formally democratic, highly heterogeneous society, a shift can be noted in the kind of actors that work to counter state power and authority in Mexico. This transition can be understood as a shift from "popular movements" to "civil society." *Popular movements* are large mass-based movements (e.g., union, peasant, neighborhood movements), often inspired by an explicit socialist ideology, whose specific aim was to transform the state. On the other hand, *civil society* denotes smaller movements or institutions within civil society. Organized civil society tends to be identified with the language of democracy, rights, and citizenship. This transformation was highlighted in several of the interviews I conducted. For example, a woman working with a feminist organization in Nuevo León described this shift in the following manner:

> "The 1980s are the years of the awakening of the citizen because they gave such a brutal blow to the union movements, the teachers' movement, and the movement at Fundidora.[1] . . . In the case of Nuevo León, well, part are small previous struggles that were called NGOs, part are from civil associations, yes. Some are from the ecclesiastical base communities, and some are from radical groups that were linked with the guerilla movement. It is also important to say, Monterrey had very good leaders in the urban popular guerrilla movement. So, there were also people, and now everyone fits in the bag of NGOs. . . . "

This view provides an alternative understanding of the process of democratic transition. For not only is democratization linked to a struggle to define new forms of political practice, it is also linked to the decline of other political actors—a decline that this individual claims occurred through repressive means.

A distinct perspective was presented by an individual working with the Catholic Church in Monterrey. Though certain popular actors are still politically active, as he describes, they have less access to the political arena then they had previously. He links this transition to the rise of media institutions in Monterrey:

> "Tierra y Libertad[2] is still capable of bringing together 20,000 people to protest in the streets of the city. But all of a sudden a group of gays can schedule a press conference, where they are going to bring an image of Marilyn Monroe instead of the Virgin of Guadalupe, and that will receive greater coverage than the protest of 20,000, and it can be five gays. Yes. And, well, the media, public opinion, is more receptive to those kinds of smaller initiatives."

These narratives of political transformation challenge the notion that the process of democratic transition is somehow "natural" and that certain forms of politics simply fade away while others emerge. Rather, when viewed from within civil society, "democratic transition" implies a complex process of destruction and recomposition of political actors in relationship to broad institutional structures.

The following two individuals explain, nonetheless, why, in their estimation, the emergence of autonomous civil society organizations has been important. In the first quote a Oaxacan woman, who has had experience working in a variety of civil society organizations, emphasizes the importance of a highly autonomous notion of civil society, oriented toward developing decision-making capacities. She argues that organizations within civil society have to create their own initiatives, and draw on their own capacities to construct "a good life" without depending on "the authorities" (the government).

"We consider it important that civil society has its own initiatives and that implies thinking about ourselves from ourselves. Not in reference to the other, but from our own capacities, our own possibilities. In the end, we are the ones who comprise this place. We are the ones who live it; we are the ones who work it. In principle, we think that we have all of the possibilities of having a good life without depending on the authorities who really haven't bothered to listen to the community, and that means all of the political parties, even those that pretend to be very much on the left. . . . Indigenous organizations exist with a clear position of autonomy from the government. That means that their organization is sovereign, but among political organizations we cannot find one that can be identified with popular interests. . . . "

The following individual voices a similar perspective. In the long transition to democracy, he argues that the quality of formal political actors has deteriorated. As a result, both the government and political parties are unresponsive to the needs and interests of the community. This has resulted in a turn toward civil society organizations. This individual was working with a small independent magazine in Oaxaca at the time I interviewed him.

"I think that there have been some changes, but they are minimal. . . . What is true, what we have felt, is that the credibility of different political actors has deteriorated, primarily among the political parties. Well, we have realized that people no longer believe in political parties. . . . During the electoral seasons there are a lot of promises, a lot of talk, but then the process finishes and everything remains the same. And it comes again, since there are elections every 2 years, every year, every 3 years . . . and there are promises that remain. So the people, they stop believing in them, and there is more of a tendency toward nongovernmental organizations and toward civil associations because they are the ones that respond in some way at that moment to the realities and the needs of the community."

The emergence of autonomous public spheres in Mexico is embedded in this long and difficult transition, and the concomitant emergence of an autonomous civil society. The emergence of civil society organizations has, nonetheless, a diverse set of origins. Below I provide examples of two different trajectories that demonstrate the ways in which personal experiences and local networks have become politicized. In both instances, such mobilization was linked to the reconstruction of a public and social identity. The first example comes from an interview with a feminist journalist in Oaxaca. The second example comes from an individual working with an NGO promoting alternative development and indigenous rights.

> "We want to inform women about what is happening with us. We also want to visualize women because they do not appear in the media. There is a movement now. . . . In 1994 we created a group of women reporters . . . there was a female reporter who came from Mexico, and she invited us to reflect back to see what we had written as journalists about women. As it turned out, a lot of us followed her suggestion and we went to the library to see our own work or to see the work of others. And we found out that women did not appear in the news, except in the police report, as victims or perpetrators. . . . [*Interviewer: Yes, well many talk now about civil society, of an autonomous space.*] Yes, because it changes your form of being. When you are an individual, alone in your house, with problems, reflecting and reflecting about yourself, it is very difficult, and it does not go from there. . . . But when you share what is bothering you . . . as is the case when this woman came from Mexico City and asked us, 'What have you done, as journalists? What have you done to democratize information? So that women appear in the media?' And so we reflected together, we took the work home, and we pushed forward action."

The interview excerpt from this individual indicates that public spheres within civil society emerge through dialogue, reflection, and action. Democratization is, in this sense, a struggle. She also explains the importance of autonomous movements within civil society. Such autonomy provided a place in which both identity and political action could be rethought, in both a collective and an individual manner. Her description points to the process of constructing an autonomous public sphere engaged in reformulating the identity of women, particularly in how they appear in the local media. The goal of this effort was, in her words, "to make information more democratic" and to shift the boundaries with regard to what kinds of issues and whose voices are considered part of the public.

The interview excerpt below presents a distinct set of origins for the emergence of a civil society organization linked to the reformulation of a public identity:

> "The most important motivation was based in Christian principles, but in 1992 with the 500-year. . . . Well, some call it—in a very elegant fashion—the encounter of two worlds, others call it the conquest, and we say it was the Iberian invasion. . . . With the 500-year anniversary there is a critical juncture at the international level, within which we began to rethink our cultural identity, our roots, and with the Zapatista movement this only becomes stronger. So, our organization is in the process of recuperating our cultural identity. It has not been easy. Imagine people who, with a lot of effort, study for a career in a very hostile environment, in which indigenous peoples are marginalized, they are laughed at. We have friends whose parents speak Zapotec, and we, almost as a measure of security, we deny this and lose the language. So this process wasn't easy, to find yourself all of a sudden, to say that indigenousness has value."

The origins of this civil society organization can be found at the intersection of an international event (the 500-year anniversary of the Conquest), a national event (the Zapatista uprising), and local experience of cultural marginalization. This example indicates, therefore, that international political moments can spark local political transformation. As in the case above, one of the major interests of this organization is reinterpreting the meaning of a social identity, that of indigenousness. As this interviewee described later on, the organization had the additional goals of maintaining both autonomy from the government and organizational transparency. They also articulate a particular notion of democracy shaped by indigenous political practices.

Producing Information

Whether the origins of an organization are local or international in scope, the primary focus of many of the civil society organizations I interviewed was local in scope (either at the municipal or the state level).

As suggested above, organizations within civil society have been working to promote the construction of autonomous public spheres in a way
that introduced new identities and needs into the public arena. To do so,
one of the primary political strategies they use is the production of new
forms of information aimed at a local public audience (partial examples
of this have already been seen above). For example, a woman working
with a feminist organization in Monterrey emphasized that, in her perspective, the strategic production of information is more important than
mobilizing a mass based movement.

> "How many people belong? For example, in the Plural Pact there are
> about 40 people; in Feminist Millennium, Nuevo León, there are
> five people; and Catholics for the Right to Decide, well, there is just
> one in Nuevo León. So, there are a few people, but because of that
> we haven't been interested in massive work. One of these persons,
> for example, is a local congressional deputy, and that deputy can
> bring together 3,000 people, but that doesn't interest us. Now, if you
> have someone who writes in the newspaper, for me that is enough.
> Or if suddenly you are invited to give a lecture. Recently, we just
> gave a lecture to 600, 700 gynecologists, talking about emergency
> contraceptives. Why would you want a lot of people? And the gyne-
> cologists committed to support emergency contraceptives. Why
> would you want more people? And one person coordinated all of
> that. A doctor. So, it is not really a mass movement, it's a movement
> based in specific strategy."

Indeed, the production of public information from within organizations
in civil society took on a variety of forms. These included lobbying, public "performances,"[3] conducting opinion polls, and *consultas* (referen-
dums). Thus my research indicated that organizations within civil society place a high premium on tracking, constructing, and diffusing
information. I now turn to a few more specific examples of this political
strategy.
   The following excerpt comes from an individual working with a
feminist organization in Monterrey. She places emphasis on diffusing information and lobbying the state congress on certain issues:

> "Well, I think everything has a political aspect . . . all work that is de-
> veloped is political. Generally, we diffuse information about certain
> social issues that we would like to be seen. We 'lobby'[4] in the Con-
> gress so that certain initiatives become concrete."

Here is a second example. The following interview excerpt came from a
discussion that I held with a group that serves as an umbrella organiza-

tion for civil organizations in Monterrey. This organization conducts surveys to evaluate local municipal services. The organization then carries this information to local municipal governments, to provide a "citizens' analysis" of how the municipal government is doing. Under the auspices of this organization one can also see, therefore, a direct and public flow of information from organization to government.

> "Look, we have a series of commissions. Groups of people who participate. We have a health commission, a safety commission. . . . And these commissions do evaluations. Conduct surveys. And we approach the different offices of government in order to tell them, 'Look, according to the latest survey, this is what the people think of you.' "

Because access to the mainstream media remains tenuous in some instances, civil society organizations also construct alternative media in order to express their opinions and ideas. The following interview excerpt comes from an individual working with just such a project in Oaxaca. He begins by describing the appearance and general contents of the magazine he helps to edit:

> "There is no particular 'line' in this project. I mean, there is nothing that says this is leftist and against the Right. Here we include a series of opinions and concerns, and points of view. . . . Perhaps one of the requirements is quality, but there is no ideological limitation. The people that participate in each number are different. . . . We might say that this is political in the sense that we diffuse information, but not in the sense of following a particular person, someone who wants to be a deputy in congress. [*Interviewer: Is it part of the process of democratization?*] I think the objective is to become aware of different activities, different forms of thinking, expressions that allow consciousness to be generated."

The editors of this magazine see their role as one of diffusing information and ideas. This creates the possibility, in their perspective, of transforming politics from a practice of personality to one of generating consciousness. They also see the project as not being defined by ideology (left or right) but rather by debate and opinion. This is a project, therefore, that is suggestive of the construction of an alternative public sphere, directed in particular toward the Oaxacan society.

Imagined Communities

Thus far, I have explored the origins of autonomous civil society organizations in Mexico that promote the construction of autonomous public spheres. These spheres serve as sites for identity construction and for the production of information that makes alternative needs and interests public. A final dimension of the construction of autonomous public spheres that should be discussed is the imagined communities upon which the individuals working within these arenas draw. Modern political theory has suggested that there is a territorial congruence between the imagined communities in which we live and the scope of political action in which we engage. One of the most striking outcomes of the research that I conducted in Mexico was the lack of this kind of congruence. Here, in particular, a turn toward a postnational, and at times global, sense of political community can be seen. This move can be linked to the imagined and material networks within which civil society organizations are immersed and it also reflects a decisive turn away from a strictly national identity and community.

I will highlight four examples to demonstrate a range of perspectives. The first perspective, provided by a man who works with a consumer NGO in Monterrey, affirms a national sense of identity. The following three examples, however, turn away from a strong articulation of a national identity, and toward a postnational sense of belonging in which local and global identities are more salient than a strictly national one.

> "I think at the bottom, it is authentic, that feeling, that identity of being Mexican. I think what has hurt the expression of 'Mexicanness' a lot is the government that we have had—government that has not moved human development forward, the growth of people. I think that we have a lot of values about which we can be proud . . . a rich history, artistic values, profound roots—I mean I believe it and I feel it. I am convinced that it exists."

In contrast to the perspective above, the individual quoted below views a concern with a strictly national sense of identity as part of the authoritarian history of Mexico, in which a homogeneous mestizo identity was imposed at the expense of local place-based and ethnic differences. In this light, engagement with local differences forms part of a shift toward a more democratic culture. The importance of acknowledging and affirming local place-based identities was a common theme in the interviews I conducted. The following individual, who expressed such an opinion, worked with a newspaper in Monterrey at the time I interviewed him.

> "I think that the discourse of identity in Mexico is part of the authoritarian discourse. When we are the country that has the most living

languages in the world, right? It's amazing, and where are they? We are a cultural mosaic more complicated than a jigsaw puzzle and we are worried that everyone be the same, I don't know, that discourse of identity. Furthermore, it's a centralist discourse."

Instead he argues:

"We should be concerned with all of the prejudices we have. Certainly, the search for diversity in terms of culture is something that did not happen before midcentury because the thumb was placed on the issue of identity. . . . Oaxaca and Guadalajara are cities that are completely different. We should look for and live alongside that mosaic, and really allow different cultures to blossom and participate."

Another common theme, which ran in quite the opposite direction, was an engagement with a global sense of identity and community. This perspective is eloquently depicted in the quote below:

"I dream that that idea of nations as differentiated groups, let's say, will be destroyed some day. I think that the tendency at the world level is going in that direction. Now one speaks of 'the global village,' of interconnection through the Internet, of technology. Science has been evolving rapidly and is taking human beings to confront that kind of challenge, of what it means to be Mexican, what it means to be North American. I think that those are differences marked by groups in the government that have to do with economic questions. But little by little and with the pressure of the same technological advance, but also with political groups in the world that are so diverse, so dissimilar, they will have to become diluted. It has to do with the exploitation of resources, but also with artificial borders and distances. So I consider myself a citizen of the world, where I can talk perfectly, and at the same level with another citizen from another country."

This individual worked with an AIDS-advocacy NGO in Monterrey and thus was concerned about the needs and interests of a particular group, that of sexual minorities. Although this group was enmeshed in networks at the national level, its primary arena of action and advocacy was the local state. Nevertheless, the imagined community in which he places himself is a global one. A similar perspective was provided by a person who works with an organization that defends the rights of people who have been subject to practices of detention-disappearance by Mexican authorities. Her perspective reinforces the notion that the processes of globalization have profoundly influenced political subjectivities within civil society in Mexico.

> "I think that more than talking about nationalisms, now that we are in an era of globalization in which the peculiarities of each country tend to disappear, I think that we should fight for an identity, for the good of humanity. I believe that it is respect for the human being. I believe that, more than talking about nationality, it is to speak of an identity built out of our similarities, the respect for our similarities."

What emerged, therefore, in many (though not all) of the interviews I conducted with organizations in civil society was a turn toward both local and global senses of community and belonging. What is significant about this recourse to imagined communities at both the global and the local scale is not that either a global or a local sense of identity is, in an abstract sense, "better" than a national identity. Rather, these alternative imagined communities serve as crucial political resources for rethinking and reimagining political life. In this sense, the emergence of autonomous public spheres in Mexico may be at least partially embedded within a "postnational" democratic imagination.

CONCLUSIONS

In this chapter I have examined the process of democratization in Mexico in relationship to the construction of autonomous public spheres. This view has outlined a path of political struggle, of the articulation of new kinds of identities, and the production of new kinds of information. This process was embedded in the long transition to democracy of the 1980s and 1990s. During that time period, as the legitimacy of the Mexican national government continued to erode, organizations within civil society sought distinct arenas through which to construct a democratic politics. These arenas were both local and transnational in scope. The origins of some organizations within civil society can be traced to transnational moments; others are much more local in origin. The audiences for who autonomous public spheres are constructed are often decidedly local, and yet they are embedded within imagined communities that reside at both the subnational and the supranational scales. Thus the geography of this alternative "postnational" political order is both subnational and transnational in scope.

Yet, as formal democratization unfolds in Mexico, there have been multiple ways in which the newly elected Mexican government has been embraced on the international stage, particularly after the 2000 elections. As this has occurred, the legitimacy and authority of the nation-state in Mexico vis-à-vis democracy has been reestablished. Signs of this trend include the fact that Mexico was elected to a spot on the United Nations Security Council in 2001. In addition, two prominent international meetings have been hosted in Mexico (the United Nations International Conference on Financing for Development, held in Monterrey in 2001, and the World Trade Organization Ministerial Meeting held in Cancun in 2003). Finally, Vicente Fox received the International Medal of Democracy from the Center for Democracy in 2000. This reinscription of democracy within the nation-state suggests that a mismatch may exist between the spaces of democracy found at the formal political level and the spaces of democracy found at the informal political level. In other words, democratic transition in Mexico may be caught between a "national constellation" at the formal level and a "postnational constellation" embedded within civil society. Furthermore, as the national government gains increased legitimacy by reasserting the national boundaries of politics, this may marginalize and delegitimize alternative geographies of democracy that have arisen in contemporary Mexico, weakening movements to expand and deepen the process of democratization.

NOTES

1. "Fundidora" refers to a foundry in Monterrey that was closed in the 1980s. The workers at Fundidora were reputed to be radical and autonomous from government control.
2. Tierra y Libertad is an urban-popular movement in Monterrey that founded an expansive communal neighborhood organization after a large squatter land invasion in 1973.
3. The word "performance" was spoken in English.
4. The word "lobby" was spoken in English.

REFERENCES

Agnew, J. (2002). *Place and politics in modern Italy.* Chicago: University of Chicago Press.

Alvarez, S., Dagnino, E., & Escobar, A. (1998). Introduction: The cultural and the political in Latin American social movements. In S. Alvarez, E. Dagnino, & A. Escobar (Eds.), *Culture of politics, politics of culture: Re-visioning Latin American social movements* (pp. 1–29). Boulder, CO: Westview Press.

Benhabib, S. (1998). Models of public space: Hannah Arendt, the liberal tradition, and Jürgen Habermas. In J. B. Landes (Ed.), *Feminism: The public and the private* (pp. 65–99). Oxford, UK, and New York: Oxford University Press.

Brenner, L., & Aguilar, A. G. (2002). Luxury tourism and regional economic development in Mexico. *Professional Geographer, 54*(4), 500–520.

Brooks-Higgenbotham, E. (1993). *Righteous discontent: The women's movement in the black Baptist Church, 1880–1920.* Cambridge: Harvard University Press.

Cabildo, M., Montes, R., & Vera, R. (2000). Fox hacía la presidencia: Mil 91 días de marketing. *PROCESO, 1235,* 18–21.

Canclini, N. G. (1999). *La globalización imaginada.* México, D.F.: Editorial Paidos Mexicana.

Cason, J., & Brooks, D. (2000, June 30). Un millon de dólares, gasto de EU en observadores electorales. *La Jornada* (México, D.F.), p. 11.

Castañeda, J. (1996). *The Mexican shock: Its meaning for the United States.* New York: New Press.

Cheah, P., & Robbins, B. (Eds.). (1998). *Cosmopolitics: Thinking and feeling beyond the nation.* Minneapolis: University of Minnesota Press.

Cohen, R., & Arato, A. (1992). *Civil society and political theory.* Cambridge, MA: MIT Press.

Cohen, R., & Rai, S. (Eds.). (2000). *Global social movements.* London and New Brunswick, NJ: Athlone Press.

Collins, P. H. (1998). *Fighting words, black women and the search for justice.* Minneapolis and London: University of Minnesota Press.

Commission on Global Governance. (1995). *Our global neighborhood.* Oxford, UK: Oxford University Press.

Dagnino, E. (1998). Culture, citizenship and democracy: Changing discourses and practices of the Latin American left. In S. Alvarez, E. Dagnino, & A. Escobar (Eds.), *Culture of politics, politics of culture: Re-visioning Latin American social movements* (pp. 33–63). Boulder, CO: Westview Press.

Devetak, R., & Higgott, R. (1999). Justice unbound?: Globalization, states, and the transformation of the social bond. *International Affairs, 75*(3), 483–498.

Dussel Peters, E. (2000). *Polarizing Mexico: The impact of liberalization strategy.* Boulder, CO, and London: Lynne Rienner.

Eley, G. (1992). Nations, publics, and political cultures: Placing Habermas in the nineteenth century. In C. Calhoun (Ed.), *Habermas and the public sphere* (pp. 289–339). Cambridge, MA: MIT Press.

Eschle, C. (2001). *Global democracy, social movements, and feminism.* Boulder, CO: Westview Press.

Ferree, M., Gamson, W., Gerhards, J., & Rucht, D. (2002). Four models of the public sphere in modern democracies. *Theory and Society, 31,* 289–324.

Fox, J. (1994). The difficult transition from clientelism to citizenship: Lessons from Mexico. *World Politics, 46*(2), 151–184.

Fraser, N. (1995). Politics, culture, and the public sphere: towards a postmodern conception. In L. Nicholson & S. Seidman (Eds.), *Social postmodernism: Beyond identity politics* (pp. 287–312). Cambridge, UK: Cambridge University Press.

Frohling, O., Gallaher, C., Jones, J. P. III. (2001). Imagining the Mexican election. *Antipode, 33*(1), 1–16.

Garcelon, M. (1997). The shadow of the Leviathan: Public and private in communist and post-communist society. In C. Calhoun (Ed.), *Public and private in thought and practice* (pp. 303–332). Chicago: University of Chicago Press.

Gupta, A., & Ferguson, J. (1992). Beyond "culture": Space, identity, and the politics of difference. *Cultural Anthropology, 7*(1), 6–23.

Gwynne, R. (1999). Globalization, neoliberalism and economic change in South America and Mexico. In R. Gwynne & C. Kay (Eds.), *Latin America transformed: Globalization and modernity* (pp. 68–97). London: Edward Arnold.

Habermas, J. (1989). *The structural transformation of the public sphere.* Cambridge, MA: MIT Press.

Habermas, J. (2001). *The postnational constellation.* Cambridge, MA: The MIT Press.

Held, D. (1991). Democracy, the nation-state and the global system. In D. Held (Ed.), *Political theory today* (pp. 197–235). Cambridge: Polity Press.

Held, D. (1995). *Democracy and the global order.* Cambridge, UK: Polity Press.

Holston, J., (Ed.). (1999). *Cities and citizenship.* Durham, NC: Duke University Press.

Ibarra, M. E., & Elizalde, T. (2000, July 3). Avalan observadores el proceso electoral. *La Jornada* (México, D.F.), p. 14.

Klak, T. (1998). Thirteen theses on globalization and neoliberalism. In T. Klak (Ed.), *Globalization and neoliberalism: The Caribbean context* (pp. 3–23). Lanham, MD: Rowman & Littlefield.

Laclau, E., & Mouffe, C. (1985). *Hegemony and socialist strategy.* London: Verso.

Landes, J. B. (1998). The public and the private sphere: A feminist reconsideration. In J. B. Landes (Ed.), *Feminism: The public and the private* (pp. 135–163). New York: Oxford University Press.

Lomintz, C. (2001). *Deep Mexico, silent Mexico.* Minneapolis: University of Minnesota Press.

McGrew, A. (1997). Democracy beyond borders?: Globalization and the reconstruction of democratic theory and practice. In A. McGrew (Ed.), *The transformation of democracy?* (pp. 231–266). Cambridge, UK: Polity Press.

Monsiváis, C. (1996). Will nationalism be bilingual? In E. McAnany & K. Wilkinson (Eds.), *Mass media and free trade: NAFTA and the culture industries* (pp. 131–141). Austin: University of Texas Press.

Mouffe, C. (1992). Preface: Democratic politics today. In C. Mouffe (Ed.), *Dimensions of radical democracy* (pp. 1–14). London: Verso.

Nagar, R. (1997). The making of Hindu communal organizations: Places and identities in postcolonial Dar es Salaam. *Environment and Planning D: Society and Space, 15,* 707–730.

Olvera, A. J. (2001, September 6–8). *Sociedad civil, esfera pública y gobernabilidad democrática en México.* Paper presented at the annual meeting of the Latin American Studies Association, Washington, DC.

Radcliffe, S. A. (1999). Civil society, social difference and politics: Issues of identity and representation. In R. Gwynne & C. Kay (Eds.), *Latin America trans-*

formed: Globalization and modernity (pp. 203–223). London: Edward Arnold.

Robles Gil, R. R. (1998). *Abriendo Veredas: Iniciativas públicas y sociales de las redes de organizaciones civiles.* México, D.F.: Convergencia de Organismos Civiles por la Democracia.

Rodríguez, V. (1997). *Decentralization in Mexico.* Boulder, CO: Westview Press.

Ross, J. (2000). Zapping free trade. *Texas Observer.* Article available online at *http://www.texasobserver.org.* Accessed on October 21, 2003.

Ross, J. (2003). Mexican data grab. *The Progressive, 67*(8), 30–33.

Ryan, M. P. (1998). Gender and public access: Women's politics in nineteenth-century America. In J. P. Landes (Ed.), *Feminism: The public and the private* (pp. 195–222). New York: Oxford University Press.

Saenz, M. (2002). Introduction: Periphery at the core. In M. Saenz (Ed.), *Latin American perspectives on globalization* (pp. 1–21). Lanham, MD: Rowman & Littlefield.

Sassen, S. (1996). *Losing control?: Sovereignty in an age of globalization.* New York: Columbia University Press.

Smith, M. P., & Guarnizo, L. E. (Eds.). (1998). *Transnationalism from below.* New Brunswick, NJ: Transaction.

Smith, R. C. (1998). Transnational localities: Community, technology and the politics of membership within the context of Mexico and U.S. migration. In M. P.Smith & L. E. Guarnizo (Eds.), *Transnationalism from below* (pp. 196–238). New Brunswick, NJ: Transaction.

Uribe-Urán, V. M. (2000). The birth of a public sphere in Latin America during the age of revolution. *Comparative Studies in Society and History, 42*(2), 425–457.

Urla, J. (1997). Outlaw language: Creating alternative public spheres in Basque Free Radio. In L. Lowe & D. Lloyds (Eds.), *The politics of culture in the shadow of capital* (pp. 280–300). Durham, NC: Duke University Press.

16

Global Institutions and
the Creation of Social Capital

WILLIAM MUCK

The presence or absence of social capital can be detected in nearly every facet of public life. Its presence can mean the difference between waiting in an orderly line to board an arriving bus, or having to push, shove, and wrestle your way through an unsympathetic mass of humanity to find your way onto what is inevitably an overcrowded bus. At the community level, social capital can be plainly observed in the equitable and efficient delivery of local services. The mail gets delivered on time and the trash is regularly picked up. Government officials, whether they are health care providers, police officers, or teachers, actually show up for their job and take pride in their work. At the international level, social capital fuels the open exchange of vital information about technology, global markets, and various strategies of development. All told, its presence enables citizens to rise above their own narrow and immediate self-interest and embrace the benefits that come from collective engagement.

Inspired by the foundational research of James Coleman (1990) and Robert Putnam (1993), social capital—understood here as the norms of civic engagement that foster collective action—has become a hotbed of academic interest over the last decade. Countless scholars have become captivated by a quest to understand the macroimplications of a very simple idea: that everyone would be better off if we could all trust each other and work together. In fact, much of social capital's appeal appears to stem from the simplicity of the idea. For instance, as Woolcock (1998) points out, on some basic level one expects that communities that have frequent social interactions and openly trust one another would be

> safer, cleaner, wealthier, more literate, better governed, and generally "happier" than those with low stocks, because their members are able to find and keep good jobs, initiate projects serving public interests, costlessly monitor one another's behavior, enforce contractual agreements, use existing resources more efficiently, resolve disputes more amicably, and respond to citizens' concerns more promptly. (p. 155)

The empirical evidence to emerge from the growing body of social capital literature has reinforced this notion, indicating that communities with an active, engaged, and trusting civil society are able to overcome collective action problems and are thus better positioned to solve the economic, political, and social problems that plague all communities. All told, the evidence suggests that we should think of social capital as being akin to the oil that enables the smooth operation of any high-performance machine. Just as a Mercedes Benz will find itself broken down on the side of the road if it is not infused with the proper oil, a community endowed with significant material resources will have difficulty achieving economic and political success without the requisite stock of social capital to lubricate the development process. Therefore, whether it is fueling democratic growth (Putnam, 1993), improving public school performance (Coleman, 1990; Schneider, Teske, Marschall, Mintrom, & Roch, 1997), enhancing the banking industry (Jackman & Miller, 1998; van Bastelaer, 2000), providing local health care (Tendler, 1997), creating an irrigation or sewer system (Lam, 1996; Ostrom, 1996), fostering industrial or agricultural growth (Heller, 1996; Reid & Salmen, 2000), or fueling the "East Asian Miracle" (Evans, 1995; Stiglitz, 1996), social capital has repeatedly been shown as an indispensable factor in economic and political development.

The implication of the findings of this substantial body of research is that greater attention needs to be devoted to finding ways to artificially produce the social capital that Putnam (1993) and others have deemed so critical to proper economic and political development. In other words, because the causal relationship between social capital and economic and political development appears now to be sufficiently established, scholars ought to redirect their effort toward identifying the institutional conditions conducive to the emergence of social capital. Whereas many theories of social capital initially implied that civil society sprang organically from long-standing communities (Putnam, 1993; Fukuyama, 1995), more recent evidence suggests that under the right institutional conditions social capital can in fact be cultivated (Schneider et al., 1997; Skocpol, 1996).

In light of the above discussion, this chapter considers the role that intergovernmental organizations (IGOs) and nongovernmental organizations (NGOs), working in tandem with the state, can play in developing civil society and social capital as part of a broader effort to promote development in a globalized world economy. The above organizations have, with increasing regularity, attempted to intervene in the political and social relationships and practices within communities in ways that will enhance their abilities to work together to address collective action problems. Traditionally, scholars have looked to the state as the primary source for the mobilization of civil society. While the state remains a principal motivator, the ensuing analysis will demonstrate that IGOs and NGOs have emerged as vital new resources for the mobilization of civil society and social capital. In fact, nonstate actors like the World Bank have assumed a leading role in the globalization of ideas about the crucial role that civil society and social capital play in the development process.

In order to accurately review the ability of IGOs, NGOs, and the state to cultivate social capital, it is necessary that one first recognize that social capital is a multidimensional concept that can govern a whole host of social interactions. The prominence of Putnam's (1993) work on Italy and his exclusive focus on horizontal networks of civic engagement (churches, clubs, and civic groups) may have contributed to a false assumption that social capital refers exclusively to social trust existing at the communal level. In contrast, recent research has suggested that social capital can be applied to a wide variety of issue areas and can exist in a multitude of networks that govern social relations extending all the way to the global level (Evans, 1996; Gittell & Vidal, 1998; Grootaert & van Bastelaer, 2002; Woolcock, 1998; Woolcock & Narayan, 2000). Globalization has tied the world together, and in doing so has dramatically increased the number and intensity of interactions among previously isolated populations. It is important that we expand our understanding of social capital to take these different, more global, social networks into account. Woolcock (1998), for instance, details three social relationships that are governed by three very different forms of social capital; intracommunity ties (integration), extracommunity ties (linkage), and state–society relationships (synergy). This chapter makes use of Woolcock's three categories and reviews successful efforts made by IGOs, NGOs, and the state to produce these three distinct types of social capital.

Substantively, the chapter is organized into five sections. I begin by briefly addressing the debate surrounding the possibility of creating social capital. While a significant amount of evidence exists to suggest that social capital can in fact be created, this has yet to become the universally accepted position within the literature. The next three sections detail each of the three dimensions of social capital (integration, linkage, and synergy), but, more importantly, review attempts by IGOs, NGOs, and the state to nurture these three dimensions of social capital. The concluding section summarizes the findings and suggests that in order to fully reap the economic and political benefits of social capital efforts must be made to simultaneously cultivate all dimensions. Promoting only one or two elements of social capital will undercut efforts to advance economic and political development.[1]

CAN SOCIAL CAPITAL BE CREATED?

While the stated purpose of this chapter is to detail the efforts of IGO, NGOs, and the state to produce social capital, one should not assume that there is universal agreement among scholars that social capital can be manufactured. In fact, a tense debate has emerged between scholars who see social capital as an enduring cultural value that is extremely difficult to produce artificially, and those who reject the idea that social capital is culture-based and instead see it as the product of an enabling institutional environment. The outrage expressed by an Italian regional president, in response to Putnam's assertion that the prospects of creating social capital in southern Italy appeared bleak, is representative. Declaring Putnam's findings "a counsel of despair!," the regional president was obviously troubled to hear that "the fate of the reform was sealed centuries ago" (Putnam, 1993, p. 183). Putnam (1993) and Fukuyama (1995) have become most widely known for treating social capital as culture. For those who see social capital in these cultural terms, the idea that one can instantaneously create a culture of reciprocal trust is preposterous. Developing the necessary networks of civic engagement and proper norms of reciprocity is a terribly slow process that requires centuries to emerge. Fukuyama (1995), for instance, has suggested that one should think of social capital as "a ratchet that is more easily turned in one direction than another" (p. 62). In other words, it is much easier to dismantle existing stocks of social capital than to build them up again.

While thinking of social capital as a largely cultural variable has become popular of late, Jackman and Miller (1998) note this "was certainly not the intent of the original social-capital theorists" (p. 50).

Citing some of the earliest social capital research, Jackman and Miller critique the exclusively cultural conception of social capital and point out that the founding social capital research was embedded in a rational choice framework. Furthermore, this early research placed a priority on identifying the institutional conditions conducive to the emergence of social trust. Recently, a growing number of scholars, frustrated with the determinism of the cultural understanding of social capital, have revisited the institutional argument and endeavored to find whether the right institutional environment could not trigger a blossoming of social capital. This research has produced a number of interesting and optimistic findings (Bebbington & Carroll, 2000; de Souza Briggs, 1997; Schneider et al., 1997). My intent in this chapter is to build upon these recent findings by highlighting the ability of nonstate actors, working in tandem with the state, to promote three distinct types of social capital.

PROMOTING INTRACOMMUNITY TIES: INTEGRATION

Integration, the most widely discussed type of social capital, looks at the number and degree of intracommunal ties that exist within a given society. Popularized by Putnam (1993), this type of social capital emerged when scholars were unable to account for the drastically different levels of economic and political success experienced by states endowed with very similar amounts of human and physical capital. These markedly different levels of economic and political performance led scholars to ask whether there was not some internal societal factor that influenced the ability of a community to productively act upon its natural resource base. Putnam's (1993) well-known finding for Italy was that different levels of civic engagement and communal trust explained the drastic differences in economic and political performance one saw between the northern and southern regions of Italy.

The logic behind integration is that an extensive network of horizontal civic associations can foster norms of trust and communal reciprocity. These norms of reciprocity allow a community to overcome collective action problems and in turn, generate better economic and political outcomes. In other words, the more individuals interact in sports leagues, church clubs, music or theater groups, study groups, or any other type of local association, the more they learn to trust one another. When citizens trust one another, they are more likely to help out those who have fallen on hard times, devote time to a school or community project, or simply keep their yard clean. However, this collective action can only emerge when individuals within the community have overcome their fear of the "free rider" and are sufficiently convinced that everyone within the community will contribute to the collective (Axelrod, 1984). The free rider is someone who benefits from the collective good without contributing to it—for instance, someone who enjoys and utilizes a community's park system yet refuses to pay the taxes that sustain the park. Only if individuals are convinced of others' good intentions will they put themselves at greater risk by contributing to the collective good.

The State: Reforming the Institutional Context

The question then becomes, How does one promote or encourage the development of this type of social capital? The obvious answer is simply to create an environment conducive to the emergence of a large number of horizontal associations. If one increases the number of organizations that the average individual participates in, intracommunal trust (integration) will soon follow. However, it is important to remember that individuals are not predisposed to join any and every organization put in front of them. An individual needs a rational reason to organize collectively. Consequently, our focus needs to be on the role IGOs, NGOs, and the state can play in arousing interest in joining these horizontal associations. Research on group membership has suggested that communities characterized by serious power imbalances are unlikely to produce a significant number of horizontal associations (Das Gupta, Grandvoinnet, & Romani, 2003). Absent an adequate stake in the economic or political proceedings, the average individual will not have much interest in organizing, other than possibly to protest lack of access. However, if an individual is empowered by being granted an economic or political stake (e.g., he or she has something to lose), it is very likely that he or she will become vastly more interested in the affairs of the community.

For instance, Whang (1981) and Das Gupta et al. (2003) show how efforts taken by the state to redistribute land within South Korea, Taiwan, and China provided local peasants with a tangible incentive to pursue organizational engagement. In the case of South Korea, Whang (1981) argued that land reform enacted by the state "provided the social and psychological preconditions necessary for co-operation among farmers" (pp. 97–98). In these cases, the state took the first step by reconfiguring the local institutional conditions (land reform), and, in doing so, provided members of the community with a greater stake in the economic and political proceedings. Horizontal organizations rapidly emerged in these communities as local farmers endeavored to pool their funds in order to acquire irrigation, plowing, and other agricultural necessities (Foner, 1989; Das Gupta et al., 2003).

Analysis of the political success achieved in the Novgorod region of

Russia suggests that reconfiguring the local political context through electoral reform can similarly generate increased political participation and horizontal organization. For instance, in his analysis of Novgorod, Petro (2001) details the remarkable political success that this region has enjoyed and the social capital it has been able to produce despite the fact that it resides in a state struggling through a very painful economic and political transition. Novgorod has transformed itself from being gravely deficient in civic activism to being "among the top quarter of all Russian regions in number of clubs and cultural associations per capita" (Petro, 2001, p. 234). According to Petro, the opportunity to have a real say in the political process has given the people of Novgorod a tangible incentive to organize and participate in the governing process. It appears that by altering the economic and political context and giving people a say in these processes, the state can encourage participation as well as the horizontal associations needed to organize these political interests.

I provide these examples not to suggest that all one needs to do in order to generate horizontal associations (integration) is to institute land and electoral reform, but rather to suggest that in efforts to promote integration it is critical to look at the root causes of group membership. The land and electoral reform examples suggest that the state can encourage organizational participation that alters the institutional context so that previously nonparticipating individuals find themselves with a tangible incentive to join. The challenge is to find ways, other than through land and electoral reforms, to give individuals a sufficient reason to seek out others for collective action. A perfect example of this is the Anning Valley Agricultural Development Project occurring in China. A joint effort of the Chinese government, the World Bank, and local NGOs, this project attempts to promote local development and civic engagement by empowering local women's groups (World Bank, n.d.-a). The $120 million project covers 15 rural communities and has incorporated local women's federations in all aspects of the agricultural development project, from design and planning to the actual implementation. By targeting a previously isolated group, the Anning Valley Project has promoted local organization, reduced the gender gap within these communities, and—most importantly—generated economic growth.

Sustaining Horizontal Integration

Although the first step in attempting to encourage the development of integration needs to focus on setting the preconditions for the emergence of civic engagement, the second crucial step to consider is the role these organizations can play in helping to preserve and develop these horizontal organizations once they emerge. While establishing the right institutional environment for the emergence of these organizations is critical, if these organizations are not subsequently provided with the proper organizational training and financial assistance, their survival is often uncertain. These efforts need not always be overly complex or financially taxing. For instance, Fox (1996) points out that the state in Mexico has supported local peasant organizations by providing transportation to peasants that otherwise might not be able to get to meetings. This very simple and low-cost effort is just one way that the state can help sustain budding local organization. In Sweden, the government has attempted to sustain organizational engagement and to develop social capital by funding local study groups. These study groups, which 75% of the adult population has attended at some point, are small groups that meet about once a week to talk and educate themselves on various subjects ranging from foreign languages, cooking, and computers, to the European Union (Rothstein, 2001, p. 216). While the education these study groups provide is important, it is their social component that significantly improves the overall feeling of intracommunal trust throughout the society and makes them a sound investment for the Swedish government.

Using the comparative advantage it has in finance, management, and training skills, the World Bank has funded a number of programs that support the development of civic organizations. Two such programs are the "Small Grants Program" and the "Civic Association Outreach and Training Program." The former provides grants directly to local civil society organizations hoping to strengthen the capacity of these small indigenous organizations. The latter program also aims to strengthen local organization by providing local civic associations with the "analytical and strategic tools that they need to participate more effectively in the public debate" (Monico, 2002, p. 27).

NGOs, working in tandem with the World Bank, have played a similar role in helping to sustain budding local organization. For instance, in the "Benin Food Security Project," the World Bank and international NGOs jointly endeavored to promote the long-term development of local organization by "providing informational, technical and logistical support and facilitating technology transfer" to emerging local organizations (World Bank, n.d.-b). Similarly, the World Bank's "Improved Environmental Management and Advocacy Project" in Indonesia teamed an international NGO up with 12 local organizations in a training effort designed to improve the capacity of local NGOs to address the community's environmental dangers (World Bank, n.d.-c). Finally, in India, the World Bank's "Uttar Pradesh Land Reclamation Project" has provided funding to local NGOs to organize and train local farmers' water management groups on reclaiming salt-affected lands. Despite the fact that many of these programs are in the very early stage of development, they have nevertheless initiated widespread communal interaction. Moreover, once these intracommunity ties are established, they open up countless other opportunities for community development.

In summary, it appears that the development of integration is dependent upon the state taking a proactive role in establishing the institutional preconditions for the emergence of intracommunal cooperation. However, once the conditions for cooperation are created, IGOs and NGOs play an indispensable role in the development of horizontal organization through financial, logistical, and organizational training. Thus while the state continues to play a leading role in the mobilization of civil society, nonstate actors like IGOs and NGOs have emerged as equally important resources for the development of intracommunal relationships.

PROMOTING EXTRACOMMUNITY TIES: LINKAGE

While few doubt the importance of intracommunal relationships, a growing number of scholars (Coleman, 1990; Gittell & Vidal, 1998; Portes, 1998) have found that extracommunal ties (linkages) that transcend local loyalties are equally critical to promoting economic and political development. Dense horizontal networks, while useful for promoting cooperation at the local level, can come with a price and can hinder the development of critically important vertical relationships that extend beyond the communal level. Communities with exclusively horizontal networks of association tend to develop what has been called "negative social capital." Negative social capital comes about when group loyalties are so robust that they effectively segregate the group from anyone not affiliated with the group. This has proven to be extremely hazardous in the era of globalization because without the proper vertical relationships, a community is unable to participate in the extracommunal (i.e., global) networks that share vital information about technology, international markets, or various strategies of development. Gittell and Vidal (1998) describe the difference between horizontal and vertical networks as the difference between "bonding" and "bridging." The former refers to strong communal loyalties that enable local collective action by reducing uncertainty, whereas the latter focuses on the weak associations that facilitate information sharing and market interactions. Economic and political development appears to be dependent upon finding the right balance between the two. Reinforcing this point, Woolcock (1998) suggests that for development to proceed normally, "the initial benefits of intensive intra-community integration, such as they are, must give way over time to extensive extra-community linkages: too much or too little of either dimension at any given moment undermines economic advancement" (p. 175).

A good example of a community plagued by an overabundance of "bonding" and a scarcity of "bridging" was detailed by Pantoja (2000) in his analysis of efforts to rehabilitate the coal-mining industry in Orissa, India. Despite the fact that Pantoja found extremely high levels of interpersonal trust in the areas immediately surrounding the Orissa mines, the coal-mining industry continued to flounder. By delving more deeply into the social fabric of Orissa, Pantoja found that even though high levels of trust existed, these bonds were exclusive and generally fragmented by gender, caste, or class (Pantoja, 2000; Grootaert & van Bastelaer, 2001). The rejection of extracommunal ties in Orissa translated into a deficiency of the cross-network linkages that support the open exchange of information and problem-solving techniques, and thus explains the industry's continued deterioration

International Organizations: Promoting International Discourse

The lack of extracommunity linkage in the case of Orissa, India, demonstrates the importance of thinking about social capital as a multidimensional concept that encompasses a number of distinct social relationships. Strong community ties are critical to local collective action; however, if they are not supplemented with weaker extracommunity ties, overall economic and political growth will inevitably be stunted. So how does the development community encourage this type of extracommunal linkage? As the example of Orissa, India, demonstrated, "bridging" needs to begin at the local level. If a state is to excel in the current globalized economy, these extracommunal linkages need to extend to the global level. While the state is in an excellent position to promote intracommunal dialogue within its borders, the international standing of IGOs and NGOs place them in a unique position to provide the environment for extracommunal global engagement.

One of the best examples of the role that international organizations can play in creating extracommunal linkages was the World Bank's International Forum on Capacity Building (IFCB). The World Bank's goal for this project was to provide "an environment in which nongovernmental organizations from developing regions can engage other Southern NGOs and donors about approaches, practices, and policies on capacity building" (World Bank, n.d.-d). However, beyond simply facilitating multistakeholder dialogue among Southern NGOs, the forum also promoted dialogue between Southern and Northern NGOs, donors, and government representatives on various issues relating to development (Monico, 2001, p. 26). The ability to engage each other as well as experts from the developed world has dramatically augmented the amount of information available to Southern NGOs and government representatives on the development process. In contrast to the dense intracommunal ties, the relationships established at the IFCB are more focused on acquiring skills, resources, and information that will enhance a state's ability to generate economic growth in today's globalized economy.

The World Bank, working in tandem with NGOs, has taken the same approach when constructing its development strategies for specific countries by attempting to increase the dialogue that occurs between local civic associations, the state, and donor agencies. For instance, in project consultations with Honduras, Argentina, and Brazil, the World Bank engaged a number of local NGOs, including local women's and faith-based groups, labor unions, universities, private firms, and the state and local government leaders about various strategies of development (Monico, 2001, p. 26). In Morocco the World Bank organized a series of outreach meetings and workshops in the hopes of creating greater extracommunal dialogue between civil society, the state, and international donor agencies. These meetings produced a regulatory framework to facilitate the frequent exchange of ideas and information between these groups. In Cairo, the World Bank made a similar effort during a number of workshops that focused on detailing how civil society was a critical partner in sustainable development. The World Bank saw two advantages to this multistakeholder dialogue. First, by increasing the participants in the dialogue, the World Bank intended to promote extracommunal engagement within the countries themselves. Second, the World Bank has suggested that the more extensive dialogue has helped it to better understand local development challenges and how to best target these issues.

Taking a more global view of development, it appears that without the proper extracommunal dialogue, a state essentially isolates itself from the insights and economic opportunities that other countries and international institutions can provide. Reversing the perspective and looking at development from a top-down perspective, it is clear that without the inputs of the local community, international organizations lack the contextually sensitive information needed to construct effective development projects. Thus we find ourselves in a situation where too little intracommunal trust (integration) promotes uncertainty and prevents collective action, but too much isolates the community and prevents the acquisition of the skills needed to successfully participate in the extracommunal networks of the mainstream global economy (Woolcock & Narayan, 2000). International organizations (both IGOs and NGOs) find themselves in a unique position to promote the extracommunal engagement necessary for successful economic development in an increasingly globalized world economy.

CREATING STATE–SOCIETY SYNERGIES

The emphasis of the two previous sections was on horizontal and vertical relationships between individuals and the proclivity of those individuals toward intra- and extracommunal engagement. However, both of these types of social capital neglect a critical relationship that exists between the state and society, synergy. Popularized by Evans (1996), *synergy* refers to a mutually constitutive relationship between the state and society where, according to Evans, "civic engagement strengthens state institutions and effective state institutions create an environment in which civic engagement is more likely to thrive" (p. 1034). Citizens, because they are more actively engaged in public affairs, demand and consequently receive better performance from their government. This improved governance results in better school systems, improved health care, increased economic opportunities, and a generally more organized and efficient state. The synergy literature openly challenged the prevailing neoliberal belief that state–society relationships must be thought of in zero-sum terms. An increased role for the state invariably led to a decrease in civil society activity (for a critique of this view, see Skocpol, 1996). In challenging this view, Evans, and the other contributors to the special issue of *World Development* (1996) on synergy, looked at cases from all over the world where a symbiotic trust between the state and civil society played a critical role in fueling economic and political development (Burawoy, 1996; Lam, 1996; Ostrom, 1996).

State Institutions and the Creation of Synergy

Creating a mutually supportive and trusting relationship between state and society is not something that occurs overnight. History has shown that governments, which have abused or neglected their constituencies, are unlikely quickly to recapture the trust of the masses. So then how do developing countries, which tend to be defined by poor state–society relations, begin the process of creating synergy? Building on the argument developed in the earlier section on integration, the state appears to be in a unique position to manipulate the local institutional context so as to create an environment supportive of the development of synergy. As Uphoff (1992) suggests, "paradoxical though it may seem, 'top-down' efforts are usually needed to introduce, sustain, and institutionalize 'bottom-up' development" (p. 273). For instance, Rothstein (2001) has suggested that in constructing the social welfare system, the Swedish state opted for a neocorporatist structure largely because it believed this particular institutional structure would help foster greater interaction and trust between state and society. According to Rothstein, there was widespread acceptance among Swedes that this type of democratic corporatism was "the most politically effective way to handle social and economic problems, arguing that it would generate trust between the parties involved and would make it possible to secure both functioning compromises in the process of policy formulation and a smooth implementation" (2000, pp. 214–215). This is not to suggest that a tripartite organization of capital, labor and the state is the answer to all the developing world's problems, but to point to an example where the state created an environment conducive to the long-term development of synergy by reconfiguring the local institutional context.

Tendler's (1997) frequently cited analysis of the "Ceará Miracle" is one of the most prominent examples of a state enacting institutional reform in the attempt to promote a more synergistic relationship between state and society. Ceará, a state within Brazil, was able to drastically improve its health care in a matter of 4 years; vaccination coverage increased, infant death rates fell, and access to medical care drastically improved. Prior to the 1987 program, the state was in complete financial disarray, unable to pay public servants and using what little medical supplies it had for political favors. Five years later, the corruption was eliminated and citizens of Ceará had developed a sense of collective responsibility for their government health program as well as for fellow members of the community. The effect of this state-led program was that community perceptions began to change. Health agents began to see their work as more than simply a job and community members recognized the impact they could have on their collective well-being. The following statement made by a health agent illustrates the transformation that has occurred in Ceará and the synergy that quickly developed. "This town was nothing before the health program started. I was ready to leave and look for a job in São Paulo, but now I love my job and I would never leave—I would never abandon my community" (in Tendler, 1997, p. 28). The Ceará example suggests that by implementing the appropriate institutional reform the state can begin the process of engaging the local community and making them a vital component of their own development.

Providing a Forum for Synergy: International Organizations

The above examples suggest that by manipulating the institutional environment the state can create state–society engagement and ultimately promote a synergistic relationship. Creating a positive state–society relationship appears dependent upon the proper forum for dialogue. However, we should not assume that the state is the only agent capable of promoting state–society synergies. Organizations like the World Bank have also endeavored to promote synergy. While the World Bank lacks the sovereign power to alter the local institutional context, it is favorably positioned to provide a forum for open state–society dialogue. For instance, in Morocco, the World Bank organized a number of outreach meetings between community organizations and the government in the hope of promoting a dialogue and better understanding of the "impediments" to greater state–society interactions. Focusing on education, these World Bank meetings between the Moroccan state and local NGOs led to an agreement where 53 local NGOs have agreed to carry out literacy training programs throughout the country (Monico, 2001, p. 13). Similarly, in Belarus, the World Bank held 14 consultation meetings involving over 1,500 representatives from civil society organizations as well as representatives from local government. The far-reaching dialogue led to the creation of two government–civil society task forces on health and sustainable development. Thus we see that international organizations can provide a neutral forum where state and society can engage one another and with a bit of luck promote social capital.

NGOs: Building Bridges and Enabling Rational Behavior

This chapter has so far looked at the role that states and international organizations can play in altering the local institutional context in order to increase the chances for producing various dimensions of social capital (integration, linkage, or synergy). The overarching argument holds that if the institutional context can be manipulated in the proper way, the state can artificially produce the various dimensions of social capital thought to be so critical to modernization and democracy. While this intuitively makes sense, it is important to note that rationality is a contextually sensitive issue, and what appears rational to one group may not appear that way to another. In terms of development programs, a state program intended to empower the local community by placing them in charge of a local development project may appear at the community level to be an attempt by the state to cut costs and avoid responsibility. Instead of empowering local individuals and encouraging local organization, a negative community perception of a state-funded program may have the contrary effect of pushing individuals further away from any form of positive civic engagement.

It is at this point that we need to consider the role that NGOs can play in helping overcome these dilemmas through issue framing. Communities that are deficient in horizontal and vertical networks of engagement also tend to lack any real trust between the community and the state (synergy). Without proper legitimacy, a state will find it increasingly difficult to successfully implement a development program. However, this problem can be overcome if a local NGO is present to help rationalize the utility of the program to the local community. Due to its close proximity to the community, a local NGO would be much more likely to garner the trust of the local community than would the state. For instance, in his analysis, Brown (1996) found that "in government-initiated programs, NGOs helped reframe initial definitions of problems to emphasize the interests and the participation of grassroots groups" (p. 223).

A concrete example of this grassroots model was the role the Orangi Pilot Project (OPP), a local Pakistani NGO, played in reframing a program to build sewage systems throughout Pakistan. Initially seen by local communities as an attempt to shirk responsibility, OPP was able to reframe the sewage project as an opportunity for local empowerment and control. The OPP's efforts to reframe the issue and detail how the program benefitted the community fueled the emergence of thousands of urban neighborhood organizations, which helped build the sewage systems and latrines in Pakistan. Brown also notes that after completing the project these neighborhood organizations "took on a variety of other self-help projects" (p. 237). Brown goes on to detail how an immunization program in Bangladesh, an irrigation program in Indonesia, and a village savings program in Zimbabwe were all successful because of the role local NGOs played in facilitating a productive and trusting relationship between the state and society. In each case, the local NGO played a vital role in reframing a development program so as to emphasize how and why it would be beneficial to the local community.

In terms of manipulating the local institutional context so as to create an economic or political incentive for organizational engagement and state–society synergy, we see that NGOs play a vital role in enabling civic engagement and social capital. Without a legitimate and trusted broker to frame the issue in terms of the immediate and long-term interests of the local community, the successful implementation of any government program will be in doubt. Without proper framing, a local community is unlikely to be able to see or to trust that a specific government program is actually in its own interest. Yet, because of their access and legitimacy, local NGOs allow communities to become aware of their own and others' interests, and thus confidently to pursue those interests through collective engagement. Consequently, we need to think of local NGOs as actors that enable rational collective behavior. We should think of them as bridge builders, which can transcend the existing mistrust between state and society and facilitate a more trusting relation-ship. The long-term development of the various dimensions of social capital appears to be highly dependent upon the ability of the state to re-construct the local institutional context and on the ability of local NGOs to enable local communities to see and act upon the opportunities the state has created for them.

CONCLUSION

My original intent for this chapter was to draw attention to a handful of successful efforts at promoting social capital. However, as I delved more deeply into the chapter, it soon became apparent that globalization and the fundamental changes it had ushered in had greatly complicated my task. Globalization has drastically transformed the nature of the interna-tional system and made the world an increasingly economically, politi-cally, and socially interconnected place. Civic interactions are no longer restricted to the community level and are occurring globally with a de-gree and intensity never seen before in history. Individuals from every re-mote corner of the globe can now quickly reach around the world and access others who have vital information or know-how. All told, global-ization has made it clear that intracommunity ties, Putnam's (1993) orig-inal focus, are but one of many forms of civic engagement that must be simultaneously nurtured in order to ensure proper economic and politi-cal development. As a result, this chapter reviewed efforts by IGOs, NGOs, and the state to promote three distinct forms of social capital.

Why such an emphasis on nonstate actors? The argument pursued in this chapter is that global nonstate actors (IGOS and NGOs) have emerged as a vital new resource for the mobilization of these various forms of civil society and social capital. The examples from the chapter suggest that while the state remains the primary motivator of social capi-tal, IGOs and NGOs have a unique comparative advantage in the pro-duction of social capital.

I can make a few concluding generalizations about the comparative advantage that state and nonstate actors have in promoting social capi-tal. First, because of its sovereign control, the state is in a unique posi-tion to manipulate the local institutional context to promote engagement between individuals (integration) as well as between state and society (synergy). The section on integration examined the root cause of group membership and found that the state can encourage organizational en-gagement by altering the local environment so that previously nonpartic-ipating individuals find themselves with a tangible economic or political incentive to engage in collective behavior. Similarly, the section on synergy detailed how the political reorganization that occurred in Sweden and Ceará helped foster a more mutually trusting relationship between state and society. Consequently, social capital development, specifically integration and synergy, seems dependent upon the state taking a proactive role in establishing the institutional preconditions for the emergence of intracommunal and state–society cooperation.

Second, evidence from the World Bank suggested that while international organizations lack the capacity to directly manipulate the local environmental, they are in a favorable position to provide an impartial forum for extracommunal (linkage) and state–society dialogue (synergy). The section on linkage detailed the World Bank's ability to foster vertical relationships and how the establishment of these extracommunal relationships significantly improved the delivery of information to developing countries. However, beyond fostering extracommunal and synergistic engagement, the preceding analysis showed that international organizations are well-positioned to provide financial support and organizational training for emerging community organizations (integration). Thus international organizations have a distinct comparative advantage in advancing constructive dialogue among distrustful parties. The long-term development of social capital appears to be directly linked to the ability of international organizations to continue in this capacity.

Third, NGOs obviously have a critical role to play in social capital development. However, unlike the state or intergovernmental organizations, NGOs' comparative advantage stems from the close ties they can establish with local communities. The trusting relationship between NGOs and local communities assures affected populations that their participation with the state (synergy) or other actors outside the immediate community (linkage) is truly in their long-term interests. In that sense, NGOs act as bridge builders, enabling the rational behavior necessary for social capital development. Through the process of issue framing, NGOs allow local communities to become aware of their own and others' interests, and thus to confidently pursue those interests through collective engagement.

The preceding analysis provided a few insights into the comparative advantages that these organizations have in promoting the different dimensions of social capital. Moreover, it only makes sense that in order to fully reap the benefits of social capital, all three dimensions require simultaneous nurturing. Scholars therefore should not be surprised when isolated and short-term efforts to improve social engagement have only weak effects on social capital formation (Gugerty & Kremer, 2000). Only when all aspects of social capital are jointly produced over an extended period of time should we expect to see significant improvement in economic and political performance.

Finally, the World Bank has taken center stage for much of this chapter. I want to be very clear that my intent was not to glamorize the World Bank, as a global institution that can do no wrong. A plethora of research has detailed the many inadequacies that continue to plague the World Bank as well as the other Bretton Woods institutions (for a good example, see Stiglitz, 2003). Even accepting these limitations, the World Bank has taken the global lead when it comes to social capital initiatives. In many ways, the World Bank has been the only IGO to have so whole-heartedly incorporated the development of social capital into its development plan. For instance, 68% of World Bank projects in 2001 had planned involvement of civil society (Monico, 2001). Yet the World Bank's success in promoting social capital should not detract attention from its many other problems. My hope for this project is that the other global institutions devoted to global development will follow the World Bank's lead in terms of social capital development and embrace the unique comparative advantage they posses. A concerted effort made by the world's many other nonstate actors (both IGOs and NGOs) could dramatically improve the overall prospects for global development.

NOTE

1. Because of time and space constraints, my discussion of international organizations will focus exclusively on the efforts of the World Bank to create social capital. While other intergovernmental organizations have made efforts to promote social capital, the World Bank has been the leading advocate of the promotion of civil society and social capital.

REFERENCES

Axelrod, R. (1984). *The evolution of cooperation*. New York: Basic Books.
Bebbington, A., & Carroll, T. (2000). *Induced social capital and federations of the rural poor* (Social Capital Initiative Working Paper No. 19). Washington, DC: World Bank, Social Development Department.
Brown, D. (1996). Creating social capital: Nongovernmental development organizations and intersectoral problem solving. In *Promoting civil society—Government cooperation: A selection of IDR reports*. Boston: Institute for Development Research.
Burawoy, M. (1996). The state and economic involution: Russia through a Chinese lens. *World Development, 26*, 1105–1117.
Coleman, J. (1988). Social capital in the creation of human capital. *American Journal of Sociology, 94*, 95–120.

Coleman, J. (1990). *Foundations of social theory.* Cambridge, MA: Harvard University Press.

Das Gupta, M., Grandvoinnet, H., & Romani, M. (2003). *Fostering community-driven development: What role for the state?* (World Bank Policy Research, Working Paper No. 2969). Washington, DC: World Bank

de Souza Briggs, X. (1997). Social capital and the cities: Advice to change agents. *National Civic Review, 86,* 111–117.

Evans, P. (1995). *Embedded autonomy.* Princeton, NJ: Princeton University Press.

Foner, E. (1989). *Reconstruction: America's unfinished revolution, 1863–1877.* New York: Harper & Row.

Fox, J. (1996). How does civil society thicken? *World Development, 24,* 1089–1103.

Fukuyama, F. (1995). *Trust: The social virtues and the creation of prosperity.* New York: Free Press.

Gittell, R., & Vidal, A. (1998). *Community organizing: Building social capital as a development strategy.* Newbury Park, CA: Sage.

Granovetter, M. (1973). The strength of weak ties. *American Journal of Sociology, 78,* 1360–1380.

Grootaert, C., & van Bastelaer, T. (2001). *Understanding and measuring social capital: A synthesis of findings and recommendations from the social capital initiative* (Social Capital Initiative Working Paper No. 24). Washington, DC: World Bank, Social Development Department.

Gugerty, M. K., & Kremer, M. (2000). *Does development assistance help build social capital?* (Social Capital Initiative Working Paper No. 20). Washington, DC: World Bank, Social Development Department.

Heller, P. (1996). Social capital as a product of class mobilization and state intervention: Industrial workers in Kerala, India. *World Development, 24,* 1055–1071.

Jackman, R., & Miller, R. (1998). Social capital and politics. *Annual Review of Political Science, 1,* 47–73.

Lam, W. F. (1996). Institutional design of public agencies and co-production: A study of irrigation associations. *World Development, 24,* 1039–1054.

Monico, C. (2001). *World Bank–civil society collaboration: Progress report for fiscal years 2000 and 2001.* Washington, DC: World Bank.

Ostrom, E. (1996). Crossing the great divide: Coproduction, synergy and development. *World Development, 24,* 1073–1087.

Pantoja, E. (2000). *Exploring the concept of social capital and its relevance for community-based development: The case of coal mining areas in Orissa, India* (Social Capital Initiative Working Paper No. 18). Washington, DC: World Bank, Social Development Department.

Petro, N. (2001). Creating social capital in Russia: The Novgorod model. *World Development, 29,* 229–244.

Portes, A. (1998). Social capital: Its origins and applications in contemporary sociology. *Annual Review of Sociology, 24,* 1–24.

Putnam, R. (1993). *Making democracy work: Civic tradition in modern Italy.* Princeton, NJ: Princeton University Press.

Reid, C., & Salmen, L. (2000). *Understanding social capital, agricultural exten-sion in Mali: trust and social cohesion* (Social Capital Initiative Working Pa-per No. 22). Washington, DC: World Bank, Social Development Department.

Rothstein, B. (2001). Social capital in the social democratic welfare state. *Politics and Society, 29*, 207–241.

Schneider, M., Teske, P., Marschall, M., Mintrom, M., & Roch, C. (1997). Institu-tional arrangements and the creation of social capital: The effects of public school choice. *The American Political Science Review, 91*, 82–93.

Skocpol, T. (1996). Unraveling from above. *The American Prospect, 6*, 20–25.

Stiglitz, J. (1996). Some lessons from the East Asian miracle. *World Bank Research Observer, 11*, 151–177.

Stiglitz, J. (2003). *Globalization and its discontents*. New York: Norton.

Tendler, J. (1997). *Good government in the tropics*. Baltimore: Johns Hopkins Uni-versity Press.

Uphoff, N. (1992). *Learning from Gal Oya: Possibilities for participatory devel-opment and post-Newtonian social science*. Ithaca, NY: Cornell University Press.

van Bastelaer, T. (2000). *Does social capital facilitate the poor's access to credit?* (Social Capital Initiative Working Paper No. 8). Washington, DC: World Bank, Social Development Department.

Whang, I.-J. (1981). *Management of rural change in Korea*. Seoul: Seoul National University Press.

Woolcock, M. (1998). Social capital and economic development: Toward a theo-retical synthesis and policy framework. *Theory and Society, 27*, 151–208.

Woolcock, M., & Narayan, D. (2000). Social capital: Implication for development theory, research, and policy. *World Bank Research Observer, 15*(2), 225–249.

World Bank. (n.d.-a). Anning Valley Agricultural Development Project. Available at *http://www.worldbank.org.cn/English/content/149h1213169.shtml*.

World Bank. (n.d.-b). Benin Food Security Project. Available at *http://www.worldbank.org/wbi/sourcebook/sbxb0501.htm*.

World Bank. (n.d.-c). Improved Environmental Management and Advocacy Pro-ject. Available at *http://www.worldbank.org/wbi/sourcebook/sb0404t.htm*.

World Bank. (n.d.-d). International Forum on Capacity Building. Available at *http://www.ifcb-ngo.org/about/background.phtml*.

17

Dreaming of "Liberation" by Riding on Globalization
Oppositional Movements in Okinawa

TAKASHI YAMAZAKI

Is globalization become a way to liberate a local ethnic group feeling oppressed by a nation-state? Can riding on globalization, or opening itself to the world economy, provide such a group with a new opportunity to rid itself of the structural constraints of a nation-state? This study is an attempt to answer these questions in the case of the territory of Okinawa Prefecture, Japan. As Taylor and Flint (2000, pp. 42–46) argue, the scale of the nation-state typically intervenes in the interaction between the capitalist world economy and localities. The reality of the world economy tends to be distorted through the scale of the nation-state and experienced as such by people living at the local scale. States can adopt liberal economic policies to allow imports or they can adopt more protectionist policies to support local enterprises under threat from imports. This ideological choice is increasingly forced by growing economic globalization, and the outcomes of the choice for localities are significant. The vertical three-tier world systems model of scale, however, does not explain how localities can "detour" around the scale of the nation-state and negotiate directly with the scale of the world economy, its institutions, and its operatives in the form of transnational corporations. Localities are not a static given but an active field where different forces (ideological and economic) from different scales (nation-state and world economy) come together and achieve their most evident impacts.

Theses such as the retreat of the state (Strange, 1996) and the end of the nation-state (Ohmae, 1995) suggest that we are in the age when, whether they wish it or not, people living in localities have to confront the reality of the world economy. In this time–space context, how local people act toward the reality of the world economy can become a crucial matter since their action itself reconstructs both the reality and their subjectivity. According to Giddens (1979, 1984), the dynamism of the world economy constitutes a structure that enables and constrains the actions of agents. Therefore, while people living at the local scale may strategically detour around the scale of the nation-state and attempt to negotiate directly with the actors of the world economy such as transnational corporations, the liberation of people from the constraints of a nation-state may simply lead to the substitution of a new oppression wrought by the world economy. Additionally, people's assessments of their relation to the nation-state or the world economy may be split and reveal internal conflicts or cleavages among themselves. Most localities that chafe at the strictures of the nation-state are defined by their unique ethnic or cultural characteristics. This interaction between a minority ethnic group and a state typically constitutes a center–periphery relation (Rokkan & Urwin, 1983; Flora, 1999). The internal dynamism of the group is also affected by the nature of the relation. Whether the relation is harmonious or hostile differentiates the reaction of each segment of the group.

Using the interactional perspectives of local–nation-state–global, this study examines the case in which a local ethnic group faces the opportunity to rid itself of the political economic constraints of a nation-state by riding on globalization or opening itself directly to the world economy. One way to attempt this strategy is to set up free trade zones to attract international investment, without the involvement of the central state. The case of Okinawa, Japan, offers an example of new realities brought to a locality by increased economic globalization at a time of reenergized ethnic mobilization and identification in a locality. Group solidarity splintered in the face of important economic choices generated by a new relationship with the Tokyo government, consequent to the changing geopolitical situation in northeast Asia. While the strong presence of the U.S. military in Okinawa is obviously important to the story, the bidirectional nature of global–nation-state–local relations in this case can offer some insights into other regional claims for autonomy and redress of grievances against the nation-state.

OKINAWA IN THE 1990S

Okinawa (Okinawa Prefecture of Japan) is a group of 108 islands stretching between the Japan Islands and Taiwan (Figure 17.1). It has a population of 1.3 million and an area of 2,267 square kilometers. Their location on small scattered islands far away from Japan proper has not only allowed Okinawans to maintain their distinctive ethnic traits but has also maintained Okinawa's status as the poorest prefecture in Japan (a little more than 70% of the national average). Other than the attractions of its semitropical climate as an important tourist resource, Okinawa's industry is not self-supporting. What makes Okinawa unique is the 38 U.S. military bases and installations located there, which occupy approximately 20% of the area of Okinawa Island (Figure 17.2). Al-

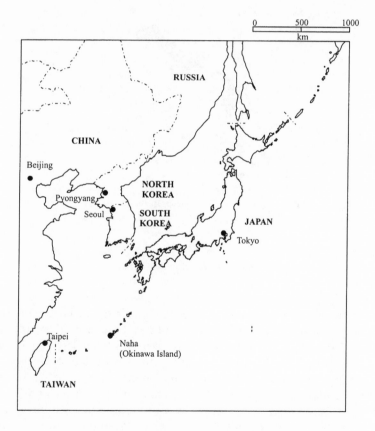

FIGURE 17.1. Northeast Asia.

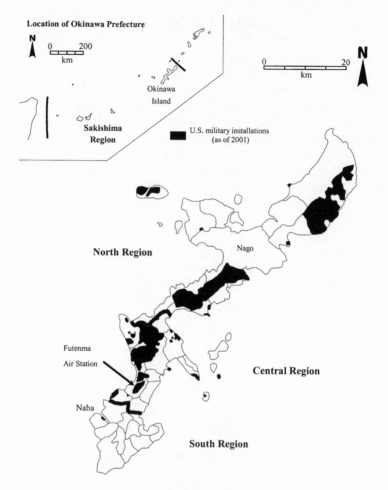

FIGURE 17.2. Okinawa Island.

though the area of Okinawa is only 0.6% of that of Japan, Okinawa contains 75% of the area of the U.S. military bases within Japan, a legacy of U.S. military administration of Okinawa from 1945 to 1972.

The militarization of the islands after 1945 inevitably changed Okinawan society and its economy. The objective of the U.S. administration of Okinawa was clearly defined in the early 1950s as land for new military bases. This resulted in a "base economy" that employed landless Okinawans. Consequently, the 27-year U.S. military administration of Okinawa transformed Okinawan society and the economy in such a profound way so that Okinawans have come to hate the U.S. but at the same time need U.S. military bases. This ambivalence toward U.S. bases caused a large number of antibase protests by Okinawans and also generated important internal conflicts (cleavages) among locals over the necessity of the bases. For Okinawans, U.S. military bases have not only been the sources of problems such as forcible land seizure, obstacles to effective development, physical and environmental damage, and homes of violent criminals, but also the sources of jobs and rent income. Okinawan interests related to the bases have not at all been straightforward but instead are multifaceted and contradictory.

When Okinawa reverted from U.S. military control to Japan in 1972, the U.S. bases remained almost intact according to the agreement between the Japanese and U.S. governments that maintained the U.S. military presence in Okinawa as necessary for the security of Japan and East Asia. In exchange for the military burden that Okinawa continued to carry for Japan and as a "compensation" for its long-term separation from Japan, the Japanese government has provided a large amount of developmental subsidies for Okinawa over the past 30 years, amounting to 700 billion yen (about $6 billion). As the management of U.S. bases was rationalized and as the Okinawan economy improved in the 1980s through subsidized public works, Okinawa's dependence on the base economy decreased. Instead, Okinawa became more dependent on the public works subsidized by the Japanese government. In this complex context, the U.S. military presence, Okinawan economic dependency, and Okinawa's sociocultural integration into Japan have been the major stimuli in the continuation of protest against the U.S. and Japanese governments (Nakano & Arasaki, 1976; Arasaki, 1996) and controversies about the future of Okinawa (Maeshiro, Makino, & Takara, 1998; Arakawa, 2000; Idaka, 2001).

In the 1970s and 1980s the economic recovery and development to reduce the gap between Okinawa and Japan proper were prioritized in the agendas of the Okinawa Prefecture government (akin to state governments in the United States). Such policies were successful in increasing the prefecture's products and population. The conservative prefecture government from 1978 to 1990 was able to provide Okinawans with material gains (*Okinawa Taimusu*, 11/24/90: 3) and Okinawans became more satisfied with the result of the reversion (the return of governmental control of Okinawa to Japan).[1] Although protest actions against U.S. military bases continued during these years, they were fragmented, localized, and attracted fewer participants for each action compared to the previous decades of U.S. military control (Figure 17.3). The era of Okinawa's economic recovery, led by a strong Japanese economy, seems to have ended when the Japanese economy peaked in 1990 (Figure 17.4). Furthermore, the end of the cold war about the same time

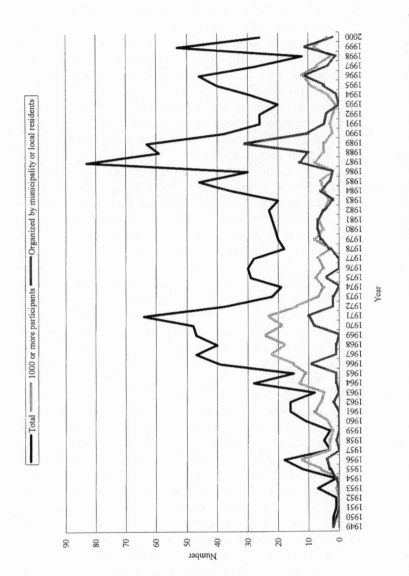

FIGURE 17.3. Number of collective actions, 1949–2000. "Collective action" includes rallies, demonstrations, and strikes. Data from the *Okinawa Times*, 1949–2000.

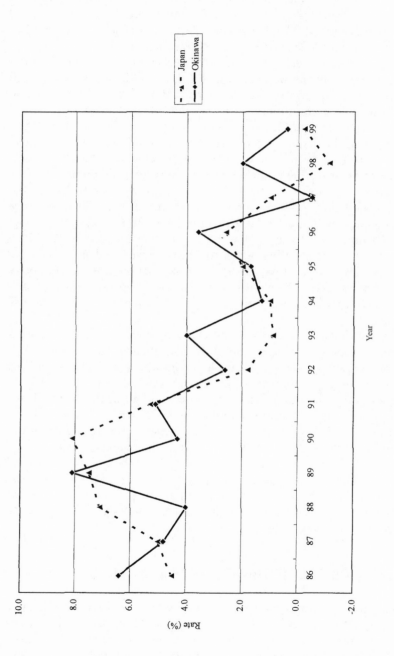

FIGURE 17.4. Economic growth rates. Data from *Okinawa ken tokei ka* (1999) and *Tokeikyoku* (2000).

provided an opportunity for Okinawans to rethink the political status of their own islands. A shift of Okinawan political preference was reflected in the result of the 1990 gubernatorial election, in which a new reformist (leftist) governor defeated the conservative incumbent.

The period between 1990 and 1998 became one of the most important historical epochs in Okinawa due to two significant contextual changes. First, the end of the cold war increased the expectation that Okinawa's long-term military burden would finally end with the withdrawal of the nuclear threats of the former Soviet Union. It was also expected that the demise of the Soviet Union would promote the democratization of East Asia. Second, the end of the cold war shifted U.S. policy priorities providing from military security to promoting economic prosperity in the world. Anticipation rose in Okinawa on the belief that riding on globalization would create a new stage of economic prosperity. For example, the Bogor Declaration of the Asia-Pacific Economic Cooperation Conference (APEC, of which Japan is a member) in 1994 requested developed member states to liberate trade and investment by 2010. In this new context, economic globalization was viewed as an alternative for Okinawa to overcome excessive dependence on Japan's national economy and public finance and establish a local self-supportive economy.

In addition to these contextual changes, there were also local factors that promoted the orientation toward globalization. First, in 1995, U.S. servicemen raped a 12-year-old Okinawan girl. This violent act drastically increased Okinawan grievances and led to a series of mass protests against the U.S. military in the second half of the 1990s. These political uprisings underpinned a new stage of reformist politics seeking pacifism in Okinawa. Second, Masahide Ota was elected as governor in November 1990. He was the third reformist governor in postreversion Okinawa. In his second term, from 1994 to 1998, he actively attempted to promote antiwar, pacifist policies by taking advantage of the political uprisings after the rape in 1995. In addition, the prefecture government led by Ota attempted to frame unique policies linking globalization to Okinawa's economic self-supportiveness.

THE POLITICS OF GOVERNOR MASAHIDE OTA

In order to understand the nature of the interaction between Okinawa and Japan proper and the internal dynamism of Okinawan society in the 1990s, we must examine the political processes during the terms of Governor Masahide Ota. Ota had been a professor of journalism at the University of the Ryukyus in Okinawa and was well known for his leftist opinions. His predecessor, Junji Nishime, who served from 1978 to 1990, was renowned as a conservative.[2] Seemingly tired of "the logic of development based on public works" promoted by the conservative governor (*Okinawa Taimusu*, 11/24/90: 3), Okinawan voters expected the new reformist governor Ota to change the course of prefecture politics (*Okinawa Taimusu*, 11/19/90: 1). The years around 1990 were the peak of recent economic prosperity in Okinawa as well as in Japan. For the previous two decades, Okinawan growth rates were the highest and unemployment the lowest of the postwar period, paralleling developments in Japan as a whole. Okinawan voters, relieved of the worry of economic underdevelopment, expressed their political preference for the more pacifist, antiwar politics that Ota pledged to embody.

Following the end of the cold war, Ota's first term from 1990 to 1994 had few political issues resulting in his reelection victory by a huge margin in 1994. Voters' interest in prefecture politics was shown in the lowest voter turnout (62.54%) in the gubernatorial elections since 1972 (*Okinawa Taimusu*, 11/21/94: 1). Ota's second term, however, was dramatically different from his first term after the rape of the 12-year-old in 1995. This incident drastically changed the course of prefecture politics. It reverberated in a significant impact on Japan–United States security relations. Political rallies on October 21, 1995, attracted approximately 85,000 participants on Okinawa Island and 3,000 on the Sakishima Islands (*Okinawa Taimusu*, 10/22/95: 1). Not only protesting the rape, speakers at these rallies blamed the Japanese and U.S. governments for their continuing unfair treatment of Okinawa, and demanded that both governments revise the Japan–United States Status of Forces Agreement (SOFA)[3] and reduce the number and size of U.S. military bases (*Okinawa Taimusu*, 10/22/95: 1–3). Unable to ignore the explosion of Okinawan grievances, the Japanese and U.S. governments began to revise the SOFA and to develop a plan to return (the land used for) U.S. military bases to Okinawa.

In a second development, Ota promoted a series of pacifist policies in 1996 by building on the antibase protests. He refused to sign the land lease contracts for U.S. military bases.[4] Responding to public petitions, Ota carried out a prefecture referendum in September 1996 asking whether the SOFA should be revised and whether U.S. military bases on the island s should be reduced. Although the result of the referendum was nonbinding, the revision of the SOFA and the reduction of bases won easily.[5] The growing collective movement supported Ota and protested against the Japanese and U.S. governments.

Finally, this tense political environment motivated Ota to outline a "grand design" for Okinawa's in January 1996, which was embodied in the Action Program for the Reduction of U.S. Military Bases (the Action

Program) and the Cosmopolitan City Formation Concept. These policies stimulated debates over Okinawa's subordinate situation and internal political economic structure in conjunction with the increasing pressure of globalization in the form of economic liberalization in the Asia-Pacific region.

OKINAWA'S "GRAND DESIGN" AND GLOBALIZATION

The Action Program and the Cosmopolitan City Formation Concept of January 1996 were framed after the Japanese and U.S. governments formed the Special Action Committee on Okinawa (SACO) in November 1995 to develop a plan to consolidate U.S. military bases and installations and return 11 of them to Okinawa. Departing from SACO's plan, the prefecture government drafted its Action Program for the phased removal of all U.S. military bases by 2015, arguing that piecemeal return of the bases according to the convenience of the U.S. military forces would prevent the prefecture and municipal governments from devising a long-term, consistent developmental plan.

The Cosmopolitan City Formation Concept was a concrete, if unrealistic, measure for the development. As the title of the policy implies, it was an attempt to develop Okinawa and locate it in the context of globalization in the Asia-Pacific region, following the model of Hong Kong. The implementation of the Cosmopolitan City Formation Concept required the use of the large land area then occupied by the U.S. military. Since the removal of bases would result in a loss of job opportunities and income sources for many Okinawans,[6] it was evident that new industries would need to be created to counterbalance the loss. Thus, these two policies complemented each other to constitute "Okinawa's Grand Design for the 21st Century" (Okinawa Prefecture Government, 1997: Preface).

The Cosmopolitan City Formation Concept was embodied in the New Industrial Promotion Policy for Cosmopolitan City Formation in November 1997 by extending an FTZ. Even though the term "globalization" appears only twice in the text of the policy statement, the idea of economic and cultural globalization in the Asia-Pacific region was set against the militarization of the islands. As the original English text states:

> The Cosmopolitan City Formation Concept is Okinawa's grand design for the 21st century and its goal is the promotion of regional characteristics which will contribute to the self-supportive economic development of Okinawa, and continuous development of the Asia-Pacific Region, as

well as assisting in maintaining peace. The concept also aims to transform a military-based island into a peaceful island, and to positively promote various policies based on three basic principles: peaceful exchange, technological cooperation, and economic/cultural exchanges. Above all, it proposes to implement decisive measures based on a thorough review of the rigid economy of Okinawa today and its progress toward globalization inside and outside the territory to achieve the goal of "creating and promoting new industries suitable for the 21st century."

The Policy statement, based on the Cosmopolitan City Formation Concept, has three general orientations: the advancement of the FTZ, the integration and enhancement of the information and communication industry, and the formation of a hub for the international tourism and destination industry. Among the three policy directions, the advancement of the FTZ was the core strategy to promote Okinawa's economic self-supportiveness and competitiveness in the Asia-Pacific context by encouraging Japanese and foreign direct investments and opening Okinawa to the world economy. Thus it was most closely related to the idea of globalization. As the text states:

> Okinawa's problems in relation to the promotion of industry include the limited availability of land due to the existence of vast U.S. military bases, delayed improvement of locations for industrial development as a direct result of having been excluded from various post-war industrial promotion policies, and comparatively high transportation costs because of its geographical distance from other parts of Japan and being a prefecture of many relatively small islands.

Japan's present weak economic condition, including industrial migration to foreign regions and competition with rapidly growing neighboring economies in East Asia, should be taken into consideration. As the document states:

> Thus, it is necessary to develop new industrial promotion policies which utilize Okinawa's regional characteristics and resources in a positive way. Improvements designated to attract domestic as well as foreign industries need to be expanded by introducing a free trade zone with tax incentives, focusing on deregulation, and constructing an infrastructure which includes an international airport, harbors, and information and communication facilities. (Okinawa Prefecture Government, 1997: Chapter I)

From the contents of the Policy statement, as well as the Action Program, it seems clear that the prefecture government was attempting to "liberate" Okinawa from the U.S. military bases and from Japan's national economy and public finance at the same time. The prefecture government sought to weaken the long-time yoke of the Japanese nation-state and to promote the economic autonomy of Okinawa.

DEBATES OVER THE FREE TRADE ZONE

In order to implement Okinawa's "grand design," the prefecture government began to examine possible measures to improve its political economy. One of the most widely debated measures was the establishment of the prefecture-wide FTZ, probably because the economic liberalization of Okinawa contained a symbolic meaning—the complete transformation of the territoriality of the islands by the prefecture itself in the face of globalization. Moreover, the tendency of debates to focus on economic liberalization was conditioned by two structural constraints: the survival of the cold war in East Asia and Okinawa's innate economic weakness. Before examining the debates over the FTZ, let us look at these two premises.

The Survival of the Cold War?

Although the ideal of the removal of U.S. bases and the advancement of an FTZ was stated in the two policies, the geopolitical and geoeconomic context of Okinawa did not necessarily allow the prefecture to rid itself of the structural constraints of the nation-state. With regard to U.S. bases in Okinawa, the Japanese and U.S. governments were unable to accept a rigid time limit for the removal of the bases since they believed that potential security threats (i.e., China and North Korea) still existed in East Asia and that the military value of Okinawa was still high. For that reason, it was unlikely that U.S. military bases located in Okinawa would be drastically reduced. Such an interpretation was clearly expressed in the redefinition of the Japanese–U.S. security relations, contained in the New Guidelines for Japan–U.S. Defense Cooperation of 1997. Although the Japanese government, led by the Liberal Democratic Party (LDP) in 1996, basically supported the Cosmopolitan City Formation Concept, it did not make any clear statement regarding whether the Action Program could be implemented on time. If the Japanese government and/or the U.S. government disagreed about the Action Program, the Cosmopolitan City Formation Concept could not become a truly "pacifist" policy since such an economic policy could be promoted regardless of the existence of U.S. bases. Therefore, it was doubtful that Ota's policies to link the removal of the bases to the promotion of new industries could be implemented as planned. It was instead predicted that the contents of the Cosmopolitan City Formation Concept might be absorbed into, or hijacked by, the existing national developmental plans for Okinawa.

Innate Economic Weakness of Okinawa

In terms of the prospect of Okinawa's economy, Ota's policies had another weakness. Due to his strong reformist-leftist stance against the Japanese and U.S. governments, the conservative LDP-led Japanese government stopped providing regional subsidies. Eventually, this antagonistic relationship began affecting the material life of Okinawans.[7] In addition, even though neither the Action Program nor the Cosmopolitan City Formation Concept addressed Okinawa's political independence, these policies obviously suggested an independent Okinawa modeled on Hong Kong (i.e., two political–economic systems within one state). This prospect, however, was too optimistic.

As the text of the New Industrial Promotion Policy for Cosmopolitan City Formation recognized, the decline of Japan's economy was clear in the second half of the 1990s. Consequently, Okinawa's economy was weakened since it was firmly integrated into Japan's national economy. The society of Okinawa is often said to be dependent on three K's: *kichi* (military bases), *kokyo-jigyo* (public works), and *kanko* (tourism). While the ratio of base-related income to the gross prefecture expenditure (GPE) is decreasing, the amount of public financial transfer received by Okinawa and the income of the tourist industry occupy a substantial part of the GPE.[8]

After Okinawa reverted to Japan in 1972, the U.S. base-related income was replaced by subsidies from the Japanese government. Since the local economy had long been dependent on such external resources, the growth of the manufacturing industry was hindered while the construction industry as a recipient of the subsidies was promoted.[9] In such circumstances, state laws and institutions have protected many sectors of Okinawa's economy, especially its primary industries. In addition, the loss of the land for production and overdevelopment has damaged Okinawa's natural environment.[10] Such a distortion of Okinawa's economy has prevented Okinawa from becoming self-supportive and has reproduced its underdevelopment.

Tanaka Committee Report

In order to implement the Cosmopolitan City Formation Concept, the prefecture government established the Committee of Industrial and Economic Promotion and Deregulation, led by a well-known economist, Naoki Tanaka (hereafter the Tanaka Committee). It examined desirable options for Okinawa in the face of globalization and published a final report in July 1997 (*Okinawa Taimusu*, 7/25/97: 13).

The report recommended that the prefecture government employ a "decisive" measure for industrial promotion such as "two systems within one state" (the Hong Kong model) and proposed the introduction of a prefecture-wide FTZ in 2001 to attract foreign direct investment. Following the Bogor Declaration of the APEC, which expected developed member countries to liberalize trade and investment by 2010, the Tanaka Committee emphasized that the establishment of the prefecture-wide FTZ would become a model that could be extended to Japan proper. The report also stated that the people of Okinawa Prefecture needed to end the reversion programs[11] by themselves and to tackle "the creation of a new Okinawa" based on the principles of self-determination and self-responsibility. Although what the Tanaka Committee recommended was straightforward liberalization, it sparked heated debates.

Governor Ota expressed his approval of the report by stating that, even though large state subsidies had been very important and necessary for Okinawa, whether one liked it or not, liberalization would proceed based on the APEC agreement (*Okinawa Taimusu*, 7/23/97: 1). Many sectors of Okinawan industry, however, were concerned about or objected to the "decisive" report of the Tanaka Committee. A survey about the influence of the prefecture-wide FTZ showed that such concern was shared among the corporations protected by the special reversion measures or operating within the closed prefecture market (*Okinawa Taimusu*, 7/23/97: 3; 7/24/97: 3). While the tourist industry and ambitious smaller businesses welcomed the prefecture-wide FTZ, the manufacturing, agricultural, and food processing industries expressed serious concern about the damage that liberalization might cause (*Okinawa Taimusu*, 7/31/97: 11; 8/1/97: 9; 8/2/97: 11; 8/5/97: 9; 8/6/97: 11; 8/7/97: 13). These industries feared that the cutthroat competition that Japanese and foreign corporations would bring into Okinawa might destroy their businesses.

The Prefecture Plans for the FTZ

Based on the Tanaka Committee Report, the prefecture government began to devise its own plan for the FTZ. The draft for the plan was published in September 1997 (*Okinawa Taimusu*, 9/2/97: 3). The main points of the draft adhered to the Report but the draft was significantly different from the Report in two areas. First, the draft included the exemption of customs on imported products but excluded the items that would be negatively affected by the import of foreign products. The draft also listed the introduction of duties preferential for Okinawa. Second, the draft included the reduction of the national corporate tax, which was higher than those in other countries. The Report avoided inclusion of such a measure so as not to attract the corporations that were escaping taxation and not to neglect the constitutional duty of tax payment. Compared to the Report, the draft had a more protectionist nuance and demanded further preferential treatment of Okinawa from the Japanese government, indicating the increased pressure of concession. After issuing the draft, the prefecture government published the plan titled "The Development of the Industrial Promotion Policies for Cosmopolitan City Formation" (*Okinawa Taimusu*, 10/28/97: 2–3). This plan accepted public objections to the introduction of the prefecture-wide FTZ in 2001 and stated that it was appropriate to attempt the phased expansion of subprefecture free trade districts while introducing the prefecture-wide FTZ in 2005. With regard to the favorable treatment of taxation, the plan listed the reduction of the corporate tax as well as the investment tax and the exemption of the local tax. The prefecture government accepted the objections from the businesses protected by existing special measures and demanded the continuing favorable treatment of Okinawa from the Japanese government.

The Responses of the Japanese Government

With regard to the idea of the prefecture-wide FTZ, the Okinawa Small Committee of the LDP Tax Commission in Tokyo decided to reject it (*Okinawa Taimusu*, 11/18/97: 3). The LDP held a protectionist stance toward Okinawa that was in line with its local party position in the prefecture[12] Another obstacle to Okinawa was the Ministry of Finance. The Tanaka Committee Report attempted to avoid measures creating problematic "two systems within one state" such as the reduction of the corporate tax and visa exemption because of their effectiveness and/or constitutionality (*Okinawa Taimusu*, 7/27/97: 1). The Ministry of Finance already had a negative attitude regarding the inclusion of favorable tax systems in the prefecture's draft when it was published in September 1997 (*Okinawa Taimusu*, 9/3/97: 2).

The response of the Japanese government to the final prefecture plan was revealed in a speech given by Prime Minister Ryutaro Hashimoto at the ceremony for the 25-year anniversary of Okinawa's reversion (*Shusho kantei*, 1997). Although Hashimoto promised that the Japanese government would help to promote Okinawa's self-supportiveness with a new economic promotion plan, he mentioned neither the prefecture-wide FTZ nor the reduction of the corporate tax. Instead, he proposed the establishment of a regionally limited FTZ and taxation measures for investment.[13] In sum, the Japanese government acted in order to protect local Okinawan businesses and the integrity of the state, indicating that the developmental plan for Okinawa resulted in the amalgamation of liberalization and protectionism. One of the cores of Okinawa's "grand design" was thus watered down by both internal and external pressures (Hook, 2003).

OKINAWAN RESISTANCE AND POLITICAL ECONOMIC PROCESSES

The other core of Okinawa's "grand design" was the Action Program. Following the rape in 1995, Rengo Okinawa (the Okinawa Prefecture branch of the Japanese Trade Union Confederation) initiated a signature-collection campaign in January 1996. This campaign was for the enactment of a prefecture ordinance to carry out a nonbinding prefecture referendum, which would question the necessity of U.S. bases in Okinawa (*Okinawa Taimusu*, 1/23/96: 23). After the successful campaign, the ordinance was enacted in June and the referendum was to be carried out in September 1996, about a year before the Cosmopolitan City Formation Concept was embodied. When the prefecture referendum was held, the prefecture government had not yet presented any concrete plan to counterbalance the economic losses that the removal of the U.S. bases would cause. For this reason, military landowners and base employees objected to or expressed concerns about the referendum. The prefecture LDP opposed it for the same reason. Furthermore, the Japanese government regarded the program as a denial of the Japan–U.S. Security Treaty (*Okinawa taimusu sha*, 1996: 79).

The result of the referendum showed that more than half (53.04%) of the voters approved the revision of SOFA and the reduction of U.S. bases. However, an analysis of the referendum reveals the following interesting facts.[14] First, voters in the municipalities with a higher per capita income or with a higher proportion of state subsidies tended to approve the revision and the reduction. Second, these proposals were opposed by people in municipalities with larger areas occupied by U.S. bases (implying a larger amount of base-related subsidies), with a higher proportion of employees in the construction industry (i.e., a higher proportion of public work recipients), with a lower proportion of employees in the commercial industry (i.e., more agricultural workers), and with a higher proportion of developmental expenditure. These relationships indicate that the Action Program was supported in wealthier municipalities or those with higher administrative demands,[15] but that factors such as U.S. bases, protected industries, or larger developmental expenditure may have induced voters to oppose the program. The analysis clearly shows that the base problems in Okinawa were not only political issues but also economic ones. In this sense, the holding of the referendum in 1996 without presenting any alternative economic plan could be half-successful at most.

Meanwhile, the Japanese and the U.S. governments agreed in 1996 on the phased reduction and removal of U.S. bases, but this agreement gave rise to new problems. The most surprising response from both governments was the complete return of the Futenma Air Station, but its removal to Japan proper was strongly refused by possible host municipalities there. As a result, the air station had to be removed to some place within Okinawa. The Japanese government mostly neglected the intentions of the Action Program. Okinawans ridiculed this situation by calling it *"tarai mawashi"* of bases, indicating irresponsibly passing around the bases within Okinawa (*Okinawa Taimusu*, 12/22/96: 1).

Many protest actions over the removal of U.S. bases, especially the Futenma Air Station, took place from 1996 to 1997.[16] In particular, Nago City was chosen as a candidate recipient of the removed (i.e., newly built) base and became a center of protest. The removal of existing U.S. bases to the northern region illustrated another ambivalent aspect of Okinawan life. As the economy of Okinawa improved after the reversion, socioeconomic disparities between the north region and the rest of Okinawa became larger.[17] For the northern region, these disparities needed to be narrowed. Therefore, the removal of bases to that region meant not only transferring the source of problems but also providing a new opportunity for economic promotion. Local economies in Okinawa, affected by the decline of Japan's economy, needed such promotion measures.

Nago City carried out a municipal referendum in December 1997. The issue was whether the city should accept a new offshore base as a substitute for the Futenma Air Station. Although 52.8% of the votes were cast against the new base, the mayor of the city announced the acceptance of the new base for the reduction of U.S. bases and the further development of the city in exchange for his resignation (*Okinawa Taimusu*, 12/26/97: 1). In the mayoralty election of February 1998, the candidate supporting the new base was elected in spite of Governor Ota's objection to the intraprefecture removal. The Japanese government welcomed the result because U.S. bases could be maintained in Okinawa. In November 1997 the Japanese government already included the plan for the promotion of the north region in the Okinawa Promotion 21st Century Plan.

U.S. bases have been resources for political and economic deals between Okinawa and the Japanese government. The existence of U.S. bases in Okinawa and the provision of state subsidies for Okinawa's development have been inseparably interwoven. As long as the bases were used for such deals, building a new link between bases and globalization was not an easy task. While many Okinawans have accumulated strong grievances against U.S. bases and the Japanese government, Okinawan political behavior has not been consistent overall. It is reasonable to assume that economic considerations have driven the vote choices, both overall and at the local levels. Okinawans have been swinging between resistance against and integration into the state (Arakawa, 2000; Idaka, 2001).

CONCLUSION: THE END OF THE REFORMIST ERA

In the highly tense relationship between Okinawa and the Japanese government from 1995 to 1998, Governor Ota sometimes made concessions to Tokyo in order not to lose financial resources for Okinawa. An attitude such as this, however, induced criticism from Okinawan protesters for being inconsistent, thus weakening group solidarity. Being concerned about long-term economic stagnancy, business circles as well as the prefecture LDP decided to select Keiichi Inamine as their candidate for the 1998 gubernatorial election. Ota also decided to run for his third term.

The issues for the gubernatorial election were quite clear. In their electoral campaigns, Ota emphasized the implementation of the Action Program and the prefecture-wide FTZ, while Inamine pledged to restore the relationship between Okinawa and the Japanese government and to overcome economic stagnancy using state promotion measures for Okinawa. Inamine's side tactically used the strategy that ascribed economic stagnancy to the failure of Ota's policies. In truth, economic stagnancy was not necessarily brought about by Ota's local policies but by the overall downturn of the Japanese economy. However, by calling the stagnancy "kensei fukyo" (the recession caused by the prefecture politics), Inamine's side implied that Ota overemphasized an antibase and pacifist ideology and neglected effective economic promotion measures. This negative campaign succeeded in transforming the national problem into a local one. Structural constraints Okinawa had been experiencing, such as dependency and subordination in the center–periphery relations, were effectively hidden from the voters. The result was Ota's prefecture-wide defeat, indicating that the majority of Okinawans were worried about the decline of Okinawa's economy, which was dependent on external resources. After this result, the Japanese government promptly resumed state economic promotion measures for Okinawa. The 8-year reformist era was over.

As detailed in this study, globalization was a key emphasis from 1996 to 1997 for the reformist Okinawa Prefecture government. It represented a new space in which Okinawa could be located for new prosperity and self-supportiveness as opposed to the constraining territory and national rules of Japan on which Okinawa had been dependent. Okinawa's "grand design" and related policies depicted a new geopolitical and geoeconomic context for Okinawa in the post-cold war era. However, Okinawa's dependence on Japan through economic and financial flows was so profound that the political and economic cleavages there finally hindered its "liberation" from state economic and financial control. In this center–periphery relation, political issues over the removal of U.S. bases were transformed into economic ones, and political decision-making tended to be an outcome of the "rational" choice based on relatively short-term utility economic calculation. Although the problematic of globalization was sometimes presented in the political–economic arguments examined in this study, "Okinawa more open to the world" or "Okinawa situated in the international trade network" were represented as ideals substituting for the closeness and fixity of the nation-state. The prefecture-wide FTZ embodied such an idealistic vision and, for this very reason, induced a great deal of realistic opposition from the sectors protected by the Japanese government.

Part of the Okinawan population continued to need the protection of the nation-state, and therefore significant cleavages remained among the local people. The results of the elections and referenda mentioned in this study basically followed such cleavages. As long as these cleavages are reproduced through the center–periphery relation between Okinawa and Japan proper, a new consensus over the self-supportiveness of Okinawa will be difficult to construct. Since the concept of economic globalization failed to promote such a consensus, riding on globalization was denied as a way to "liberate" Okinawa. However, it is noteworthy for a locality to make an attempt to detour around the structural constraints of the nation-state by opening itself directly to the world economy. The case of Okinawa shows that the concept of economic globalization can become a way to mobilize, if not liberate, a locality against the pressure of the nation-state. Whether the practical potential of globalization for localities can be anything more than a mere ideological strategy remains to be seen.

NOTES

1. According to the public opinion survey in 1989, 84.7% of the Okinawan respondents showed satisfaction with the reversion (*Naikaku soridaijin kanbo kohositsu*, 1989: 41).
2. The terms (years) of successive governors (or a chief executive) are 8 (Yara, reformist, 1968–1976), 2 (Taira, reformist, 1976–1978), 12 (Nishime, conservative, 1978–1990), 8 (Ota, reformist, 1990–98), and 6 (Inamine, conservative, 1998 to present).
3. SOFA was concluded in 1960 to determine the status of the U.S. military force stationed in Japan. What SOFA provides is Japan's obligation to offer facilities and areas to the U.S. military force, U.S. authority to control its military installations, and the legal jurisdiction over U.S. military servicemen. The jurisdiction over the U.S. servicemen who commit a serious crime has been a focal point of protest actions in Okinawa. From Okinawa's point of view, SOFA unfairly protects the status of the U.S. military force and U.S. servicemen.
4. According to the law, in order to allow the U.S. military force to use the land of Japanese citizens, the Japanese government leases the land and offers it to the U.S. military force. If Japanese landowners refuse to rent their land, the Japanese government has legal power to force them to do so. In this coercive process, the head of the pertinent local government must sign the land lease contracts instead of the landowners. Since Ota refused to sign as the head of Okinawa Prefecture, the Japanese government sued him for violating the law. This lawsuit was finally brought to the Supreme Court, which decided against Ota in August 1996.
5. The results were as follows: approval votes amounted to 88.0%, objection votes to 8.2%. Voter turnout was 59.5%.
6. The base-related economy in Okinawa consists of base-related employment (8,400 workers in 1999), construction, rent, national subsidies, and other economic impacts. The ratio of base-related consumption, employment, and rent in the gross prefecture expenditure was estimated as 5.2% in 1997 (*Okinawa ken*, 2000: 45).
7. Because of Ota's objection to state policies toward Okinawa, the negotiations over state promotion measures between Okinawa and Japan were often suspended.
8. The proportions of the three external incomes to the GPE in 1994 were as follows: base-related, 4.8%; public sector, 31.8%; and tourist industry, 10.1%. Those in 1974 were 11.3%, 36.7%, and 6.7% respectively (Kurima, 1998: 31; *Okinawa ken*, 2000: 45).
9. In 1997, the proportions of the manufacturing and the construction industries to the GPP were 5.5% and 12.3%, respectively, while the national counterparts were 24.4% and 9.8% (*Shusho kantei*, 2000a).
10. The erosion of red clay, which is facilitated by heavy rainfall and public works, causes serious environmental damage on coral reefs as fishing grounds and tourist resources.

11. "The reversion programs" refer to postreversion developmental measures for Okinawa to fill the socioeconomic disparities between Okinawa and Japan proper.

12. Parties in Okinawa first expressed mixed feelings toward the proposal of the prefecture-wide FTZ (*Okinawa Taimusu*, 7/25/97: 3). Among the governmental parties, only the Komei Party overtly supported the proposal, while the Communist Party objected to it because it might weaken smaller businesses and primary industries. On the other hand, the LDP and the New Frontier Party showed a positive understanding of the proposal. Even though the prefecture government was a reformist one, centrist or conservative parties first supported such a neoliberal policy as the FTZ. However, as the FTZ was publicly debated, the LDP as well as the Communist and Social Mass Parties (one of the governmental parties) began to show overt objections to the prefecture-wide FTZ, and other parties remained cautious about the proposal (*Okinawa Taimusu*, 9/26/97: 2).

13. The final version of the Japanese government's plan in 2000 did not mention any tax system (*Shusho kantei*, 2000b).

14. Full results are reported in Yamazaki (2004).

15. A closer examination of the financial balance of each municipality indicates that municipalities with higher expenditures for education and/or social welfare tended to receive more state subsidies.

16. According to the *Okinawa Taimusu* published during the period, 18 rallies took place, most of which were in the north region such as Nago City.

17. According to Okinawa ken's *Okinawa ken tokei nenkan*, per capita annual income by region in 1995 was as follows: Naha 2.34, south 1.90, central 2.06, north 1.77, Sakishima 1.84, and isolated islands 2.06 million yen. The figure for the north region is the lowest.

REFERENCES

Arakawa A. (2000). *Okinawa: Togo to hangyaku* [Okinawa: Integration and rebellion]. Tokyo: Chikuma shobo.

Arasaki M. (1996). *Okinawa gendaishi* [The contemporary history of Okinawa]. Tokyo: Iwanami-shoten.

Giddens, A. (1979). *Central problems in social theory: Action, structure and contradiction in social analysis.* Berkeley and Los Angeles: University of California Press.

Giddens, A. (1984). *The constitution of society: Outline of the theory of structuration.* Berkeley and Los Angeles: University of California Press.

Flora, P. (Ed.). (1999). *State formation, nation-building, and mass politics in Europe: The theory of Stein Rokkan.* Oxford, UK: Oxford University Press.

Hook, G. D. (2003). Responding to globalization: Okinawa's free trade zone in microregional context. In G. D. Hook & R. Siddle (Eds.), *Japan and Okinawa: Structure and subjectivity* (pp. 39–54). London: RoutledgeCurzon.

Idaka H. (2001). *Soto no Okinawa: Aidentiti kuraishisu* [Bicephalous Okinawa: Identity crisis]. Tokyo: Gendai kikakushitsu.

Kurima Y. (1998). *Okinawa keizai no genso to genjitsu* [Illusion and reality of Okinawa's economy]. Tokyo: Nihon keizai hyoron sha.

Maeshiro, M., Makino, H., & Takara, K. (1998). *Okinawa no jiko-kensho: Teidan, "jonen" kara "ronri" e* [Self-examination of Okinawa: A three-man talk, from "emotion" to "logic"]. Naha: Hirugi sha.

Naikaku soridaijin kanbo kohositsu. (1989). *Okinawa-kenmin no ishiki ni kansuru yoron-chosa* [Public opinion survey on the consciousness of the people of Okinawa Prefecture]. Tokyo: Author.

Nakano Y., & Arasaki, M. (1976). *Okinawa sengoshi* [Postwar history of Okinawa]. Tokyo: Iwanami-shoten.

Ohmae K. (1995). *The end of the nation-state: The rise of regional economies.* New York: Free Press.

Okinawa ken. (2000). *Okinawa no beigun oyobi jieitai-kichi* [The bases of U.S. military force and the Self Defense Forces in Okinawa]. Naha: Author.

Okinawa ken. (Various years). *Okinawa ken tokei nenkan* [The Okinawa Prefecture statistical yearbook]. Naha: Author.

Okinawa ken tokei ka. Heisei 11 nendo kenmin keizai keisan [Prefecture people's economic account in Heisei 11]. Naha: Author.

Okinawa Prefecture Government. (1997). *New industrial promotion policy for cosmopolitan city formation.* Available online at *www.pref.okinawa.jp/97/ FTZ/kokutoshi/index-e.html.* Accessed March 2, 2002.

Okinawa taimusu sha. (Ed.). (1996). *50 nenme no gekido* [Turmoil in the 50th year]. Naha: Author.

Rokkan, S., & Urwin, D. (1983). *Economy, territory, and identity: Politics of West European peripheries.* Beverly Hills, CA: Sage.

Shusho kantei. (1997). *Okinawa fukki nijugo-shunen kinen shikiten naikaku soridaijin shikiji* [Prime minister's speech at the 25-year anniversary of Okinawa's reversion]. Available online at *www.kantei.go.jp/jp/hasimotosouri/speech/ 1997/1121soriokinawa.html.* Accessed July 3, 2003.

Shusho kantei (2000a). *Sangyo-betsu soseisan no taizenkoku-hikaku* [Comparison of gross products by industry to the national average]. Available online at *www.kantei.go.jp/jp/singi/okinawa/21century/siryou09.html.* Accessed July 3, 2003.

Shusho kantei. (2000b). "Okinawa keizai shinko 21 seiki plan" saishu-hokoku-sho ["The 21st Century Plan for Okinawa's Economic Promotion," Final Report]. Available online at *www.kantei.go.jp/jp/singi/okinawa/21century/ 21plan.html.* Accessed July 3, 2003.

Statistics Bureau. (Various years). *Nihon tokei nenkan* [Japan statistical yearbook]. Tokyo: Nihon tokei kyokai.

Strange, S. (1996). *The retreat of the state: The diffusion of power in the world economy.* Cambridge, UK: Cambridge University Press.

Taylor, P. J., & Flint, C. (2000). *Political geography: World Economy World-economy, nation-state, and locality* (4th ed.). Harlow: Prentice-Hall.

Tokeikyoku. (2000). *Dai 50 kai nihon tokei nenkan* [The 50th edition Japan statistical yearbook]. Tokyo: Nihon tokei kyokai.

Yamazaki, T. (2004). *Political space of Okinawa: Geographical perspectives on ethno-regional integration and protest.* Unpublished doctoral dissertation, University of Colorado, Boulder.

ARTICLES IN THE *OKINAWA TAIMUSU* (MONTH/DAY/YEAR: PAGE)

11/19/90: 1. Soko no zen yato kyoryoku [The successful joint struggle of opposition parties].

11/24/90: 3. Nishime kensei 12 nen no kiseki [The twelve-year tracks of Nishime's prefecture politics].

11/21/94: 1. Ota-shi taisa de tosen [Mr. Ota won the election by a huge margin].

10/22/95: 1-3. Kichi shukusho shi kyotei minaose [Reduce military bases and revise the agreement].

1/23/96: 23. Jumin-tohyo de kichi tou [Questioning bases with the referendum].

12/22/96: 1. Kichi no tarai-mawashi kyudan [Accusing the rotation of the base].

7/23/97: 1. Zenken furi zohn: 128 mannin no sentaku 2 [The prefecture-wide FTZ: The choice of 1.28 million people 2].

7/23/97: 3. Okinawa keizai tokubetsuku ga kennai sangyo ni oyobosu eikyo chosa, jo [The survey on the influence the Okinawa special economic district will have on prefecture industries, first part].

7/24/97: 3. Okinawa keizai tokubetsuku ga kennai sangyo ni oyobosu eikyo chosa, ge [The survey on the influence the Okinawa special economic district will have on prefecture industries, second part].

7/25/97: 3. Zenken furi zohn: 128 mannin no sentaku 4 [The prefecture-wide FTZ: The choice of 1.28 million people 4].

7/25/97: 13. Sangyo to keizai no shinko to kisei-kanwa tou kento iinkai hokoku zenbun [Full text of the Report of the Committee of Industrial and Economic Promotion and Deregulation].

7/27/97: 1. Zenken furi zohn: 128 mannin no sentaku 6 [The prefecture-wide FTZ: The choice of 1.28 million people 6].

7/31/97: 11. Zenken furi zohn: Keizaikai no shiten 1 [The prefecture-wide FTZ: Perspectives of economic circles 1].

8/1/97: 9. Zenken furi zohn: Keizaikai no shiten 2 [The prefecture-wide FTZ: Perspectives of economic circles 2].

8/2/97: 11. Zenken furi zohn: Keizaikai no shiten 3 [The prefecture-wide FTZ: Perspectives of economic circles 3].

8/5/97: 9. Zenken furi zohn: Keizaikai no shiten 4 [The prefecture-wide FTZ: Perspectives of economic circles 4].

8/6/97: 11.Zenken furi zohn: Keizaikai no shiten 5 [The prefecture-wide FTZ: Perspectives of economic circles 5].

8/7/97: 13. Zenken furi zohn: Keizaikai no shiten 6 [The prefecture-wide FTZ: Perspectives of economic circles 6].

9/2/97: 3. Zenken jiyu boeki chiiki no ken soan zenbun [Full text of the prefecture's draft for the prefecture-wide FTZ].

9/3/97: 2. Zenken furi zohn: 128 mannin no sentaku 32 [The prefecture-wide FTZ: The choice of 1.28 million people 32].

9/26/97: 2. Zenken furi zohn: 128 mannin no sentaku 40 [The prefecture-wide FTZ: The choice of 1.28 million people 40].

10/28/97: 2–3. Kokusai toshi kesei ni muketa sangyo shinkosaku no tenkai [The prefecture's plan for the Development of the Industrial Promotion Policies for Cosmopolitan City Formation].

11/18/97: 3. NIRA chukan hokoku [The NIRA interim report].

12/26/97: 1. Higa shicho jinin wo teishutsu [Mayor Higa submitted his resignation].

18

The "War on Terrorism" and the "Hegemonic Dilemma"

Extraterritoriality, Reterritorialization, and the Implications for Globalization

COLIN FLINT

The United States has been the dominant, or hegemonic, world power since the end of World War II. The current period of "globalization" is a moment in the United States' cycle of hegemonic power. It is a period when the United States must simultaneously promote a global politics of open borders, while also maintaining its sovereign control over "domestic" policies. It is the tension between these two politics that I have termed the "hegemonic dilemma"—securing domestic politics from external shocks while promoting a global regime of economic networks and flows. To illustrate the argument, I analyze quotations from the Bush administration in the aftermath of the terrorist attacks of September 11, 2001 (9/11). These quotations illustrate the United States' need to maintain a global presence, but also the imperative to define and defend its own borders. Though the United States, as hegemonic power and victim of terrorist attacks, may experience this dilemma most intensely, the dilemma is also indicative of the tension between a global outlook and domestic stability facing all countries.

I use a world-systems approach to make my argument. World-systems theory contends that the tensions between global flows and domestic politics are not unique to the contemporary period of globalization, but have been in existence since the establishment of the capitalist world-economy approximately 500 years ago. In the first section of the chapter, the world-systems interpretation of globalization is discussed and the concept of hegemony is defined. Next, the global and domestic politics of the United States, as hegemonic power, are outlined, leading to the definition of the hegemonic dilemma. The empirical section of this chapter uses Bush administration quotes regarding the "War on Terrorism" to illuminate one manifestation of the hegemonic dilemma. Finally, in my concluding remarks, I discuss the implications of the hegemonic dilemma for all countries.

GLOBALIZATION AND WORLD-SYSTEMS THEORY

Globalization has been defined in numerous ways (see Mittelman, Chapter 2, this volume). The definition chosen determines the identification of the geopolitical processes that underpin globalization, as well as their geopolitical implications. Santos's (1999) concentration upon multifaceted processes plus the important component of diffusion allows for an understanding of both the content of globalization and its geographical expression. For Santos, *globalization* is "the process by which a given local condition or entity succeeds in extending its reach over the globe and, by doing so, develops the capacity to designate a rival social condition or entity as local" (p. 216). This definition provides insight into the geopolitics of imposition, on the one hand, and resistance, on the other. Thus globalization is the competitive diffusion of economic, political, and social practices from one locality over a significant portion of the globe.

But the geographical component is not enough to understand globalization; the timing is of utmost importance too. Globalization is seen as a qualitative progression in the developmental trajectory of capitalism that has ushered in a new form of social organization (Hardt & Negri, 2000). We are also warned, however, that globalization is not necessarily a one-way process, as it was created, in part, by the actions of states (Sassen, 1996, p. 22); hence states could decide to reimpose a regime centered upon the sovereignty of nation-states (Block, 1987). A more historically sensitive interpretation of globalization places it within the cycle of the rise and fall of great, or hegemonic, powers that has been a sporadic feature of the capitalist world-economy since its inception in the mid-1400s (Arrighi, 1994).

World-systems theory provides a framework to understand contemporary globalization as a form of the diffusion of economic, political, and cultural practices established and promoted by one state, the United States. To make this argument, I must first explore how globalization relates to the features of the capitalist world-economy and then describe the special role of the United States.

World-systems theory interprets globalization as a particular moment in the history of the capitalist world-economy. The capitalist world-economy is a historical social system that first emerged in Europe in the mid-1400s in the wake of feudalism. Since that time, it has expanded to encompass the whole globe and is currently the only form of social organization. In relation to the globalization debate, two interrelated features of the capitalist world-economy are of importance: a single economy and multiple states (Taylor & Flint, 2000). Since the inception of the capitalist world-economy, commodities have been priced and traded in a market that extends beyond single states or countries. As a contemporary example, the prices of oil, coffee, uranium, and so forth are established in a "global" market. Hence, the capitalist world-economy is partially comprised of economic networks channeling flows of commodities, money, and people across the globe. Economic networks, however, operate over a political terrain fragmented into separate units or entities; these entities are states, or countries. States attempt to manipulate economic flows to their own advantage either by opening their borders to trade if they can be strong in the global market, or by closing their borders if they are threatened. For example, different countries will argue for varying degrees of protectionism over the trade of, say, steel, depending upon the relative competitiveness of their national steel industry. In this sense, for the past 500 years the politics of the capitalist world-economy has been all about the tension between state policies and global flows and power—the very stuff that is the focus of the globalization debate and its concentration upon our times as being unique.

Despite the world-systems identification of tension between global flows and state sovereignty over the last 500 years, there is still room for agreement that something important is happening with this balance now. In that sense, this is a period of globalization, but it is an intensification of historic processes rather than a unique phenomenon. For world-systems theory, the current period is one of "financialization," or the need to intensify global investment in the wake of overproduction. This process is a feature of the hegemonic cycle. So first I must explain hegemony.

In the 500-year-long history of the capitalist world-economy, there have been times when one state has dominated the whole system: the Dutch in the 1600s, the British through the 19th century, and the United

States during the 20th century. In world-systems theory, these states are termed "hegemonic powers." The world-systems understanding of hegemony has become more complete and complex over time. It began with an initial concentration upon economic prowess (Wallerstein, 1984), through a connection with the establishment of geopolitical world orders (Taylor, 1996; Taylor & Flint, 2000), to the important inclusion of the role of social and cultural practices defined and disseminated by the hegemonic power (Taylor, 1999). Hegemony is founded upon the clustering of dominant production processes and technological innovation within the borders of one state (with intrastate uneven development) that allows for dominance in commerce, which ultimately provides for global financial domination. Economic hegemony allows for, and is facilitated by, political domination of the world that is reflected in the establishment of periods of geopolitical stability, otherwise known as "hegemonic geopolitical world orders." But the power and dominance of the hegemonic power is not just a product of economic strength or political might. The power base is based upon a subtler tactic: the definition of a modern way of life that is, on the whole, desired and emulated by social groups within the hegemonic power and across the globe.

This understanding of hegemony provides for the multifaceted nature of globalization discussed by Santos (1999); it is economic, political, social, and cultural. It also includes the processes of diffusion. Economic influence, for example, is diffused through the necessity to establish free trade across as much of the world-economy as possible to provide relatively easy market access for the goods produced by the hegemonic power. As the "home" of the most innovative products and production processes, the hegemonic power has the goods that others covet, and the ability to make them quickly and cheaply. Hence, it is to the advantage of the hegemonic power to create a regime of global free trade to maximize the geographical extent of the market for such goods.

The hegemonic power also plays the role of world leader (Modelski, 1987), establishing a period of relative geopolitical calm that simultaneously is fueled by, and itself facilitates, economic growth.[1] Political influence is felt in the establishment of institutions with global reach, such as the North Atlantic Treaty Organization (NATO), the United Nations (UN), and the International Monetary Fund (IMF) and the World Bank in the period of U.S. hegemony. The purpose of these institutions is to influence and alter the domestic policies of other countries in a way that best suits the hegemonic power's idea of a global economic regime. Such influence in the politics of other countries is called "extraterritoriality."

In addition, hegemonic economic prowess is built upon a social reorganization that comes to define what it means to be modern (Taylor, 1999). In the case of the United States, modernity has been centered upon the consumer society and its suburban landscape. This social reorganization is built upon a social compact—a combined social, racial, and gender division of labor that defines access to a variety of life chances (Silver & Slater, 1999). The new products and divisions of labor that undergird the hegemon's economic power helps to define, and are also driven by, changing cultural practices. For example, economic growth, the production of the automobile and other consumer durables, and gender and racial divisions of labor helped to promote the landscape of suburbia. In turn, the cultural practices of the suburban way of life promoted continued demand for economic products, such as oil. To be modern was to be suburban (Taylor, 1999). It was the suburban way of life, broadcast from Hollywood, that was sold to the world as the prime modernity to be emulated.

The *prime modernity* is the combination of the cultural and economic innovation and vitality that enables a country to reach hegemonic status. Thus, the United States can claim to be modern, efficient, and the home of universal human values (Wallerstein, 2002). And it is these attributes of the hegemonic power that other countries want to emulate (Taylor, 1999). In turn, this emulation means that the prime modernity—in terms of a way of life and cultural values—is exportable. Other countries seek to be modern too, which results in the adoption of global and domestic policies that mirror those of the hegemonic power.

Much has been made in recent years, especially since the end of the cold war, about the United States being a global power like no others that have come before it (Hardt & Negri, 2000). Especially the argument is brought forward that the United States' hegemonic rule is more complete, not just in terms of its global extent, but in its vision of what the world should be like (Hobsbawm, 2003). A world-systems perspective would take issue with this point of view. First, the process of U.S. hegemony began at the beginning of the 20th century, and was firmly established by its role in the Allied victory in World War II. Given this time frame, then, it can be acknowledged that U.S. hegemony has been severely challenged by the counterideology of international socialism, the first cold war, and the revolt of the periphery, or "third world," in the second cold war (Halliday, 1986). The defeat and demise of the Soviet Union was indeed a partial victory for the hegemonic power, but resistance to its policies across the globe have increased to such an extent that it has become the new geopolitical agenda with the identification of the "axis of evil." Challenge exists domestically too, in the form of extreme religious and conservative groups critical of the "values," a key component of prime modernity, of secular U.S. government (Juergensmeyer, 2000).

But, still, it may be argued that the United States' military capacity and reach is unprecedented. Maybe so, but a historical approach can raise further words of warning. The establishment of the British Empire came toward the end of Britain's hegemonic reign; the need to exert territorial control through physical presence is a sign of hegemonic weakness rather than of strength (James, 1994; Kennedy, 1987). In the past, great powers have entered a spiral of decline as they have become "overstretched" in the need to police challenges to their power (Kennedy, 1987). Though the U.S. economy is in a position of relative strength compared to those of its main competitors (i.e., the European Union and Japan), there still remain questions regarding its ability to take on global policing responsibilities that may be interpreted as responses to challenges to its hegemonic role. As the *Economist* (2003) remarked:

> Surely, the world's hyperpower can replenish its 147,000 strong Mesopotamian garrison by itself? Apparently not. Of the army's 490,000 men and women, 362,000 are deployed in 120 countries, including Afghanistan and South Korea. On July 23rd [2003], the Pentagon announced a complicated "rotation" for Iraq stretching into next year. It will need to deploy the part-time national guard and various untested troops. It also confessed it needed "more infantry . . . more military police . . . and more civil-affairs [specialists]." (p. 28)

It is not a simple relationship between the relative strength of a country's economy and its army that matters, but the actual ability to maintain a global military presence over a sustained period. Instead of a realist fixation upon military strength, the ability to exert power across the globe requires the political vision to construct a geopolitical project that will gain international support as well as the political will to carry it out.

The *Economist* (2003b) supports my opinion that the United States does not possess the political ability to maintain global power, despite its military strength. In addition, sustained military activity requires sustained domestic and international political support; it will take more than a response to the terrorist attacks of 9/11 to produce domestic political quiescence and stable international coalitions. It will also require belief in the need and the value of a global civilizing mission. The beginning of a new period of American hegemony would be best signaled by a global political project that is acclaimed by the rest of the world, rather than by the current military incursions that are dominated by the self-interests of the United States. It is quite possible that the United States could define such a mission and combine it with economic and political strength to forge a new period of hegemony. However, contemporary U.S. military strength and political commitment do not seem strong enough at the moment to dismiss the possibility that we are seeing military adventurism that is the rearguard action of hegemonic decline, rather than the construction of a new period of hegemonic rule.

A clue to the likelihood that the global military presence of the United States does not reflect the establishment of a new period of hegemony is found in the "emptiness of administration policy" (Hobsbawm, 2003). In an interesting contradiction, the United States as a global force with a universal project is also, for Hobsbawm, a country that has a vacuous foreign policy. What is, perhaps, more accurate is that the contemporary ingredients of U.S. foreign policy are transparently self-serving, stemming from a need for national security in the wake of 9/11. Though U.S. foreign policy is couched in the promise of "freedom" and so forth, there is none of the universal vision that drove the United States' rise to hegemonic power in the first half of the 20th century. Indeed, the practice of U.S. foreign policy contradicts some of the universal benefits offered by U.S. hegemony (Prestowitz, 2003). For example, the promise of national self-determination that is a component of U.S. hegemony's universal message is clearly being countered by its military's eviction of regimes it does not like. As the history of previous great powers demonstrates, military expeditions not only weaken them in an economic sense, they also serve to expose the self-serving lies behind their promises of acting for all humankind. Boulding (1990) claims that power backed by military might alone is unsustainable without an integrative message—in other words, that carrots are more effective than sticks. Similarly, Hirsh (2003) argues that the world is no longer receptive to U.S. ideals and goals. If true, it points to a decline of U.S. hegemony rather than to its reassertion.

As social scientists, we are armed with theory and historical interpretation, not a crystal ball. Whether we are seeing the process of U.S. hegemonic decline or the establishment of not just a new period of hegemony, but a new form of global rule, is a matter for conjecture and competing perspectives. But it is through examination of the balance between the United States' global power and its domestic strength that the trajectory can be best understood.

EXTRATERRITORIALITY, THE SOCIAL COMPACT, AND THE HEGEMONIC DILEMMA

The practice of hegemony, that is, the diffusion of economic, political, and sociocultural practices and influence, requires the geopolitics of *extraterritoriality*, namely, the imposition of power and influence by one nation-state into the sovereign spaces of other nation-states (Flint, 2001). Many understandings of international politics begin with the assumption that all states have an equal ability to exercise sovereign control over their own territory. While it is known that some countries have larger militaries and economies than others, an axiom of the legalistic interpretation of world politics is that any country, despite its level of "power," has the right and ability to be sovereign in its own territory. But this is a myth; some countries have the ability to exert varying degrees of influence within others. Such influence is called "extraterritorial" in that the power of one state extends beyond its own territory into the territorial sphere of sovereignty of another. Though we can see extraterritoriality in a number of state-to-state relations—Syrian influence in Lebanon, for example—the major proponent of this form of geopolitics is the hegemonic power.

The extraterritoriality of the hegemonic power is the geopolitical manifestation of contemporary globalization, a set of diffusion processes that is primarily the product of the hegemonic nation-state. These practices and their dissemination began in the middle of the last century, and have become most manifest during a particular period in the hegemonic cycle: the period of financialization (Arrighi, 1994; Arrighi & Silver, 1999). "Financialization" refers to a period of intense flows (especially with regard to investment) in order to compensate for weakening markets for—and the economic surplus of—some goods. Its roots are found in the economic success of the hegemonic country, as it produced and exported goods. As other countries emulated the hegemonic power, an economic crisis resulted as the global market was awash with products. In the light of this crisis, new sites for investment had to be found, which required the opening of borders to foreign investment. This is a similar interpretation of the establishment of the North American Free Trade Agreement (NAFTA) and other "open-border" policies offered by scholars with a more traditional Marxist focus (Harvey, 2000). The difference is that world-systems theory offers an explanation for the timing of these processes within the rise and fall of hegemonic powers.

Extraterritoriality is only one side of the coin, though. As noted, the economic growth that is the required underlying basis of hegemonic power is based upon the creation of a social compact (Silver & Slater, 1999). The social compact includes and excludes particular groups from the benefits of a particular hegemonic project. The production processes of the United States that were the basis of its economic power included particular gender and racial divisions of labor. After World War II, the employment opportunities experienced by women and blacks during the war were retracted, as skilled manual jobs and management positions were dominated by white males. The New Deal provided benefits for blue- and white-collar workers that accepted new work relations. More radical workers movements and African Americans were not awarded the same levels of participation and reward. The social compact has not been static over time, though. The civil rights movement and the feminist movement challenged its exclusionary and discriminatory form. The result has been the gradual erosion of male dominance in the workplace, though not its extinction. In addition, economic competition from abroad forced the United States to reevaluate the privileges it had been able to award its workforce: wages, benefits, vacation time, pensions, and so forth. In other words, the social compact was attacked through both domestic and international pressure.

The hegemonic social compact also entailed differential social rewards outside of the United States. For the purpose of this chapter, however, it is apt to concentrate upon one aspect of the social compact that has not been emphasized enough. The social compact is based upon an understanding of *territorialized citizenship*, that is, belonging to a recognized and territorially bounded nation-state, and the two-way avenue of rights and responsibilities between state and citizen (Taylor, 1991, 2000). The political competition for a privileged position within the social compact of U.S. hegemony was a matter of domestic politics, and was cemented with the ability of governments to distribute benefits to different members of society. But "society" here refers to the sense of a political community defined and demarcated by state borders. The politics of the creation, maintenance, and challenge to the social compact of U.S. hegemony was a matter of "domestic" politics.

World-systems theory points us to two complementary, but at times antagonistic, politics. On the one hand, is the politics of extraterritoriality, magnified during financialization; this is characterized by the erosion of state sovereignty and the ability of countries to conduct their own politics. Political issues such as these are usually the focus of globalization studies. On the other hand, the hegemonic power's ability to exert extraterritorial influence is dependent upon a social compact that is housed within the assumptions of territorial sovereignty, namely, citizenship rights and the domestic politics of the distribution of wealth, power, and life chances.

It is this tension that I have called here the *hegemonic dilemma*—or the geopolitical need to promote extraterritoriality and infiltrate the sovereignty of other nation-states while maintaining one's own territorial sovereignty. Emphasizing that hegemony is a process of establishment, rule, and relative decline suggests that the balance between territoriality and extraterritoriality will change over time. The extraterritorial presence of U.S. hegemony has been evident since the mid-20th century. In a macrosense this can be seen in the role of U.S. ideology and capital in the IMF and World Bank, on the one hand, and its global military presence through NATO and other alliances, on the other. From the Marshall

Plan and the imposition of a political constitution in postwar Japan to the powerful cultural influence of Hollywood, the United States has exerted its influence into other sovereign spaces, sometimes as a matter of conscious political acts and in other instances as a result of entrepreneurial agency. All of these processes and institutions entwine "domestic" concerns such as employment and stock share values, for example, with activity in the sovereign spaces of other states. It follows that the domestic and international aspects of hegemony are intertwined, and that their separation in this chapter is for purely organizational and heuristic purposes. But things have changed since the terrorist attacks of 9/11. Awareness that the porosity of borders not only exists, but also has attached risks, is part of mainstream political consciousness. In this political climate, the Bush administration has had to address such public concern without derailing the hegemonic project. Thus political statements regarding 9/11 provide insights into the existence of the hegemonic dilemma, as well as how politicians have tried to negotiate it.

RHETORIC OF EXTRATERRITORIALITY
AND RETERRITORIALIZATION POST-9/11

The twin notions of hegemony (Wallerstein, 1984) and prime modernity (Taylor, 1999) suggest that the United States has a role in diffusing particular economic practices as well as a particular way of life across the spatial extent of the capitalist world-economy. Economic leadership and dominant social practices were at the forefront of comments from U.S. politicians in the aftermath of 9/11. Leading the way, the president established free trade as the defining element of U.S. power, identity, and engagement with the rest of the world:

> Fearful people, people who don't trust the ability of our entrepreneurs build walls around America. Confident people tear them down. And I'm confident in the American spirit. I'm confident that the entrepreneurs of our country—Hispanic, Anglo, African American—compete with anybody, any place, any time, and let's trade freely. (President George W. Bush, January 5, 2002[2])

Bush's comment naturalizes the global economic reach of the United States as well as asserts the importance of free trade, with its extraterritorial implications. Interestingly, the inclusion of Hispanics and African Americans in the U.S. project implies that the initial racial bias of the domestic social compact has been shattered. Similarly, the treasury secretary emphasized the relationship between economic hegemony and prime modernity:

As trade flows from nation to nation, ideas of freedom, creativity, and tolerance are part of the packaging. (Secretary of the Treasury Paul O'Neill, October 24, 2001[3])

In a way that political rhetoric need never make clear, processes of free trade that are an expression of U.S. economic might are attached to a civilizing mission that will improve the social well-being of humanity. Similar sentiments were expressed, with a heart-felt belief, by British imperialists (James, 1994, pp. 145, 183).

As in previous manifestations of hegemony, the economic reach of the United States is the key engine in ensuring power, influence, and the ability to speak as if for all. With no need for explanation, the economic power of the United States is equated with a universal global commonality of interest that denies national competition. However, such a process requires U.S. businesspeople and diplomats to be "out there," in other words, practicing extraterritoriality. As stated by Secretary of State Colin Powell:

It was not hard to pull this coalition together because instantly, on the 11th of September, every civilized nation looked and said this is an attack not just against the American World Trade Center, but the World Trade Center.

And while we are waging our campaign, you will still be out there in the world doing your work. We know that for you, as for us in the State Department, staying home is not an option.

From its beginning in 1985 when Secretary Shultz met with a handful of CEOs, OSAC [Overseas Security Advisory Council] has expanded to nearly 2,000 affiliated U.S. companies and organizations. That is a wonderful, wonderful testament to the drive and spirit of America's entrepreneurs, business people, educators and others who follow their dreams beyond our borders and spearhead America's engagement with the world. (November 7, 2001[4])

These ideas were reinforced by National Security Advisor Condoleezza Rice, who again made a direct connection between U.S. economic imperatives and global human rights:

We are committed to a world of greater trade, of greater democracy and greater human rights for all the world's people wherever they live. September 11th makes this commitment more important, not less. (February 1, 2002[5])

However, the success of the hegemonic process depends upon maintaining a balance between national self-interest and the perception that the hegemonic country is bringing benefits to all. In the following quotes, Colin L. Powell is more candid about the benefits to U.S. capital that is part-and-parcel of "transforming the world as we knew it." Both global transformation and "national" capital accumulation for capitalists are the goals and means of the hegemonic power.

> Our ideas, our know-how and our culture reach every corner of the world, and is transforming the world as we knew it.
>   Our economic engagement with the rest of the world is an important part of our effort to maintain the secure international environment within which Americans and American businesses prosper. It helps spread to others the benefits that we ourselves enjoy. (October 31, 2001[6])

With regard to extraterritoriality, a (re)definition of prime modernity took center stage over economic globalization. In response to a terrorist act, an immense criminal event, tropes of civilization and justice were used to bring the United States' role in disseminating a particular way of life in to the conversation. Numerous quotes on this matter were found from each and every member of the Bush administration. The following quotes were selected because they highlighted that, though the pursuit of justice was extraterritorial in its geography, justice was uniquely American.

> A calculated, malignant, and devastating evil has arisen in the world. Civilization cannot ignore the wrongs that have been done. America will not tolerate their being repeated. Justice has a new mission—a new calling against an old evil. (Attorney General John Ashcroft, November 8, 2001[7])

The Taliban served, in an orientalist fashion (Said, 1979), not only to define the United States as the locality that defines the freedoms that are to be globalized, but also the perceived dangers that can arise if the globalization project is thwarted.

> Walker was blessed to grow up in a country that cherishes freedom of speech, religious tolerance, political democracy, and equality between men and women. And yet he chose to reject these values in favor of their antithesis, a regime that publicly and proudly advertised its mission to extinguish freedom, enslave women, and deny education. John Walker Lindh chose to fight with the Taliban, chose to train with al Qaeda, and to be led by Osama bin Laden. (Attorney General John Ashcroft, January 15, 2002[8])

The key point, reflected in the following comments by President Bush, is that the United States, as prime modernity, is the definer and disseminator of justice. Moral understandings of justice and specific forms of legal practice are assumed, by the hegemonic power, to be universally applicable and beneficial. The pursuit of justice becomes another vehicle for hegemonic extraterritoriality.

> We will rid the world of the evil-doers. We will call together freedom loving people to fight terrorism. (President George W. Bush, September 16, 2001[9])

> Terrorists try to operate in the shadows. They try to hide. But we're going to shine the light of justice on them.
> Eventually, no corner of the world will be dark enough to hide in. (President George W. Bush, October 10, 2001[10])

> The U.S. is the defensor [sic] of liberty all over the world, and that's what this attack was about. (President George W. Bush, October 15, 2001[11])

> There is no corner of the Earth distant or dark enough to protect them. However long it takes, their hour of justice will come. (President George W. Bush, November 10, 2001[12])

Significantly, the previous quotes recognize that parts of the world exist where the hegemonic power's presence is less than welcome. As a precursor to the geopolitical trope of "axis of evil," dark corners of the world are identified—"dark" in that they have erected barriers to the progressive "light" of civilization diffused by the hegemonic power by means of extraterritorial policies. Such a strategy requires not only the political might of the hegemonic country, but also the power that comes from prime modernity; the military geopolitics of "getting them" also requires a hegemonic project of defining and nurturing a particular understanding of "human freedom."

> We're going to go after these terrorists with a global reach on our own time, but as rigorously as possible. We're going to get them where they are in whatever shape they are. (Deputy Secretary of State Richard Armitage, September 2001 [exact date not given][13])

> Our mission today is not only to root out and eliminate the terrorists— we must also enlarge the circle of human freedom to include that vast majority of Muslim people who are seeking to enjoy the benefits of living in a free and prosperous society, but do not yet do so. (Deputy Secretary of Defense Paul Wolfowitz, November 14, 2001[14])

Even a "hawk" such as Wolfowitz can see that geopolitics requires a cultural component to aid global dominance. The attacks of 9/11 were perceived by the Bush administration as a catalyst for a redefinition of the United States's global role, a role in which U.S. civilization and justice were not only to be defined but also spread across as much of the globe as possible. Seemingly benign statements such as

> I think the best way to attack—to handle the attacks of September the 11th—is to fight fear with friendship; is to fight fear with hope; is to remind people all around the world we have much more in common than people might think; that we share basic values—the importance of family, and the importance of faith, and the importance of friendship. (President George W. Bush, October 25, 2001[15])

were echoed by more explicit recognition of a historic mission:

> History has called us into action, here at home and internationally. We've been given a chance to lead, and we're going to seize the moment in this country. As we've mentioned more than once, what we do here at home is going to have lasting impact for a long time. And I want to tell you what we're doing abroad is going to have lasting impact, as well. (President George W. Bush, February 5, 2002[16])

The previous quote recognizes the need to balance a global presence with domestic benefits—and such strong statements suggest that both geographical arenas require explicit attention and change at this stage of the hegemonic process. But such a mission and political changes have to be given concrete ingredients, as specified by National Security Advisor Condoleezza Rice.

> On every continent, in every land, this President, the education President at home, wants to press the goal of education for all abroad. (February 1, 2002[17])

The hegemonic commitment requires a global strategy of diplomacy, one driven by the commonsense assumptions of the unquestioned legitimacy and supremacy of the U.S. way of life. Moreover, as Secretary of State Colin Powell acknowledges, the hegemonic project requires extraterritorial reach. However, he also notes that this is not a one-way process under the total control of the hegemonic power. Global flows can also enter and impact the United States, the other side of the hegemonic dilemma.

> The terrorist attacks of 9/11 underscore the urgency of implementing an effective public diplomacy campaign.
> There is no part of the world that we are not interested in. We

are a country of countries. We are touched by every country, and we touch every country. (February 5, 2002[18])

The comprehensiveness and multifaceted nature of this project were also made clear by the secretary of state. The hegemonic project is not simply a matter of economic accumulation, but is made manifest in transformations of state and civil society practices. In addition, the following quote also recognizes the geopolitical difficulties faced by the United States in combating a network of terrorists over a geopolitical terrain of sovereign nation-states.

But the war on terrorism starts within each of our respective sovereign borders. It will be fought with increased support for democracy programs, judicial reform, conflict resolution, poverty alleviation, economic reform and health and education programs. All of these together deny the reason for terrorists to exist or to find safe havens within those borders. (Secretary of State Colin L. Powell, November 12, 2001[19])

The emphasis placed upon the charitable and developmental image of prime modernity should not prevent us from recognizing that it is extraterritorial in its geographical expression, and hence is geopolitical. History also suggests that the imposition of a way of life into other sovereign spaces cannot be achieved without military force. Secretary of Defense Donald H. Rumsfield seemed aware of this history lesson when he stated:

We talked early on, the president did, about the opportunity to rearrange things in the world in a way that would be beneficial to our country and to peace and to stability and to free systems, and how as we're doing this do we do it in a way that because it's such a fundamental shift in how people think about the world, how do we do it in a way that benefits the world after this event is over. (January 9, 2002[20])

In remembering how hegemony and prime modernity are connected, it should be emphasized that Rumsfield recognizes national interest first and assumes its global benefits. "Rearranging things in the world" has been the historic role of hegemonic powers (Taylor, 1996), and it appears that a response to 9/11 is the current vehicle for its enactment. Military force is an expensive way of facilitating such diffusion. On the other hand, "justice" is a much more neutral term as it encourages emulation rather than forced transformation.

Some will ask whether a civilized nation—a nation of law and not of men—can use the law to defend itself from barbarians and remain civilized. Our answer, unequivocally, is "yes." Yes, we will defend civilization. And yes, we will preserve the rule of law because it makes us civilized. (Attorney General John Ashcroft, October 25, 2001[21])

At the conclusion of World War II came the reckoning at Nuremburg. Former Attorney General and Supreme Court Justice Robert Jackson led the prosecution of 21 Nazi defendants for crimes against their countrymen, against their neighbors—indeed, crimes against humanity. All pleaded not guilty. Some claimed that they were merely following orders. Others disputed the jurisdiction of the court. But Jackson successfully argued their guilt with a sense of urgency born of a civilization threatened by a new force of evil. "The wrongs which we seek to condemn and punish have been so calculated, so malignant, and so devastating," said Jackson, "that civilization cannot tolerate their being ignored, because it cannot survive their being repeated." (Attorney General John Ashcroft, November 8, 2001[22])

Ashcroft's allusion to World War II is noteworthy. The United States-led victory in 1945 is generally recognized as the moment when America achieved hegemonic status. To allude to this moment allows the Bush administration to project the image that the United States is at a starting point once again, with a new civilizing mission and economic might ushering in a renewed period of hegemony. However, such assertions may be political cover when resistance to a declining U.S. hegemony is on the increase. Time will tell which of these scenarios is correct.

Of course, the reasons for such actions require, perhaps, none of this talk of hegemony and prime modernity. It is, after all, a simple matter of manifest destiny.

Watch us, we're America. We're not going to draw back behind our oceans and behind our fences. (Secretary Colin L. Powell, November 7, 2001[23])

But it is a manifest destiny that is linked to a hegemonic project, that secures ownership of the seas as previous hegemonic powers have done while speaking for the whole world. The manifest destiny is translated as a global rather than a national process.

And I know the President will consult with our friends and allies in the world because it is not just a danger to the United States; it is a danger to the whole world, to the civilized world. (Secretary of State Colin L. Powell, February 3, 2002[24])

And the hegemonic project is diffused from a national base, over a specific period of time, with perceived global benefits, reflecting both

Santos's (1999) definition of globalization and world-systems theory's understanding of the hegemonic process.

> And it is our national security, the United States of America, at this time in history, that is able to contribute to peace and stability in the world. (Secretary of Defense Donald H. Rumsfeld, February 3, 2002[25])

Furthermore, in the wake of 9/11, the tropes of justice and freedom are the key contemporary code words for the practices of U.S. hegemony that are to be encouraged across the globe. Whether such tropes will form the ideological foundation for a new period of hegemony is debatable, seeing as they so clearly reflect U.S. imperatives rather than a response to a global clamor.

> We act today to protect the lives and safety not just of Americans but all of those who believe in this idea and ideal of freedom, and all of those who have sacrificed to live amidst the blessings of freedom. (Attorney General John Ashcroft, November 14, 2001[26])

The flip-side of the hegemonic dilemma is that the necessary practices of extraterritoriality, the expansion of influence into other sovereign spaces, may decrease the security of one's own space. Tom Ridge's interpretation of his cabinet responsibilities seems to be in direct contrast with the extraterritorial mission defined by Rumsfield, Powell, and Ashcroft.

> The only turf we should be worried about protecting is the turf we stand on. (Governor Tom Ridge, October 8, 2001 [quoted in Becker & Sciolino, 2001])

The establishment of a cabinet-level Office of Homeland Security recognized that the flows of globalization are not operating just one-way, but that the sources of that insecurity are the practices of U.S. hegemony.

> . . . to deal with a 21st century environment that says the challenges to American's sovereignty and our security, which historically have been offshore, but because of the 20th century environment we find that the challenges are here. . . . (Governor Tom Ridge, October 18, 2001[27])

Or to put it another way, the geography of the current challenges is a spatiality of networks that blurs any distinction between "domestic" and "foreign" (O'Tuathail, 2000). These networks are a product of historically prior U.S. hegemonic practice, but the embedded statism of formal foreign policy requires a separation of the two (Agnew, 1999; Oas, 2002; Taylor, 2000; Walker, 1993).

It's one war, but there are two fronts. There's a battlefield outside this country and there's a war and a battlefield inside this country. (Governor Tom Ridge, October 22, 2001[28])

In the minds of policymakers the distinction between the domestic and the political arenas remains, and the key political–geographical prophylactic separating them is the international border of the United States. The border creates a spatial dichotomy between a territorialized and revered sense of U.S. justice, on the one hand, and foreign "evils," on the other:

But any time there are borders that are that open and that substantial, there are risks that people crossing the border could be individuals who are involved in very serious activities that could be troublesome. (Attorney General John Ashcroft, October 2, 2001[29])

I had an opportunity several times to testify before the Senate and the House, and to make the point that what we're dealing with here is not immigration; we're dealing with evil. (Attorney General John Ashcroft, October 31, 2001[30])

I'd like to note that the INS [Immigration and Naturalization Service] has been and continues to be a very vital player in this war on terrorism, in this investigation, as well as the ongoing process of protecting the American people from what we see as the forces of evil. (Commissioner of the Immigration and Naturalization Service Jim Ziglar, October 31, 2001[31])

These quotes reinforce the foreign/domestic polarity underlying mainstream understanding of international politics (Agnew & Corbridge, 1995, p. 86; Walker, 1993). But the balance is not as simple as keeping terrorists and criminals outside of the United States; it is also about maintaining an immigration policy that protects but does not disrupt the practices of hegemony. U.S. hegemony still requires relatively open borders, but it appears that the commitment to this openness is being questioned currently.

You know, we used to think of America the beautiful, fortress America, trusting America. And we find that perhaps we've trusted too much. (Governor Tom Ridge, October 16, 2001[32])

At the very least, policymakers are discussing a balance between openness and security, and raising the broader question of whether nation-states are seeking to redefine the balance between globalization and state sovereignty. The need to "track people" pits together two aspects of the geopolitical terrain, economic flows (in this case, human beings) with the policing of territorial states. However, this process is complicated by a recognition that the freedom of movement is, at least in principle, essential to the translation of U.S. values into the hegemonic project.

> We need to do more with respect to tracking people within a society that is an open society. And we have to do it in a way that protects us but, at the same time, does not cause us to be a closed society, be the kind of society that would not be reflective of American values. (Secretary of State Colin L. Powell, September 12, 2001[33])

The policy commentary post-9/11 served to territorialize particular notions of freedom and civilization—U.S. versions that were perceived to be for the benefit of all and so worthy of imposition by processes of extraterritoriality.

> And it should be a testimony and inspiration to every American everywhere, to understand that public safety is everybody's business, and it's our opportunity to do those things that preserve our liberty and the integrity of what it means to enjoy the freedoms we call America. (Attorney General John Ashcroft, September 20, 2001[34])

U.S. politicians have been eager to use the attacks of 9/11 as a catalyst for a politics reasserting the ability, and moral necessity, of American hegemony. Ashcroft's earlier quote invoking the justice of the Nuremburg trials offered a reminder of how World War II offered a rhetorical counterpoint to illustrate the global need for U.S. power and political leadership. The first necessity in this discursive strategy is to define a period of global chaos that requires the order that a hegemonic power may offer. The terrorist attacks of 9/11 were used to define a time of emerging global chaos requiring the reimposition of U.S. order and its associated extraterritoriality.

> British Prime Minister Tony Blair recently spoke of the fragility of our borders in the face of transnational terrorists. Conflicts, he said, rarely stay within national boundaries. Tremors in one country reverberate throughout the world. The threat, Prime Minister Blair concluded, is chaos. (Attorney General John Ashcroft, October 29, 2001[35])

The attacks upon U.S. symbols of hegemony were used to reinvigorate a hegemonic message and mission. It must be emphasized that this invigoration has both domestic and foreign components that are intertwined (Walker, 1993).

We'll be resolute in our determination to rout out terror wherever it exists—in our neighborhood or neighborhoods around the world. (President George W. Bush, January 16, 2002[36])

We still face a shadowy enemy who dwells in the dark corners of the earth. Dangers and sacrifices lie ahead. Yet, America will not rest, we will not tire until every terrorist group of global reach has been found, has been stopped, and has been defeated. (President George W. Bush, January 23, 2002[37])

The series of quotes from the Bush administration presented here have served to highlight a number of points. First, extraterritoriality is a key component of the practices of hegemony. Second, extraterritoriality is founded upon economic strength and self-interest. Third, economic reach is facilitated by the diffusion of a cultural or "civilizational" model—the prime modernity that is American in definition but, rhetorically, benefits all. Fourth, the global diffusions initiated by the hegemonic power are also disruptive of the social compact upon which hegemony is built; hence the sovereign space of the hegemonic power must be protected in a way that does not derail the hegemonic project. This is the hegemonic dilemma currently facing the United States.

Government responses to the hegemonic dilemma constitute a process of reterritorialization, as the relative porosity of borders to capital, commodities, and immigration flows is reversed. Indeed, in the wake of 9/11 immigration visa policies have become much stricter. Security provisions have been tightened along the border and at airports and other entry points. This suggests that the porosity of borders that is an essential component of globalization is quite reversible. But, as Eichengreen (2001) warns us, we should not be carried away by such attempts to restrict the movement of people. States have always imposed greater control over the movement of people compared to the movement of capital and commodities. The attacks of 9/11 are more likely to alter the composition and direction of capital flows than their total volume (Eichengreen, 2001). In addition, liberty—one of the key values of the United States—has also been under threat by the creation of military tribunals and campaigns to silence academic critique of U.S. policies (Wallerstein, 2002). Both extraterritoriality and prime modernity are open to revision post-9/11.

The twin themes of the extraterritoriality of the United States' global reach and interest and its concern for domestic security via reterritorialization and increasing border controls were found in the emphases of the Bush administration quotes. The attacks of 9/11 and their aftermath magnified the relationship between the United States' global role and the security of its citizens. Where we go from here will be the product of political agency. But it seems likely that a faltering domestic economy, immigration, the threat of terrorism, and foreign military intervention are likely to heighten the hegemonic dilemma in the near future.

CONCLUSION

Understanding globalization as a suite of diffusion processes (Santos, 1999) allows for the identification of the prime diffusers, the hegemonic powers. But diffusion is as much a temporal as a spatial process. The timing of the flows of globalization is linked to the establishment of U.S. economic and political hegemony. Previous hegemonic powers of the capitalist world economy have declined, ushering in new periods of state control of global flows (Polanyi, 1957; Schwartz, 1994). The concerns for territorial security within the hallways of Washington, DC, suggest that the globalization of U.S. hegemony is not an irreversible process.

Globalization is a form of geopolitics, led by a dominant or hegemonic nation-state. It requires the erosion of the relative ability of other nation-states to control flows across their borders, and the imposition of extraterritorial institutions. Also manifest is resistance to both of these processes by state and nonstate actors. The politics of resistance against globalization are evident in this case study of the United States in governmental concerns and policy. The geopolitics of globalization continues to be about the relative integrity of borders versus the power of state transcending networks. Perhaps history will see the contemporary "War on Terrorism" as a tide-turning moment in which nation-states reimpose some of their abilities to control global flows. On the other hand, the momentum of economic globalization may ultimately supersede the needs of the United States, and other countries for "homeland security."

In the empirical snapshot provided by this chapter, I hope to illustrate the continuous but changing tensions between extraterritoriality and territoriality that faces the hegemonic power. The chapter may also serve as a cursory and initial inquiry into two competing visions of globalization held by world-systems analysts. On the one hand, Peter Taylor (1993) and John Agnew (1993) interpret the innovations of U.S. hegemony as ushering in fundamental change in the balance between the world-economy and sovereign nation-states. The result is that the latter have been so weakened they can no longer provide the territorial haven upon which a new round of hegemony can be based. The United States is, therefore, "The Last of the Hegemons" (Taylor, 1993), and the role of nation-states within the world-economy has been changed permanently.

If this happens to be the case, we would expect to see a continued commitment to extraterritoriality by the hegemonic power as it attempts to continue to assert its comparative advantage of global reach. Alternatively, the cyclical emphasis of Arrighi (1994) and Silver and Slater (1999) argues that the recent experience of global financialization is a temporary phase in response to the global overaccumulation initiated by the hegemonic power's prior dominance in production. Hence, the reassertion of political sovereignty is not only possible but, given a historical understanding, most likely as well. If this scenario is correct, then we would expect to see political moves, even within the hegemonic power, to tilt the balance toward protecting territorial sovereignty.

The analysis of the U.S. political reaction can help us to consider the relative likelihood of these two scenarios. Of course, such snapshots regarding one political issue over such a short period of time are suggestive rather than conclusive. They do, however, illustrate that hegemonic powers must balance extraterritoriality and territoriality, and that this hegemonic dilemma creates security concerns related to the twin needs of global reach and "homeland security." The realities of globalization mean that the tension between economic openness and domestic sovereignty are faced by all the countries of the world. In this sense, the politics of the hegemonic dilemma have become pervasive and are evident in, for example, British debates over the benefits of adopting a European currency and Iran's disputes over reform or geopolitical isolation. The manner in which these disputes are resolved will go a long way in defining the political geography of the 21st century, and especially the degree to which we will live in a globalized world or not.

ACKNOWLEDGMENTS

Thanks to Catherine Adams, Sebastian Castrechini, Douglas Grane, Neeta Maniar, Drew Schaub, and Taisa Welhasch for collecting the quotes used in this chapter.

NOTES

1. Two points must be made here. First, though Modelski's (1987) theory disputes the world-systems theory identification of an economic basis of hegemony, it is useful in conceptualizing the timing of global dissent toward the hegemonic power. Second, hegemony, in the terminology of world-systems theory, is akin to Gramsci's notion, as rule is only effectively maintained by consensus. See Boulding (1990) for a discussion of the necessary role of integrative power, a role enacted by the diffusion of prime modernity.

2. *http://www.whitehouse.gov/news/releases/2002/01/20020105-3.html*.
3. *http://www.ustreas.gov/press/releases/po717.htm*.
4. *http://www.state.gov/*.
5. *http://www.whitehouse.gov/news/releases/2002/02/20020201-6.html*.
6. *http://www.state.gov/*.
7. *http://www.justice.gov/ag/speeches/2001/agcrisisremarks11_08.htm*.
8. *http://www.justice.gov/ag/speeches/2002/011502walkertranscript.htm*.
9. *http://www.whitehouse.gov/news/releases/2001/09/20010916-2.html*.
10. *http://www.whitehouse.gov/news/releases/2001/10/20011010-3.html*.
11. *http://www.whitehouse.gov/news/releases/2001/10/20011015-3.html*.
12. *http://www.whitehouse.gov/news/releases/2001/11/20011110-3.html*.
13. *http://www.pbs.org/wgbh/pages/frontline/shows/terrorism/interviews/
armitage.html*.
14. *http://www.defenselink.mil/speeches/2001/s20011114-depsecdef.html*.
15. *http://www.whitehouse.gov/news/releases/2001/10/20011025-2.html*.
16. *http://www.whitehouse.gov/news/releases/2002/02/20020205-4.html*.
17. *http://www.whitehouse.gov/news/releases/2002/02/20020201-6.html*.
18. *http://www.state.gov/secretary/rm/2002/7797.htm*.
19. *http://www.state.gov/*.
20. *http://www.defenselink.mil/news/Feb2002/t02052002_t0109wp.html*.
21. *http://www.justice.gov/ag/speeches/2001/agcrisisremarks10_25.htm*.
22. *http://www.justice.gov/ag/speeches/2001/agcrisisremarks11_08.htm*.
23. *http://www.state.gov/*.
24. *http://www.state.gov/secretary/rm/2002/7781.htm*.
25. *http://www.defenselink.mil/news/Feb2002/t02032002_t0203abc.html*.
26. *http://www.justice.gov/ag/speeches/2001/agcrisisremarks11_14.htm*.
27. *http://www.whitehouse.gov/news/releases/2001/10/20011018-1.html*.
28. *http://www.nytimes.com/2001/10/22/national/22CND-EXCE.html*.
29. *http://www.justice.gov/ag/speeches/2001/agcrisisremarks10_2.htm*.
30. *http://www.justice.gov/ag/speeches/2001/agcrisisremarks10_31.htm*.
31. *http://www.justice.gov/ag/speeches/2001/agcrisisremarks10_31.htm*.
32. *http://www.msnbc.com/news/643696.asp*.
33. *http://www.state.gov/*.
34. *http://www.justice.gov/ag/agcrisisremarks9_20.htm*.
35. *http://www.justice.gov/ag/speeches/2001/1029financialaction.htm*.
36. *http://www.whitehouse.gov/news/releases/2002/01/20020116-13.html*.
37. *http://www.whitehouse.gov/news/releases/2002/01/20020123-13.html*

REFERENCES

Agnew, J. (1993). The United States and American hegemony. In P. J. Taylor (Ed.), *Political geography of the twentieth century: A global analysis* (pp. 207–238). London: Belhaven Press.

Agnew, J. (1999). Mapping political power beyond state boundaries: Territory, identity, and movement in world politics. *Millenium, 28*, 499–521.

Agnew, J., & Corbridge, S. (1995). *Mastering space: Hegemony, territory and international political economy.* London and New York: Routledge.

Arrighi, G. (1994). *The long twentieth century: Money, power, and the origins of our times.* London and New York: Verso.

Arrighi, G., & Silver, B. (1999). *Chaos and governance in the modern world system.* Minneapolis and London: University of Minnesota Press.

Becker, E., & Sciolino, E. (2001, October 9). A nation challenged: Homeland security: A new federal office opens amid concern that its head won't have enough power. *New York Times,* p. B11.

Block, F. (1987). *Revising state theory: Essays in politics and postindustrialism.* Philadelphia: Temple University Press.

Boulding, K. (1990). *Three faces of power.* Newbury Park, CA: Sage.

The Economist. (2003a, July 26). But when will the others come back? pp. 27–28.

The Economist. (2003b, August 16). Manifest destiny warmed up? pp. 19–21.

Eichengreen, B. (2001). U.S. foreign policy after September 11th. Available online at *http://www.ssrc.org/sept11/essays/eichengreen.htm.*

Flint, C. (2001). The geopolitics of laughter and forgetting: A world-systems interpretation of the post-modern geopolitical condition. *Geopolitics, 6,* 1–16.

Halliday, F. (1986). *The making of the second cold war* (2nd ed.). London: Verso.

Hardt, M., & Negri, A. (2000). *Empire.* Cambridge, MA: Harvard University Press.

Harvey, D. (2000). *Spaces of hope.* Berkeley, CA: University of California Press.

Hirsh, M. (2003). *At war with ourselves: Why America is squandering its chance to build a better world.* Oxford, UK: Oxford University Press.

Hobsbawm, E. (2003, June 14). America's imperial delusion. *The Guardian.* Available online at *http://www.guardian.co.uk/usa/story/0,12271,977470,00.html.*

James, L. (1994). *The rise and fall of the British Empire.* New York: St. Martin's Press.

Juergensmeyer, M. (2000). *Terror in the mind of God: The global rise of religious violence.* Berkeley and Los Angeles: University of California Press.

Kennedy, P. (1987). *The rise and fall of great powers: Economic change and military conflict from 1500 to 2000.* New York: Random House.

Modelski, G. (1987). *Long cycles of world politics.* London: Macmillan.

Oas, I. (2002). *The spatial dementia of geopolitics: Online agency and U.S. hegemonic decline.* Unpublished masters thesis, Department of Geography, Pennsylvania State University, University Park.

O'Tuathail, G. (2000). The post modern geopolitical condition. *Annals of the Association of American Geographers, 90,* 166–178.

Polanyi, K. (1957). *The great transformation.* Boston: Beacon Press.

Prestowitz, C. (2003). *Rogue nation: American unilateralism and the failure of good intentions.* New York: Basic Books.

Said, E. (1979). *Orientalism.* New York: Vintage Books.

Santos, B. d. S. (1999). Toward a multicultural conception of human rights. In M. Featherstone & S. Lash (Eds.), *Spaces of culture: City, nation, world* (pp. 214–229). London, Thousand Oaks, CA, and New Delhi: Sage.

Sassen, S. (1996). *Losing control?: Sovereignty in an age of globalization.* New York: Columbia University Press.

Schwartz, H. M. (1994). *States versus markets: History, geography, and the development of the international political economy.* New York: St. Martin's Press.

Silver, B. J., & Slater, E. (1999). The social origins of world hegemonies. In G. Arrighi & B. J. Silver (Eds.), *Chaos and governance in the modern world system* (pp. 151–216). Minneapolis: University of Minnesota Press.

Taylor, P. J. (1991). The crisis of the movements: The enabling state as quisling. *Antipode, 23,* 214–228.

Taylor, P. J. (1993). The last of the hegemons: British impasse, American impasse, world impasse. *Southeastern Geographer, 33*(2), 1–22.

Taylor, P. J. (1996). *The way the modern world works: World hegemony to world impasse.* Chichester, UK: Wiley.

Taylor, P. J. (1999). *Modernities: A geohistorical interpretation.* Minneapolis: University of Minnesota Press.

Taylor, P. J. (2000). Embedded statism and the social sciences, 2: Geographies (and metageographies) in globalization. *Environment and Planning A, 32,* 1105–1114.

Taylor, P. J., & Flint, C. (2000). *Political geography: World-economy, nation-state, and locality.* London and New York: Prentice-Hall.

Walker, R. B. J. (1993). *Inside/outside: International relations as political theory.* Cambridge, UK: Cambridge University Press.

Wallerstein, I. (1984). *The politics of the world-economy.* Cambridge, UK, and New York: Cambridge University Press.

Wallerstein, I. (2002). America and the world: The twin towers as metaphor. In C. Calhoun, P. Price, & A. Timmer (Eds.), *Understanding September 11* (pp. 345–360). New York: Social Science Research Council.

Index

387

About the Editors

John O'Loughlin is Professor of Geography and Faculty Research Associate in the Institute of Behavioral Science at the University of Colorado at Boulder. His research interests include ethnoterritorial nationalism, the diffusion of democracy, democratic transitions in the former Soviet Union, Russian geopolitics, and spatial modeling of political processes. He is the editor of *Political Geography*.

Lynn Staeheli is Associate Professor of Geography and Research Associate in the Institute of Behavioral Science at the University of Colorado at Boulder. Her research interests are globalization, citizenship, and immigration.

Edward Greenberg is Director of the Research Program on Political and Economic Change in the Institute of Behavioral Science at the University of Colorado at Boulder. His research focuses on the social and political impacts of workplace and job changes on employees' well-being. He is the coauthor, with Benjamin I. Page, of *The Struggle for Democracy*, now in its sixth edition.

Contributors

George Avelino is Associate Professor of Political Science at the Fundação Getúlio Vargas in São Paulo, Brazil. His research interests include the effects of political institutions on public policies in developing countries in general and Brazil in particular.

David S. Brown is Assistant Professor of Political Science and Research Associate in the Institute of Behavioral Science at the University of Colorado at Boulder. His research interests include democratization and social spending, NGOs and their impact on politics in the developing world, and the politics of electrification in the developing world.

Colin Flint is Associate Professor of Geography at the Pennsylvania State University. His research interests include geopolitics, world-systems theory, terrorism, and the Arab world. He is the editor of *Spaces of Hate: Geographies of Hate and Intolerance in the United States* (2003) and *The Geography of War and Peace* (2004) and the coauthor (with Peter J. Taylor) of *Political Geography: World-Economy, Nation-State and Locality* (2000).

Gary Gereffi is Professor of Sociology and Director of the Markets and Management Studies Program at Duke University. His research interests include social and environmental certification in global industries, the competitive strategies of global firms, and industrial upgrading in East Asia and Latin America. His books include *Commodity Chains and Global Capitalism* (with Miguel Korzeniewicz, 1994) and *Free Trade and Uneven Development: The North American Apparel Industry after NAFTA* (with David Spener and Jennifer Bair, 2002).

Kristian Skrede Gleditsch is Assistant Professor of Political Science at the University of California, San Diego. His research interests include international conflict and cooperation, the role of international factors in democratization, and statistical methods in the social sciences. He is the author of *All International Politics Is Local: The Diffusion of Conflict, Integration, and Democratization* (2002).

Richard Grant is Associate Professor of Geography and Regional Studies at the University of Miami. His research interests include cities in the lesser developed world, especially the cities of Accra (Ghana) and Mumbai (India), economic globalization, and global trade policy. He is the editor (with John R. Short) of *Globalization and the Margins* (2002) and (with Jan Nijman) of *The Global Crisis in Foreign Aid* (1998).

Edward Greenberg (see "About the Editors").

Wendy Hunter is Associate Professor of Political Science at the University of Texas at Austin. Her recent publications have focused on democracy and social spending in Latin America and have appeared in the *American Political Science Review* and *Comparative Political Studies*. She is currently writing a book on the growth and transformation of the Workers' Party in Brazil.

Andrew Kirby is Professor of Social and Behavioral Sciences and Professor of Geography at Arizona State University, West Campus, and has been editor of *Cities: The International Journal of Urban Policy and Planning* since 1995. He is also a member of the International Advisory Board of the *Encyclopedia of the City* and series coeditor of *Society, Environment, and Place* (University of Arizona Press).

Victoria A. Lawson is Professor of Geography at the University of Washington, Seattle. She works on critical development studies, feminist analyses of globalization in relation to poverty, and inequality in the Americas. She is completing a book, *Making Development Geography*, and has published in journals such as the *Annals of the Association of American Geographers*, *Society and Space*, *Economic Geography*, and *Progress in Human Geography*.

Patricia M. Martin currently holds a Rockefeller Post-Doctoral Fellowship in Geography and Women's and Gender Studies at Dartmouth College. Her research interests include feminist political theory and the study of globalization and democratization in Latin America.

Keith E. Maskus is Professor and Chair of the Department of Economics at the University of Colorado at Boulder, and spent 2001–2002 as a lead economist at the World Bank. He is interested in international trade, investment, and intellectual property rights and is the author of *Intellectual Property Rights in the Global Economy* (2000).

Olga Memedovic is the Economist in the Strategic Research and Economics Branch of the United Nations International Development Organization (UNIDO). Before joining UNIDO in 2000, she worked at the Netherlands Economic Institute and Tinbergen Institute in the Netherlands. Her publications cover the topics of theory and measurement of comparative advantages, multilateralism and regionalism, globalization of labor markets, transition economies, and globalization of industry and the prospect for industrial upgrading by developing countries.

James H. Mittelman is Professor in the School of International Service at American University. He is the editor of *Globalization: Critical Reflections* (1996), coeditor (with Norani Othman) of *Capturing Globalization* (2001), and author of *The Globalization Syndrome* (2000), which has been translated into Chinese, Japanese, Spanish, and Malay.

William Muck is a PhD candidate in the Department of Political Science at the University of Colorado at Boulder. His research interests include American foreign policy, international relations theory, and religious identity formation. His dissertation examines the role international norms played in defining the foreign military policy of the United States during the Cold War period.

Caroline Nagel is Lecturer in Human Geography at Loughborough University, Leicestershire, United Kingdom. Her research focuses on the politics of immigration, cultural identity, and citizenship. She has a longstanding interest in Arab immigrant communities in Europe and North America and is currently working with Lynn Staeheli on a comparative study of British Arab and Arab American communities.

Michael W Nicholson is an economist with the Federal Trade Commission in Washington, DC. His professional interests include intellectual property rights, international competition policy, and multinational firms. He is currently working in Yerevan, Armenia, as a competition advisor for economic development.

Jan Nijman is Professor of Geography and Regional Studies at the University of Miami. His most recent research concentrates on the political economy and urban culture of globalizing cities, with ongoing projects in Mumbai and Miami. His work has been published in a wide range of international journals and volumes.

John O'Loughlin (see "About the Editors").

J. David Richardson is Professor of Economics and International Relations in the Maxwell School at Syracuse University and Gerald B. and Daphna Cramer Professor of Global Affairs. He is also a Senior Fellow at the Institute for International Economics (IIE) in Washington, DC. His recent work focuses on American trade and labor-market outcomes and is summarized in a forthcoming IIE book, *Global Forces, American Faces: U.S. Economic Globalization at the Grass Roots*.

Anna J. Secor is Assistant Professor of Geography at the University of Kentucky. She has published several articles on gender, political participation, and Islamism based on her field research in Turkey. Her current research project is on the urban geography of civil society in Istanbul.

Michael E. Shin is Assistant Professor of Geography at the University of California, Los Angeles. His research interests include the geographies of health, political geography, and Italian politics and society.

Lynn Staeheli (see "About the Editors").

Michael D. Ward is Professor of Political Science at the University of Washington, where he is also a member of the Center for Statistics and Social Science. He also holds a *Chaire Municipale* in the economics department at the University of Pierre Mendes France in Grenoble, France. He writes unrepentant prepost-positivist articles on the topics of globalization, political economy, and international relations in a variety of disciplines.

Takashi Yamazaki is Associate Professor in the Department of Geography at Osaka City University, Japan. His research interests are in political geography, political/social movements in Okinawa, and post-Cold War geopolitics in Asia.